ECONOMICS OF NATURAL RESOURCES AND THE ENVIRONMENT

ECONOMICS OF NATURAL RESOURCES AND THE ENVIRONMENT

David W. Pearce

and R. Kerry Turner

The Johns Hopkins University Press
Baltimore

 Published 1990
Printed in Great Britain

Originally published in hardcover and paperback, 1990
Second printing, paperback, 1991

The Johns Hopkins University Press
701 West 40th Street
Baltimore, Maryland 21211-2190

Library of Congress Cataloging-in-Publication Data

Pearce, David W. (David William)
Economics of natural resources and the environment / David W. Pearce and R. Kerry Turner.
p. cm.
Includes bibliographical references.
ISBN 0-8018-3986-6. — ISBN 0-8018-3987-4 (pbk.)
1. Environmental policy. 2. Pollution—Economic aspects.
3. Natural resources. I. Turner, R. Kerry. II. Title.
HC79.E5P37 1989 89-19855
333.7—dc20 CIP

CONTENTS

PART II: THE ECONOMICS OF POLLUTION

PREFACE AND ACKNOWLEDGEMENTS

This volume has been a long time in preparation. That is entirely our fault and we wish to thank Harvester Wheatsheaf Press, and a succession of editors, for their patience. In part the delay has been due to the fact that textbooks always go to the bottom of the list of priorities and so much else seems to happen just when it comes to drafting new material. But we have also been conscious that environmental economics and, indeed, environmental thinking in general, has been changed rapidly in the past few years. There has repeatedly been the need, therefore, to dispense with first drafts and start again.

We have aimed the text at undergraduates. The problem in doing this is that undergraduates studying environment from a social science perspective have very varied backgrounds. Many have no social science training at all, others have been raised on what we might call 'mainstream' economics. We have therefore tried to prepare chapters that will appeal to both types of audience, but we are very much aware that some readers will be irritated by the superficiality of some chapters and others will find the going in other chapters very hard. It remains our view that environmental economics as a received body of thought has a great deal to offer anyone who wants to understand modern environmental problems. We have sympathy with those who feel modern academic economics is somehow not addressing the 'real' social issues of the day, and a little less sympathy for those who want to search for a 'new' or 'alternative' economics to help solve those issues. In truth, while economists profess that they understand the narrowness of traditional economic efficiency objective as a guide to what people actually care about, many still forget the wider concerns in the practice of their subject. Economic efficiency – getting the most 'welfare' out of a given endowment of resources – remains vitally important, however, and it matters a great deal that the search for the new should not involve dispensing with the value of the old.

Where appropriate, we have indicated these wider concerns in the chapters. We have also included a chapter devoted to surveying 'environmental ethics' in order to capture some of the alternative views of the environmental problem. Since these 'deviations' from the mainstream are sign-posted fairly clearly we believe that anyone wanting a text in the conventional mode will still find it here. Those wanting an alternative economics will be disappointed, but we hope nonetheless that they will find some of their concerns addressed in this book. Moreover some, but not all, of these concerns still seem to us to be best treated by the conventional approach.

The theme through much of the book is that of 'sustainable development'. At its simplest, we take this to imply a stance which raises the political and ethical profile of natural and heritage environments as we come to the end of the twentieth century. Sustainable development thinking reflects both our greater understanding of what natural environments do for us as inhabitants and users of those environments, and stresses the uncertainty that results from our continuing ignorance of the many other ways in which we depend on life-support systems. For all we know, it may be perfectly possible to dispense with natural environments in favour of an encapsulated world of plastics and microchips. In all probability the issue is not survival in that sense. But the issue surely is one of how far down that road we wish to travel, and what regrets we think our grandchildren will have if we travel down it too far. For a great many people, we concluded, there are no substitutes for many natural assets: the oceans, the moors, the fells, the mountains, rivers and all their diverse wildlife. It is because a great many people want to experience these assets, *and* to feel reassured that they are being protected as far as seems reasonable, whether they want to experience them or not, that the social science devoted to the efficient satisfaction of their wants is relevant.

We have both spent most of our lifetimes teaching and writing about environment and environmental economics, often at times when the occupation was a very lonely one. We are sufficiently wise not to be too euphoric about the resurgence of environmental values in the late 1980s. We both witnessed, and were part of, the environmental revolution of the early 1970s. We hope this one is a sustainable development. We think it is.

As always, wives have played a vital part in getting this book on to the market. Sue Pearce has proof-read the manuscript. We are sorry it does not have much to say about African environments, her first love, but another volume is in preparation on that by one of us and she is helping with other work in the area. Merryl Turner has word-processed a number of chapters through several drafts and has offered constructive criticisms, especially of some of the 'ecological' aspects of our analysis. Our professional debts are many. Mention must be made of Anil Markandya and Edward Barbier of the London Environmental Economics Centre, Jerry Warford of the World Bank's Environment Department, Dennis Anderson of the World Bank and Shell International, Richard Sandbrook of the International Institute for Environment and Development, Tim O'Riordan of the University of East Anglia and Derek Deadman of Leicester University. We have both benefited from a long association with OECD's Environment Directorate in Paris: our thanks go especially to Michel Potier, Jean-Philippe Barde and Frank Juhasz. None of them bears the slightest responsibility for all the errors we are sure remain.

DWP, RKT

PART I

ECONOMY AND ENVIRONMENT

1 · THE HISTORICAL DEVELOPMENT OF ENVIRONMENTAL ECONOMICS

1.1 INTRODUCTION

Environmental stresses and strains are now ubiquitous phenomena appearing in all economic systems, regardless of political ideology, from the very poorest to the very rich. Despite the impression given in some of the environmental literature, environmental degradation is not the unique attribute of advanced Western industrial capitalism. Eastern bloc economies face acute water and air pollution threats, notable examples being river-water pollution in many industrial areas of Poland and declining urban air quality levels in industrial Czechoslovakia. The Soviet environment has suffered from a catalogue of pollution abuses over a long period of heavy industrialisation. Pollution there now threatens even the most precious of biospherical assets such as Lake Baikal. Among the developing economies, air pollution in cities such as Caracas, Mexico City and Sao Paulo is extremely severe and poses a significant health hazard. For the group of thirty-six (the poorest countries on earth) their very poverty is a major cause and effect of environmental problems. Poverty, which denies poor people the means to act in their own long-term interest, creates environmental stress (such as overgrazing of rangeland, soil erosion and eventual desertification) leading to resource degradation and growing population pressures.

Uncertainty still surrounds the exact nature and extent of the global interdependencies between economic growth and the supporting environmental systems. We still cannot fully quantify the risks to future human well-being posed by acid rain, ozone depletion and the greenhouse effect. Even so, half of the net natural output produced by environmental systems is now utilised by humans.

Future necessary global economic growth will further diminish that sector of nature in which self-regulating natural systems can regenerate free of human intervention. The margin for error in economic planning that has the capacity to inflict irreversible change on the natural resource base is, in the view of many, narrowing.

Environmental issues, on the boundaries of economic and natural systems, are undoubtedly complex and in many cases contain inherently uncertain outcomes. The sub-discipline of environmental economics which seeks to analyse such issues consequently sits on the boundaries of a range of social science and natural science disciplines. The current position in the economics profession is one in which mainstream analyses are firmly anchored to earlier neoclassical foundations (neoclassicism is explained on p. 10). Since any economics needs value judgements, and in the absence of agreed meta-ethical criteria for choosing between value judgements, it cannot be argued that neoclassical economics and its Paretian value judgements are 'worse' or 'better' than any other economic doctrine.

Particularly in its formative years (1960s), environmental economics encompassed a diversity of economic doctrines. A pluralistic view (i.e. one that recognises more than a single tradition in the development of economic thought) of the contribution that economics can make would guard against narrowness in economics, as well as fostering more interdisciplinary analytical linkages. We devote most of the chapters in this book to an analysis which expands the horizons of modern conventional economic thought. 'Alternative' economic paradigms are surveyed in this first chapter.

1.2 EARLY ECONOMIC PARADIGMS AND THE ENVIRONMENT

In order to appreciate the modern arguments (roughly over the period 1960 into the 1980s) both between economists themselves and between economists and other environmental analysts, the historical roots of environmental economics must be explored.

Figure 1.1 summarises some of the more important concepts and ideas that have influenced environmental economists and traces their origins back to past doctrines. Economic theories ought to be appraised within the context of their wider framework ('paradigm'). There is a complex interaction taking place as both scientific

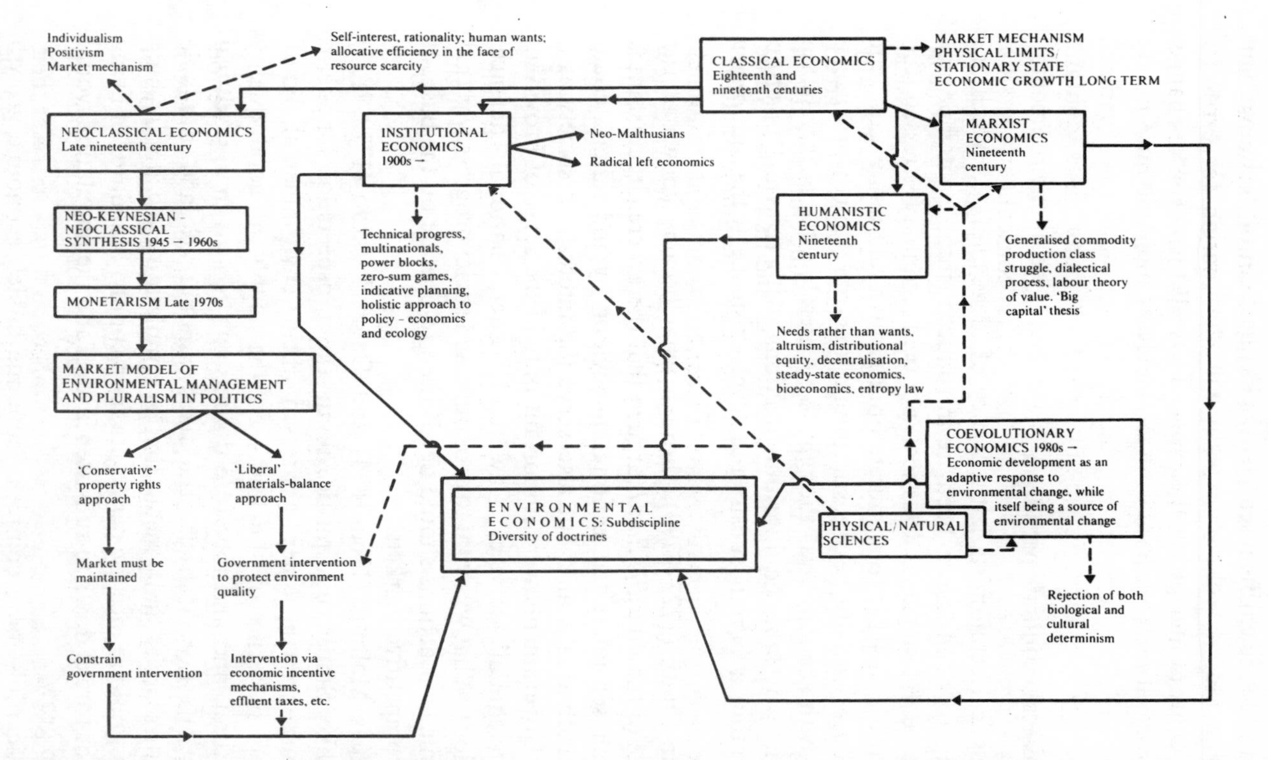

Figure 1.1 Economic paradigms and the environment. Some caveats are in order. The figure is meant to be descriptive rather than analytical. It is probably not correct to view changing economic doctrines over time in terms of Kuhnian 'scientific revolutions'. Rather it is more fruitful to think of clusters of interconnected theories or 'scientific research programmes' which compete against each other.

(natural, physical and social) theory and social order evolve. The ways in which scientific research asks its questions of the human and natural worlds it seeks to explain will at times be influenced by social, cultural and political factors. Thus attitudes toward nature and preservation/conservation will change as humanity and nature evolve.

The classical economic paradigm

The classical economists left a legacy of ideas many of which are relevant to, and have been re-introduced into, contemporary environmental debates. Classical political economy stressed the power of the market to stimulate both growth and innovation, but remained essentially pessimistic about long-run growth prospects. The growth economy was thought to be merely a temporary phase between two stable equilibrium positions, with the final position representing a barren subsistence level existence – the *stationary state*.

Adam Smith (1723–1790), through what became known as the doctrine of the *invisible hand*, argued that there were circumstances in which self-interested rational behaviour by individuals could satisfy individual wants but also serve the interests of society as a whole. Governments were important only in the sense of providing 'nightwatchman' services (law and order, national defence, education). What was vital to economic and social progress was that economic transactions should be allowed to operate on the basis of freely competitive markets.

Thomas Malthus (1766–1834) and David Ricardo (1772–1823) were, like Smith, pessimistic about the prospects of long-term economic growth. They expressed their 'environmental limits thinking' in terms of the limits on the supply of good quality agricultural land and therefore diminishing returns in agricultural production. For Malthus, the fixed amount of land available (absolute scarcity limit) meant that as population grew, diminishing returns would reduce the per capita food supply. Standards of living would be forced down to subsistence level and the population would cease to grow.

In Ricardo's more complex economic model, economic growth again peters out in the long run because of a scarcity of natural resources. Diminishing returns set in not so much because of

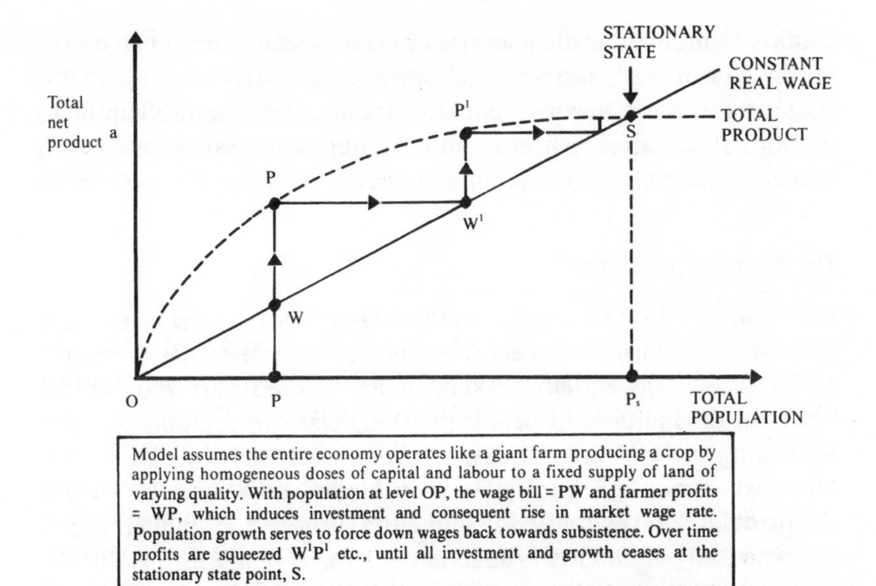

[a]Total product subject to the axiom of diminishing returns minus rent accruing to non-productive landlord class; workers are paid wages, farmers gain profits and undertake productive investment. No technical progress assumed.

Figure 1.2 Ricardo's simple commodity production model. (*Source*: M. Blaug, 1978.) Simple commodity production, i.e. production for sale by independent producers in which labour power is a commodity (a measure and a source of value); but capitalist institutions and activities treated as external to the model.

absolute scarcity but because the available land varies in quality and society is forced to move on to successively less productive land. Figure 1.2 summarises the main Ricardian analytics. Note that the lack of technical progress in the model means that the total product curve (subject to diminishing returns) remains fixed. Technical innovation (e.g. artificial fertilisers, irrigation and deep drainage, etc.) would shift the total product curve upwards, increasing output per unit of input and offsetting, but not eliminating, the tendency towards diminishing returns.

John Stuart Mill (1806–1873) conceived of economic progress in terms of a race between technical change and diminishing returns in agriculture. But unlike the other classical economists he viewed the distant prospect of the stationary state with some optimism. By then, he reasoned, technical progress would have provided for much of mankind's individualistic material wants and society would be free to pursue educational, aesthetic and other social goals.

During the nineteenth century fundamental changes in these traditional classical patterns of thought were established with Marxism, neoclassicism and humanism.

The Marxist paradigm

Karl Marx (1818–1883), adopted the *labour theory of value* from the classical economists (workers were the sole source of net economic product) and was equally pessimistic about the future standard of living for the majority of people (working class) in capitalist society. According to Marx, the classical economists had failed to place capitalist economic organisation in its historical context. He sought to formulate a generalised commodity production model which characterised commodity production as a social relationship. History was to be interpreted as a dialectical phenomenon, a process of conflicting material or economic forces out of which a synthesis, a resolution of the conflict, would emerge. Capitalist society would inevitably be beset by a class struggle (workers versus capitalist entrepreneurs) for social power. Power would be gained via control of economic resources. Marx predicted that the capitalist economic system would be faced with a falling rate of profit over time, increasing destitution for the majority working class and increased monopoly. Ultimately, the majority would overthrow the small capitalist class (existing via the exploitation of surplus value produced by workers) and seize power to create a socialist society.

Marx believed that progress was a process of natural development, inherent in human history. Progress itself was to be defined in terms of material and technological advance made possible by the exploitation ('humanising') of nature. He saw the political state as apart from nature, created as an alternative to 'natural' environment. Nature was there to be humanised via science so that inherent value could be turned into use value. But some modern Marxian writers have pointed out that Marx does emphasise the *process* of production and the fact that a viable basis for any society can only be provided if the system of production is capable of reproducing itself. There is a strong hint in this analysis that natural systems could be a limit to reproduction, as well as the economic and political make-up of society. In this sense, it can be argued that Marx took what we call today a 'materials balance' approach to the process of production over time (see Chapter 2). In terms of modern environmental issues this reproductive economic system analysis raises questions about the sources and nature of technological change. Does such change alleviate or aggravate the environmental constraints on an economy's capacity to reproduce itself? Further, is the process of reproduction consistent with reasonably stable social systems?

According to Marxian analysis, modern capitalist economic systems fail the reproduction test, i.e. capitalist systems are not sustainable, and *one* source of non-sustainability is environmental destruction. Questions of economic power, exploitation and the dialectical process encompassing the two classes in society are at the root of an inevitable environmental despoliation process, which in turn contributes to the failure of capitalism.

More formally, a Marxian economic model would have competitive capitalists seeking 'labour-saving' innovations to increase short-run labour productivity and total surplus value. This pushes up the rate of profit and capital accumulation. However, over the long run the new technologies impose a heavy cost burden on the environment as both the toxicity and durability of waste discharges increase. Pollution generates damage costs which include human morbidity and mortality. These damage costs become 'class costs' because of their uneven social incidence. Workers both at home and at work are exposed to more than their fair share of pollution. There is then an increased requirement for worker medical care in order to maintain labour productivity. If unions demand and get better compensatory health care or shorter working hours to escape occupational health hazards, their real wages will have increased at the expense of profits and capital accumulation. A familiar Marxist confrontational situation is now inevitable in the market-based economy. Just how well centrally-planned socialist economies have dealt with environmental problems is discussed in Chapter 12.

In the international economy the exploitation process manifests itself in terms of the operations of the transnational corporations. There are structural linkages between economic development in the Northern economies and the South, and these linkages radically affect the environment in the South. Changes in the environments of the South need, according to Marxians, to be understood in terms of the international redivision of labour.

Neoclassical and humanistic paradigms

Starting around 1870, neoclassical economic thought began to be developed by analysts within the mainstream of the economics profession. The labour theory of value was abandoned and a commodity's price was seen not as a measure of its labour cost but of its scarcity. The concentration on scarcity value allowed both sides of the market to be analysed simultaneously. Analysts compared the amount of a commodity that was available (supply) with the amount required (demand). The interaction of supply and demand then determined the equilibrium market price for the commodity. The economic activity that was observed in the real world was seen as the result of the interaction between productive activity (determined by technological progress) and the preferences of individual buyers constrained by the feasible range of choice and income.

The neoclassical economists also introduced a new methodology, *marginal analysis*, i.e. the study of the relationships between small or incremental changes. This type of approach was well suited to the investigation of price determination and market structures. Consequently, the classical concern with long-term growth patterns was sidelined almost completely over the period 1870–1950.

The neoclassical theory of the market was supposed to be neutral and value-free. The basic aim had been to define a set of economic laws which governs economic activity (in much the same way as physicists had done following Newton's discoveries). Rational individuals were seen in terms of seeking to satisfy substitutable wants (or preferences) and this pursuit of individual self-interest was also believed to be improving societal welfare; thus, within the 'hard core' of the neoclassical system was a particular model of human nature – the 'rational and egotistic person'. In its modern version the model has economic person holding the preference structure of indifference and operating on the basis of constrained satisfaction (utility) maximisation. The economic (instrumental) value of marketable commodities, unpriced environmental goods and services, or sympathy for future generations, is determined according to the amount of personal utility yielded. Economic person makes trade-offs at the margin to identify positions of equal personal satisfaction. The preferences of individuals are revealed by the choices they make, and efficiency and consistency of choice reflect rational behaviour.

The criterion of social desirability is usually expressed in terms of the so-called *Pareto criterion.* A Pareto optimum situation is one in which it is impossible to make any individual better off without making someone else worse off, where 'better off' means 'more preferred' and 'worse off' means 'less preferred'. Every competitive market equilibrium is a Pareto optimum and every Pareto optimum is a competitive equilibrium, as long as a set of restrictive assumptions (e.g. perfect information, absence of externalities, etc.) hold true. The 'basic theorem of welfare economics' seeks to legitimise rational behaviour as being socially desirable and also to justify some government intervention to improve the conditions under which individuals make choices. Intervention would be especially justified whenever so-called *market failures* exist, i.e. when it is clear that markets are not maximising collective welfare. The basic neoclassical view sees government as an essentially ethical agent only intervening in the market in the public interest to ease the inevitable tension between individual rationality and collective ethics. Ethical or moral obligations are not recognised at the level of the individual.

Supporters of the minority *humanistic paradigm* reject the 'rational economic person' model and instead adopt a behavioural psychology approach which emphasises a hierarchy of *needs* in place of a flat plane of substitutable wants. Humanistic analysts emphasise that preferences (tastes) are not static, independent, and determined by genetics (some of this critique is misplaced in that neoclassics does not say wants are genetically determined; it does not ask the question at all). Instead they are interdependent and can and do change over time because they are at least partially learned via the culture.

In the absence of a theory of how tastes are determined, how they differ between individuals and how they change over time, neoclassical theory treats tastes as 'exogenous'. Wants and needs are therefore not separable in the conventional analysis. Quite recently supporters of the 'human capital theory' movement have argued that all economic agents do hold exactly the same set of 'stable preferences'. It is then possible to interpret these particular preferences as beliefs about how to meet basic human needs. Needs cannot be traded off against each other or against market commodities without threatening survival. As we shall see later in this chapter, environmentalists would argue that 'high' levels of environmental quality are human needs.

Humanists, among others, have also been critical of the neoclassical theory of self-interested rationality. They argue that individuals are capable of truly altruistic acts and that an extended notion of rationality is required. *Extended rationality* could be analysed in terms of multiple preference rankings within a single individual – one self-interested and the other altruistic (group-interested). Moral consideration will then determine a 'meta-ranking' of alternative motivations, e.g. altruistic motives might be judged morally superior to ones based on self-interest. Individuals possess a sense of community which is reflected in a willingness to view assets as a common pool. This extended rationality also generates a strong obligation to abide by particular laws which are seen by the individual as promoting his/her meta-preferences, despite a potential tension between the law and narrow self-interest. The law is seen as personally beneficial not just because of its restrictive effects on others, but also because of its direct effects on the individual concerned. Given these sorts of assumptions the humanistic economy, especially in its transition phase, would be subject to central planning and direction. The government's role would not be restricted merely to correcting market failures.

The humanistic economics viewpoint would not seek to abolish the market mechanism but would seek to restrain and supplement it to a significant degree. In order to facilitate greater social system stability over the long term, increased government intervention would be required in order both to decentralise economic activity and to promote a more deliberately egalitarian distribution of income.

1.3 POST-WAR ECONOMICS AND THE RISE OF ENVIRONMENTALISM

Neoclassical economics contained the basic assumption that the economy had an in-built tendency to operate at an overall level of activity fixed by the full employment of labour. Full employment would be the norm because of the further assumption of flexible wage rates: wages would simply vary up or down until full employment is achieved. The experiences of the inter-war years (1920s and 1930s), when mass unemployment became the norm, led to the formulation of Keynesian economics with its emphasis on government intervention and deficit spending. Thus during the 1950s economic growth got back onto both the economic and political agendas. Economic growth driven by technical innovation appeared to offer limitless progress.

During the 1960s environmental pollution intensified and became more widespread. Environmental awareness was consequently heightened in some sections of industrialised societies, spawning new environmental ideologies. A number of these ideologies were basically anti-economic growth.

These events caused economists to look afresh at a central economic idea: resource scarcity in relation to possible uses. Between 1870 and 1970, mainstream economists (with some notable exceptions) appeared to believe that economic growth was sustainable indefinitely. After 1970 a majority of economists continued to argue that economic growth remained both feasible (a growing economy need not run out of natural resources) and desirable (economic growth need not reduce the overall quality of life). What was required, however, was an efficiently functioning price system. Such a system was capable of accommodating higher levels of economic activity while still preserving an acceptable level of ambient environmental quality. The 'depletion effect' of resource exhaustion would be countered by technical change (including recycling) and substitutions which would augment the quality of labour and capital, and allow for, among other things, the continued extraction of lower quality non-renewable resources.

Since 1970 a number of 'world views' has crystallised within environmentalism, providing the background to the emerging environmental economics sub-discipline. Four basic world views can be distinguished, ranging from support for a market and technology-driven growth process which is environmentally damaging, through a position favouring managed resource conservation and growth, to 'eco-preservationist' positions which explicity reject economic growth. Figure 1.3 outlines some of the main features of these different positions and a further discussion is presented in Chapter 15.

It was against this backdrop of emerging environmental ideologies that environmental economics became established as a sub-discipline. Its development within the economics profession was in one sense a reaction to the prevailing conventional paradigm. A minority of revisionists wished to alter the 'hard core' of the conventional economic research programme, in order to speed up the evolution of economics towards a paradigm that was 'relevant' to the coming zero-growth society. Others merely saw an opportunity to accommodate better the environmental systems implications of the growth economy and society within a modified, but not radically different, set of economic models. The majority mainstream view

TECHNOCENTRIC		ECOCENTRIC	
Extreme 'Cornucopian'	'Accommodating'	'Communalist'	'Deep Ecology'
Resource exploitative, growth-orientated position	Resource conservationist and 'managerial' position	Resource preservationist position	Extreme preservationist position
Economic growth ethic in material value terms	Infinite subsitution is not thought realistic but sustainable growth is a practicable option as long as certain resource management rules (e.g. for renewable resource sustainable yield management) are followed	Pre-emptive macro-environmental constraints on economic growth are required, because of physical and social limits	Minimum 'resource-take' socio-economic system (e.g. based on organic agriculture and de-industrialisation)
Maximise Gross National Product			
It is taken as axiomatic that unfettered market mechanisms or central planning (depending on the ruling political ideology) in conjunction with technological innovation will ensure infinite substitution possibilities capable of mitigating long-run physical resource scarcity		Decentralised socio-economic system is necessary for sustainability	Acceptance of bioethics (i.e. non-conventional ethical thinking which confers moral rights or interests on non-human species)
Instrumental value (i.e. of recognised value to humans) in nature	Instrumental value in nature	Instrumental and intrinsic value in nature (i.e. valuable in its own right regardless of human experience)	Intrinsic value in nature

Figure 1.3 Environmental ideologies. (*Source*: adapted from O'Riordan and Turner, 1983.)

remained optimistic about future growth prospects, with 'Ricardian scarcity' being offset by technology and compensatory market processes (see Chapter 19).

From outside economics, *ecocentrists* tried to bring to the forefront of public debate profound questions relating to the 'acceptability' of conventional growth objectives, strategies and policies. The influential Meadows Report (Meadows *et al.*, 1972) adopted a distinctive Malthusian position which implied that environmental protection policies and the promotion of economic growth objectives were incompatible (i.e. that long-run economic growth objectives were not feasible). This line of thinking led eventually to calls for steady-state (zero growth) economies, and even more radical bio-economic communities based on organic agriculture which in some people's minds ought to be guided by the ethical principles of 'deep ecology' (see Figure 1.3).

The anti-growth argument was buttressed by economic analyses which sought to highlight the social costs, especially the environmental costs, of living in a 'growth society' (i.e. desirability of economic growth and social system stability). Easterlin's 'paradox' (i.e. survey data indicating that material affluence and human happiness were not closely correlated), Hirsch's 'positional goods' concept (i.e. that the enjoyment of a range of commodities is necessarily restricted to a small group of high income earners, despite the illusion given that all sections of society might one day participate in such consumption), and Scitovsky's 'joyless economy' analysis (again emphasising human need for more than mere material affluence) are representative of 'social limits' thinking (Boskin, 1979; Hirsch, 1977; Scitovsky, 1976).

1.4 INSTITUTIONAL ECONOMICS PARADIGM

This minority economic doctrine began to emerge around the beginning of the twentieth century, although it has remained a somewhat diverse collection of views. Institutionalists have adopted what they call a 'processual paradigm' which encompasses the concept of the economy as a dynamic process. Their explanation of socio-economic change is one based on *cultural determinism*. Culture is an on-going complex of ideas, attitudes and beliefs that is absorbed by individuals ('cultural person' not 'rational economic person') in a habitual manner through institutional arrangements. Special significance is attached to scientific and technological change as factors that provide for dynamic change in the structure and functioning of the economic system.

Individual preferences are *learned* preferences which change over time and any one individual will hold both private and public preferences. The latter are thought to be important and justify an active public sector in the economy, and some analysts go as far as advocating indicative planning. Environmental problems are judged to be an inevitable result of economic growth in advanced industrial economies. Institutionalists have long accepted an approach which encompasses the notion of social costs of pollution and stresses the importance of the ecological foundations of any economic system. State intervention is required to control, as far as possible, the activities of transnational corporations and also to mediate between the interest groups (power blocks) that have emerged in modern economies. Institutionalists remain divided over the extent of intervention required in order to reach a social consensus. Some 'neo-Malthusians' believe that only an authoritarian system would be capable of bringing about the necessary changes to protect the environment, while others put their faith in decentralised socialist systems.

1.5 THE MARKET MODEL OF ENVIRONMENTAL MANAGEMENT: PROPERTY RIGHTS PARADIGM VERSUS MATERIALS BALANCE ANALYSIS

The conventional approach has generated two variants of an environmental resource management model, one more revisionist than the other in terms of the required modifications to the neoclassical blueprint. These approaches are the property rights approach and the materials balance approach.

The property rights approach

Some analysts at first maintained that pollution cost problems were non-pervasive and could be adequately mitigated via a process of re-defining the existing structure of property rights. A particular interpretation of the 'Coase theorem' (Coase, 1960) was used as the theoretical basis for this non-interventionist pollution control policy (see Chapter 5). According to Coase, given certain assumptions the most efficient solution to pollution damage situations is a bargaining process between polluter and sufferer. Each could compensate the other according to who possesses property rights: if the polluter has the right, the sufferer can 'compensate' him *not* to pollute; if the sufferer has the right, the polluter can compensate him to tolerate damage.

The non-revisionist 'property rights paradigm' approach to environmental economics has become more sophisticated. Key neoclassical assumptions about human behaviour in the marketplace (i.e. self-interested utility maximisation) have been extended to cover the activities of bureaucrats in the public sector (drawing on public choice economics literature) and notions of extended rationality (i.e. the possession of motivation other than self-interest alone) have been resisted.

Sociobiological explanations for 'rational economic man' have also been advanced. Self-interested behaviour, it is argued, is genetically programmed into humans and is therefore inevitable. At the level of the individual, economic person is still seen as making trade-offs at the margin to identify positions of equal satisfaction. The idea of 'rational ignorance' has, however, been added to the model, i.e. it is rational for individuals to obtain less than complete information before making a decision, because information is scarce and something must be given up – time, effort or money – to obtain more of it. Exactly how much information is rationally required is not specified.

It is argued that in an economy with well-defined and transferable property rights, individuals and firms have every incentive to use natural resources as efficiently as possible. Markets and prices emerge from collective economic behaviour provided exclusion is possible – i.e. any individual consuming a good can exclude other individuals from consuming the same good – and property rights exist. Environmental pollution is a form of market failure, usually because of the over-exploitation of resources held as common property or not owned at all. The market fails therefore when property rights are inadequately specified or are not controlled by those who can benefit personally by putting the resources to their most highly valued use.

According to the property rights approach, increased government intervention should be resisted because public ownership of many natural resources lies at the root of resource control conflicts: there is 'government failure'. It is assumed that the theory of the public sector should be based on the same motivational assumptions (self-interest) employed in the analysis of private individual behaviour. Thus the decision-maker will seek to maximise his own utility, not that of some institution or state, in whatever situation he finds himself. The public sector, it is argued, provides no incentives for politicians or bureaucrats to resist pressures from special interest groups. Gains to such groups often come only at a net cost to society.

The misallocation of environmental resources is not, therefore, just a question of market failure. A range of government intervention policies have themselves been the cause of environmental disruption (government failures). For example, non-integrative government policy and inefficient government intervention have resulted in 'created' land-use conflicts in wetland ecosystems and consequent sub-optimal wetland protection levels in industrialised and developing countries.

The fundamental organising concepts employed in economics and biology are strikingly similar. Some have, therefore, been tempted to claim that there is empirical proof (sociobiological evidence) of the existence of selfish economic person and of the 'optimality' of the competitive market system in a world of scarce resources. The sociobiologists interpret the findings of molecular biology as saying that human nature is fixed by our genes and the characteristics of individuals are a consequence of their biology. The Darwinian evolutionary process encourages the view that human society as well as the rest of nature progresses by the survival of the fittest in a competitive struggle. So genetically-based general forms of social organisation have been established by natural selection during the course of evolution. It is arguable then that 'selfish gene'-dominated humans (economic person) and their social organisation (the market) are a consequence of natural selection for traits that maximise reproductive fitness. For some, the competitive market process therefore represents a Darwinian process of survival. It produces exactly the same results that would ensue if all firms maximised their profits, and all consumers maximised their utility. Competitive rivalry guarantees that we will only observe survivors who actually maximise profits, and all that matters for the theory is whether it correctly predicts this outcome. It would seem that the genetically-determined competitive market is a product of natural selection and therefore must be in some sense optimal or adaptive.

The merits of this biological determinist belief will be critically examined further when sustainable development is discussed (p. 23). But there is an apparent circularity in the sociobiology/economic synthesis argument. The sociobiologists have incorporated some economic concepts such as cost-benefit analysis and game theory into their world view. It is this world view which is then cited by some economists as a justification for the continued existence of a certain social organisation.

Overall, property rights paradigm supporters would probably concede that markets are imperfect but equally they would emphasise that their failings do not automatically imply that collective action is superior. The market mechanism is then judged to be superior to any other practical alternative. Any further relinquishing of private rights and the rule of willing consent in favour of collective action will create rather than resolve environmental problems.

The materials balance approach

Revisionists have sought to incorporate materials balance models and to a more limited extent entropy limits into economic analysis (these issues are explored in detail in Chapter 2). While pollution is seen as a symptom of market failure, it is also recognised that it is a pervasive and inevitable phenomenon (because of the laws of thermodynamics) requiring government intervention via a package of regulatory and incentive instruments.

In principle, an economic optimum (efficient) level of pollution can be defined, given certain simplifying assumptions. It is that level of pollution at which the marginal net private benefits of the polluting firm are just equal to the marginal external damage costs (see Chapter 4). Because of data deficiencies and the limitations of this static approach, the optimum situation is not a practicable policy objective. Instead society sets 'acceptable' levels of ambient environmental quality, and policy instruments are directed at these standards. The analytical task is to seek out the least-cost policy package sufficient to meet acceptable ambient quality standards. Many economists favour the use of effluent taxes (per unit of pollution) but actual pollution control policy has been based on a regulatory approach often involving uniform reductions in pollution emissions across classes of industry (see Chapter 6). Because of the uncertainties involved, pollution control policy should be seen as an iterative search process based on a 'satisfising' (extended rationality) rather than an optimising principle. Pollution control instruments and policy are examined in Chapters 5, 6, 7, 8, 11 and 12.

1.6 POLICY ANALYSIS: FIXED STANDARD VERSUS COST-BENEFIT FRAMEWORK

In the face of the complexity of ecological interdependence and uncertainties surrounding resource management, two alternative approaches have been suggested. Some analysts have argued for the adoption of a *cost–benefit framework*, utilising monetary valuations but also incorporating explicit recognition of uncertainty and irreversibilities. Others urge the adoption of a *fixed standard approach*, either in selected cases or as a way of implementing a general 'macroenvironmental policy'. Macroenvironmental standards could encompass land-use zoning policy, ambient environmental quality standards for air and water, etc. In this form, such standards would operate as binding constraints and, although perhaps flexible over time (as knowledge increases), would limit the scope of cost–benefit analysis to cost effectiveness analysis.

In the policy-making context, the acceptance of the axiom of infinite substitutability, positive rates of discount and a belief in long-term ecosystem resilience capacities, would mitigate against any radical restructuring of economic growth or resource pricing policies. Revisionists, on the other hand, would caution against such optimism. The principle of infinite substitution would be rejected and a conservationist position advocated. It is now clear that in a number of developing economies severe ecosystem losses have accumulated in a matter of decades. Global pollution issues – climate warming, ozone depletion, ocean contamination – threaten more widescale problems for the future, while acid rain has generated regionally localised damage to ecosystems. Neglect of sustainability constraints could also result in irreversibility over a wide front for future generations in both developed and developing economies. The presence of irreversibility (e.g. permanent loss of unique wilderness areas and other valuable environmental resources, wetlands, productive soils, etc.) almost always favours postponement of the development options and support for resource conservation/preservation options (see Chapters 20 and 21). A safety-margin approach (for example, based on the concept of the 'safe maximum standard') to policy has been recommended in this context (see Chapter 20).

Another related idea, the 'shadow project' approach, has also been suggested in cases where locally irreversible environmental losses are likely because of economic development. The costs of the development scheme responsible for these losses (such as the destruction of a particularly valuable wetland) should be increased by an amount sufficient to fund a 'shadow' project designed to substitute for the lost environmental asset. It may be possible, for example, to restore a partly degraded wetland somewhere else in the region under consideration. We expand on these ideas in Chapter 14.

Radical re-interpretations of the cost–benefit analysis (CBA) method and technique are also thought necessary. Supporters of extended CBA have adopted a 'value sensitivity' approach and have sought to incorporate non-efficiency decision criteria into their analysis. It is argued that actual decision-makers require rational advice on both their objectives (or more strictly the implications of different objectives) and on the means to achieve these objectives. In any case, the 'positivist' attack against extended CBA is based on the curious argument that the standard value judgements that underlie the concept of a Pareto optimum (and standard CBA) command wide assent and this consensus renders them 'objective'. More convincingly, multi-criteria analysis does undoubtedly involve a trade-off of greater comprehensiveness against loss of precision. The appraisal of environmental policies that involve substantial risks and costs for future generations has also led some analysts to consider the implications of alternative systems of ethics. We survey the environmental ethics debate in Chapter 15.

1.7 ECONOMIC AND ENVIRONMENTAL VALUES

There are various interpretations of the term 'value', but economists have concentrated on monetary value as expressed via individual consumer preferences. On this basis, value only occurs because of the interaction between a subject and an object and, in terms of this explanation, is not an intrinsic quality of anything. A given object can then have a number of assigned values because of differences in the perception of held values of human valuators and different valuation contexts. Economic assigned values are expressed in terms of individual willingness to pay (WTP) and willingness to accept compensation (WTA) (see p. 125).

The environmental literature has identified three basic value relationships which seem to underlie the policy and ethics adopted in society: values expressed via individual preferences; public preference value which finds expression via social norms; and functional physical ecosystem value. Some writers argue that economic value measures are context-specific, assigned values and may therefore be inappropriate as the *sole* value measures for public resource allocation. Ecocentric ideologies seek to base policy on social norms that individuals accept as members of a community (public preferences) and that are operationalised via 'social' legislation. Deep ecology advocates place primary emphasis on a distinction between instrumental value (expressed via human-held values) and intrinsic, non-preference-related value. They lay particular stress on the argument that functions and potentials of ecosystems themselves are a rich source of intrinsic value. This value would, it is argued, exist even if humans and their experiences were extinct. Some of these issues will be examined in Chapter 15, but for now we merely raise the possibility that the intrinsic value and object–subject value distinction is not clear-cut. Humans may capture part of the intrinsic value in their preferences, e.g. valuing 'on behalf of' other species. Economists use the term 'existence value' to encompass these notions.

Economic research into the monetary valuation of environmental commodities is still in a state of flux, although considerable progress has been made. In the absence of a demand curve and market price for many environmental commodities, a number of non-market methods for estimating value has been devised. For value data collection techniques – travel cost method, participation/unit day value method, hedonic pricing and contingent valuation – have been extensively tested. A growing literature has suggested that non-use values (bequest and existence values) and option value should be counted as part of the total economic value of the natural resource. We analyse the concept of total economic value and valuation methods in Chapters 9 and 10.

1.8 SUSTAINABLE ECONOMIC GROWTH AND DEVELOPMENT

The re-birth of environmentalism in the 1960s was confined to the industrialised countries of the North. In the developing countries of the South, environmental policies, over and above a concern for basic necessities, were regarded as unaffordable luxuries. It was not until 1972, with the Stockholm Conference on the Human Environment, that a milestone was reached in the development of international environmental policy. It resulted in the establishment of the United Nations Environment Programme and the creation of national environmental protection agencies in the economies of the North. In the years that followed, developing countries, while pressing for the establishment of a new 'International Economic Order', also came to realise that the health of the environment should concern them as much as it did the industrialised countries.

In 1980, the US *Global 200 Report* (Barney, 1980) appeared to confirm environmental prophesies about the consequences of the neglect of the global 'common interest' and the over-exploitation of open-access resources. But in a re-run of the original *Limits to Growth* debate based around the Meadows report, *Global 2000* stimulated a 'cornucopian technocentrist' backlash and the publication of *The Resourceful Earth* report in 1984 (Simon and Kahn, 1984).

The rejection of the physical limits to growth thesis, the appropriate role of market forces in the development process, the role of poverty in natural resource degradation and the need to recognise and build on common interests, are all themes that reappear in highlighted form in reports such as *Our Common Future* (WCED, 1987) and *The Global Possible* (Repetto, 1985). In these documents it is accepted, in principle, that the world's resources are sufficient to meet long-term human needs. The critical issues under debate, therefore, concern the uneven spatial distribution of population relative to natural carrying capacities, together with the extent and degree of inefficient and irrational uses of natural resources.

The 1980s have also seen a re-orientation of some environmental thinking. The term *sustainability* has appeared in a range of contexts and probably most prominently in the *World Conservation Strategy* (IUCN, 1980). Underlying some sustainability thinking is an increased recognition that knowledge accumulated in the natural sciences ought to be applied to economic processes. For instance, the scale and rate of throughput (matter and energy) passing through the economic system is subject to an entropy constraint. Intervention is required because the market by itself is unable to reflect accurately this constraint. Modern economics lacks what we call an *existence theorem*: a guarantee that any economic optimum is associated with a stable ecological equilibrium (see Chapter 2). The Pareto optimality of allocation, for example, is independent of whether or not the scale of physical throughput is ecologically sustainable. There is a risk that some 'ecologically relevant' externality situations may involve damage to the ecosystem itself. Such externalities can cause false signals to regulatory authorities in such a way that sustainability conditions are not fulfilled.

A working definition of sustainable development might be as follows: it involves maximising the net benefits of economic development, subject to maintaining the services and quality of natural resources over time. Economic development is broadly construed to include not just increases in real per capita incomes but also other elements in social welfare. Development will necessarily involve structural change within the economy and in society. Maintaining the services and quality of the stock of resources over time implies, as far as is practicable, acceptance of the following rules:

(a) Utilise renewable resources at rates less than or equal to the natural rate at which they can regenerate.
(b) Optimise the efficiency with which non-renewable resources are used, subject to substitutability between resources and technological progress.

Economic development and natural resource maintenance are related in the following two broad ways:

1. Up to some level of resource base utilisation there is likely to be a trade-off between development and the services of the resource base (complementary relationship).
2. Beyond this level, economic development is likely to involve reductions in one or more of the functions of natural environments – as inputs to economic production, a waste assimilation service and recreation/amenity provision. In this

trade-off context, the multifunctionality of natural resources is a critical concept. We analyse sustainability more formally in Chapter 3.

1.9 ECOLOGICAL AND CO-EVOLUTIONARY ECONOMIC PARADIGM

Thinkers within this paradigm (Norgaard, 1984) have seriously questioned the validity of either biological determinism or cultural determinism as explanations of development and change. Reality is more complex and dynamic. There is a constant and active interaction of the organism with its environment. Organisms (especially humans) do not merely receive a given environment but actively seek alternatives or change what they find. Organisms are not simply the results but are also the causes of their own environments.

Economic development can therefore be viewed as a process of adaptation to a changing environment, while itself being a source of environmental change. From this perspective there are three distinct sources of change – the breakdown of ecological equilibrium (i.e. any combination of a method and a rate of resource use which the environment can sustain for long periods), the demands of technical consistency, and the development of new forms of need as the real costs of living are changed – none of these alone explain all change. Development is then a process of moving through a succession of ecological niches. Niche occupancy is variable, and a niche may be destroyed by means external to a society's own development process.

Over time the development process results in an increasing level of environmental exploitation. The available stock of low entropy is diminished by resource extraction and waste generation. Economic production systems become more roundabout and complex as development proceeds. Work done by natural scientists on dissipative structures is relevant to the management of complex systems. Again it appears that the evolution of such systems is neither entirely deterministic nor entirely stochastic, but a subtle mixture of both.

The *co-evolutionary perspective* has been designed to provide a link between ecological and economic analysis. Co-evolution refers to any on-going feedback process between two evolving systems.

During co-evolution, energy surpluses are generated within systems, and these are then available for stimulating new interactions between systems. If the interactions prove favourable to society the development process continues. But co-evolutionary development feedback systems frequently shift from the ecosystem to the sociosystem, i.e. production systems become more roundabout and complex and environmental exploitation increases. Since learning, knowledge and evolution are interrelated, additional co-evolutionary development potential remains untapped. However, the magnitude and extent of this development potential, which will determine how tolerable survival will be, remain uncertain.

The physical limits to growth are manifestations of the increasing complexity of the productive system. Individual subsistence requirements depend on the technology and culture of contemporary society. As complexity in the social system increases so do subsistence requirements. Preferences change because the context in which individuals learn their preferences is changed by development. This explains the 'Easterlin paradox' (i.e. survey data indicating no close correlation between material affluence and human happiness), along with explanations couched in terms of the interdependency of preferences (relative income hypothesis) and 'positional goods'. Steady-state models can be seen to be advocating the deliberate selection of a new niche of a particular kind, one which is thought would offer the prospect of prolonged occupancy (sustainability).

Some advocates of the need for more sustainable economic development strategies call for more government intervention in order to impose, for example, increasingly stringent environmental quality protection regulations. They also argue, as we have seen, that the Pareto optimality is independent of whether or not the scale of physical throughput in the economy is ecologically sustainable. But this intervention view also requires the acceptance of some global moral operative about survival, because there is the possibility that otherwise we may well be trading off survival at a lower 'quality' of life against a shorter time horizon with a higher 'quality' of life.

1.10 CONCLUSIONS

A brief overview of alternative ways of thinking about natural environments can do no more than touch on salient issues. Three in particular stand out.

First, while economists typically acknowledge the fact that there are several varied objectives possessed by individuals and by societies generally, they tend to work with only one – economic efficiency. One possibly powerful reason for this is that it may well dominate in many of the contexts that economists have typically analysed – markets in goods and services, for example. Translating that objective to the supply and demand for non-marketed goods, which is the typical context for environmental assets, permits the application of the economist's tools of the trade – optimising, marginal analysis and so on. But it may well be that finding the economically 'optimal' provision of environmental assets is not the dominant concern of individuals or societies. Issues such as fairness of access to those assets, both within a time period and over time, could be of equal and possibly greater significance. Similarly, uncertainty about the role which natural environments play in providing for the quality of our lives, and even for its very existence, might make us wary of engaging in a standard comparison of costs and benefits. Making allowance for the different objectives that people have with respect to environments could therefore alter the perspective that neoclassical economics places on those environments.

Second, some alternative views raise the issue of whether preference-based systems are relevant *at all* to the analysis of environmental issues. If habitats and their non-human occupants have some form of 'intrinsic' value, unrelated to the *act* of preferring or dispreferring by humans, then we face the problem of how to account for those values. That is, we may have values that are *of* people *for* a given environmental asset, and values that somehow reside *in* that environment and that are not *of* people. The extent to which this distinction poses a real problem depends on several factors. One is the extent to which people actually do capture some of this intrinsic value by expressing preference *on behalf* of species and habitats – we raise this issue again on p. 134 when we discuss existence value. If this is a feature of human preference expression, then the distinction between intrinsic and subject–object values may well remain valid but not be of great importance for real-world decisions. Another problem is the extent to which any prescription for social action can be based on 'rights' not possessed by humans. An animal rights campaigner would have no difficulty in affirming that human action should be guided in part by the rights of non-human beings. Others would query whether it makes sense to speak of 'rights' outside of the province of human attributes.

Third, much of the literature that questions the role of economic analysis and the environment does so, we believe, because it is not convinced that economics has come to terms with ecological conditions for sustainability. Put another way, economics does not appear to have an 'existence theorem' which enables us to be sure that whatever economy we devise will be sustainable ecologically. The only way to be sure of this sustainability is to ensure that economic models have sustainability conditions built into them. Beliefs about what would happen if we did this have clearly varied markedly over time. The 'limits to growth' school of thought would argue that such an extended economic–ecological model would show that economic growth, as traditionally understood, would not be sustainable. Others would argue that growth may be perfectly feasible but that the configuration of growth, the way in which it is achieved, would differ from the patterns we observe today. As an example, it might be growth based more on the sustainable use of renewable natural resources and less on exploitative use of exhaustible resources. But it might also be a 'hi-tech' economy in which growth is based on very low resource inputs and high technical progress: the *presumption* in much of the literature that an ecologically constrained economy is a low growth, austerity economy need not be true at all.

Our belief is that only by improving substantially our understanding of economy–environment interactions will we get a better grasp of these wider issues.

2 · THE CIRCULAR ECONOMY

2.1 NARROW AND HOLISTIC VIEWS OF ECONOMIES AND ENVIRONMENTS

Undergraduate economics textbooks now pay some attention to issues of environmental economics. But, typically, this attention is confined to supplying an 'add on' chapter illustrating how the theory in the rest of the book can be applied to environmental issues. The danger in this approach is that it obscures the fundamental ways in which the consideration of environmental matters affects our economic thinking.

Figure 2.1 shows a stylised picture of economy and environment interactions. At this stage, the diagram is deliberately vague – we make it more meaningful on p. 35. The upper square, or 'matrix', shows the economy. We consider shortly what might enter into this matrix, but the point for the moment is that economics textbooks are primarily concerned with that matrix only. For example, economics will be concerned with the way in which the various component parts of the economy interact – how consumer demand affects steel output, how the production of automobiles affects the demand for steel, how the overall size of the economy can be expanded, and so on. The lower square shows the environment. This consists of all *in situ* resources – energy sources, fisheries, land, the capacity of the environment to assimilate waste products, and so on. Clearly, there are interactions *within* this matrix as well. Water supply affects fisheries, forests affect water supply and soil quality, the supply of prey affects the number of predators, and so on. Just as within the economy matrix the relationships studied are between economic entities, so within the environment matrix the entities studied appear

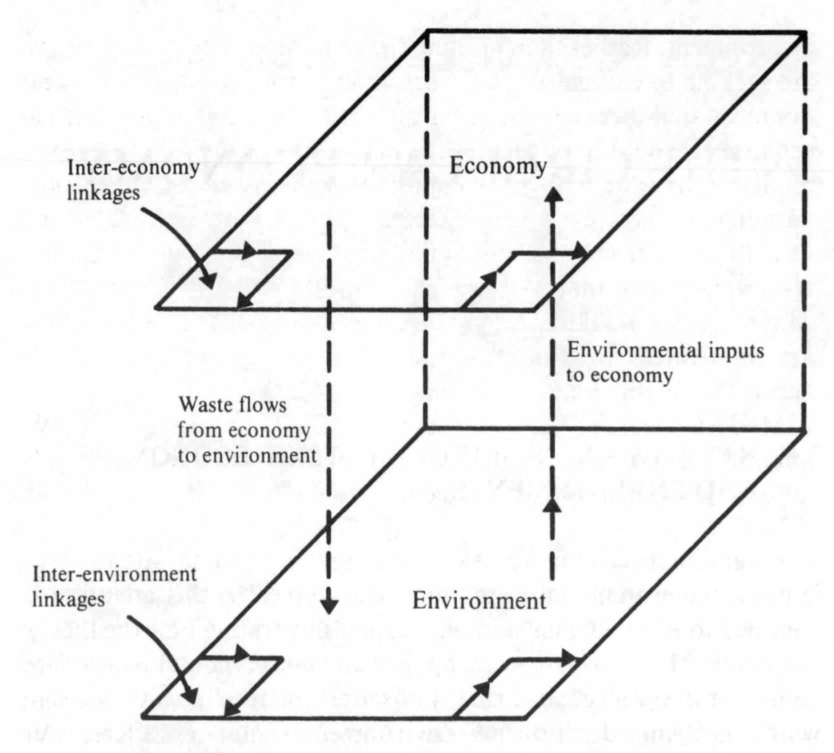

Figure 2.1 General environment–economy interaction.

to have no economic dimension.

Environmental economics is concerned with *both* matrices in Figure 2.1. Moreover, it concentrates on the interactions *between* the matrices – how the demand for steel affects the demand for water, how changing the size of the economy ('economic growth') affects the functions of the environment, and so on. Environmental economics thus tends to be more *holistic* than economics as traditionally construed – it takes a wider, more all-encompassing view of the workings of an economy.

Because it is more holistic there is a temptation to think that environmental economics is somewhat 'better' than economics as it is traditionally taught. This has led some people to think of environmental economics as an 'alternative' economics, as something that is somehow in competition with the main body of economic doctrine. This is a muddled view. In this textbook we show how we can *use* the main body of economic thought to derive important propositions about the linkages between the economy and the environment. Rather than looking for some 'different' economics, we are seeking to expand the horizons of economic thought. This does not mean that there cannot be an 'alternative' economics, but such an economics would have to alter the *paradigms* of the central body of modern economic thought. Chapter 1 has discussed such alternative paradigms. The view taken here, however, is that we have a great deal to learn from our horizon-expanding application of modern economics, and that the search for 'alternatives' is premature. Moreover, we would argue that many of the concerns of those who are motivated to find alternative ways of thinking can be accommodated within the paradigms used in this text.

Modern neoclassical economics is far from faultless, however. We attempt to show what we believe to be true and what we believe to be false in the many critiques available.

2.2 THE ENVIRONMENT–ECONOMY INTERACTION

We now need to make Figure 2.1 more meaningful since we did not specify formally what interactions take place within economies, within environments and between economies and environments. We begin with the economy and then expand the picture to include environments. Figure 2.2 pictures the economy as a set of relationships between inputs and outputs. The diagram looks a little complicated but it is fairly easy to follow. It is a big box, or matrix, made up of a series of smaller boxes or matrices. Notice that two of the categories on the vertical axis – commodities and industries – also appear on the horizontal axis.

We need to define the terms used. A 'commodity' is anything that is processed, exchanged and produced in the economy – a factory is a commodity, so is a machine, so is a TV set or take-away meal. Coal in the ground is not a commodity because it has not been processed nor yet subjected to any exchange within the economy. Industries have a familiar meaning; they are simply the institutions that undertake economic activity in the form of production or providing a service. Figure 2.2 also contains an entry for 'primary' inputs. This refers to labour and capital, but not to land which we treat separately when Figure 2.2 is developed further. 'Final demand' refers to the set of demands in the economy by final consumers, e.g. households. These demands are assumed to be determined by factors outside the

		Commodities 1,2.............N	Industries 1.............M	Final demand 1.............G	Totals
Commodities	1, 2 ... N		A	D	F
Industries	1 ... M	B			G
Primary inputs	1 ... P		C	E	H
Totals		K	L	M	J

Figure 2.2 An input–output table without environment.

model – they are said to be 'exogenous'. The numbers in each small matrix simply remind us that each matrix has a number of component parts – for example there are *M* industries, *N* commodities, *G* final demands, and so on. For our purposes we need not worry further with these numbers.

The relevant matrices have been labelled. Matrix A shows the *input* of commodities to industries. So, for a given industry, say steel, this matrix will tell us how much is required of each other commodity used in the production of steel. Matrix B shows the *output* of each commodity by each industry. Matrix C shows how much each industry spends on primary inputs – labour and capital. Matrix D shows the final demand for commodities, i.e. how much of each commodity is required to meet each type of final demand. Matrix E shows the expenditure on each primary input according to each category of final demand.

This leaves us with the column and row titled 'totals'. These are not actually matrices in the sense we have been using. For example, box F shows the total demand for commodities and this is made up of industrial demand for commodities (matrix A) and final demand for commodities (matrix D). But it will appear as a single list of demands classified by the N commodities. This list is known as a 'vector'. So, it might appear as x units of commodity 1, y units of commodity 2, z units of commodity 3, and so on. Box G shows the total outputs of each industry. It too is a vector. Vector H shows the total expenditure on primary inputs and is found by summing the elements in C and E. Vector K is the total output of commodities, vector L shows total inputs to industry, and vector M shows total expenditure on all inputs by category of final demand. The last box is J and that shows the total expenditure on all commodities and all primary inputs. It is neither a vector nor a matrix but a single number – a 'scalar'.

What use is a construct like Figure 2.2? First, we need to observe that it is a particular form of an *input–output table*. By showing the interactions within an economy, input–output tables have considerable potential value for planning purposes. If, for example, the government decides to expand final demand by inflating the economy, it is helpful to know what this will mean for the demand for labour, the demand for steel, the demand for coal and so on. Second, in ways which are beyond the scope of our interest here, it is possible to modify input–output tables in such a way that we can estimate the price impacts of changing certain key features in the economy. If we decide to raise energy prices, for example, we can show the impact on the costs of energy-using industries. This might not seem to require an input–output table. For example, if steel uses X tonnes of oil and we raise the price of oil it must surely be the case that the cost of producing steel rises by X times the increased price of oil. But we have overlooked the fact that there are other inputs to making steel, e.g. coke, which also require energy, so its price will rise too. Input–output, or I–O analysis, helps us trace these second-order effects. It is even possible to say by how much the living costs of the average family will rise, and so on.

But our interest is in the environment. Enough has been said to hint at the uses that the I–O approach might have in this context. If it were possible to introduce environmental functions into the picture then we could see how much each economic change would impact on the environment. Figure 2.3 expands Figure 2.2 in order to show this. Basically, we take Figure 2.2, and add on an extra row and an extra column. The extra row is 'environmental commodities'. This refers to

	Commodities	Industries	Final demand	Total	Waste discharge to environment
Commodities		A	D	F	N
Industries	B			G	O
Primary inputs		C	E	H	
Totals	K	L	M	J	P
Environmental commodities	Q	R		S	

Figure 2.3 An extended input–output table with environmental commodities.

all natural resources – classified here as land, air and water. In land we include natural commodities such as coal and oil, fish and forests. The environmental commodity flow will basically show us how the environment supplies *inputs* to the economy. The column that is added is the same – land, water and air – but this time it will show us how these resources act as *receiving media* for the waste products that flow from the economy. Later, we will elaborate on some important relationships between the environment as input and the environment as receiver of waste (p. 36).

We now have some extra boxes to explain. One thing to note is that all our economy boxes in Figure 2.2 were in money terms – that is, if we actually constructed such a table it would show us, for example, the money value of steel as an input to £1 or $1 of automobile output. Although major advances have been made in putting money values on some of the functions of environments, in terms of Figure 2.3, it must be recognised that the new row and column will be in *physical* terms, i.e. tonnes of sulphur oxide, tonnes of coal, etc. Matrix N now shows the amount of waste discharged as a result of the final demand for commodities measured in box F. Matrix O shows the discharge of waste products by each industry. Box P will be a vector and will show the total amount of waste discharged by the economy, classified by type of waste. Matrix Q shows the inputs of environmental commodities to economic commodities – e.g. how much water is used, how much land is used, and so on. Matrix R shows the inputs of environmental commodities to industries and box S will show the total input of environmental commodities to industrial and final demand.

Effectively, what Figure 2.3 does is to formalise the general relationships introduced in Figure 2.1. If it were possible to quantify the various relationships between environmental commodities and the economy, then we would have a clearer idea of how economy and environment interact. Some efforts have been made to do exactly this and treatment here has followed that of Victor (1972) which showed how the interactions occur in the Canadian economy. However, our purpose in introducing the idea of input-output analysis is rather different. How far one can quantify the interlinkages in detail is not our concern, although it should be evident that advances in this area could be very fruitful. The basic aim has been to show that economy and environment are linked in various ways and that, in principle at least, it is possible to model these linkages by extending a piece of analysis – input–output – that was initially developed for purposes quite unrelated to the environment. It also permits us to reflect on just what the environment does for the economy.

2.3 THE CIRCULAR ECONOMY

The previous discussion highlights some important implications of the environment–economy interaction for our conception of how economies work. If we ignore the environment then the economy appears to be a *linear system*. Production, *P*, is aimed at producing consumer goods, *C*, and capital goods, *K*. In turn, capital goods produce consumption in the future. The purpose of consumption is to create 'utility', *U*, or welfare.

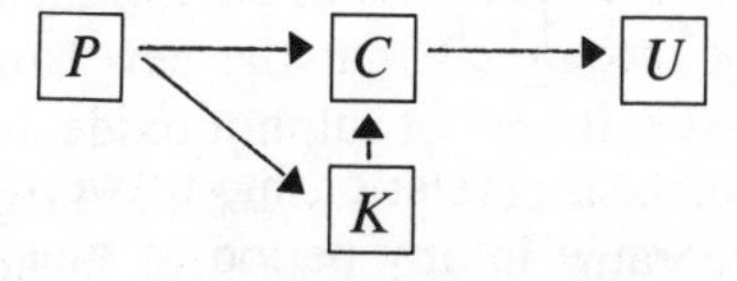

Leaving out *U* and *K*, for convenience, we can immediately add in the flow of *natural resources*, *R*, to give a more complete picture.

Resources are an input to the economic system, just as we saw in Section 2.2. Adding resource still produces a linear system:

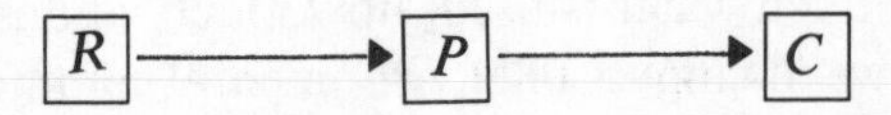

This system, however, captures the first function of natural environments, namely, to provide resource inputs to the productive system.

The picture is still incomplete because it says nothing about *waste products*. A moment's reflection will show that natural environments are the ultimate repositories of waste products: carbon dioxide and sulphur dioxide go into the atmosphere, industrial and municipal sewage goes into rivers and the sea, solid waste goes to landfill, chlorofluorocarbons go to the stratosphere, and so on. Waste comes from the economic system but we should not be led into believing that natural systems do not have their own waste. Trees dispose of their leaves, for example. This is waste. The basic difference between natural and economic systems, however, is that natural systems tend to recycle their waste. The leaves decompose and are converted into an organic fertiliser for plants and for the very tree creating the waste in the first place. Economies have no such in-built tendency to recycle. It seems fair therefore to concentrate on wastes from the economy in extending our picture of economy–environment interaction.

Waste arises at each stage of the production process. The processing of resources creates waste, as with overburden tips at coal mines; production creates waste in the form of industrial effluent and air pollution and solid waste; final consumers create waste by generating sewage, litter, and municipal refuse. So, we might take the linear system and expand it a little further:

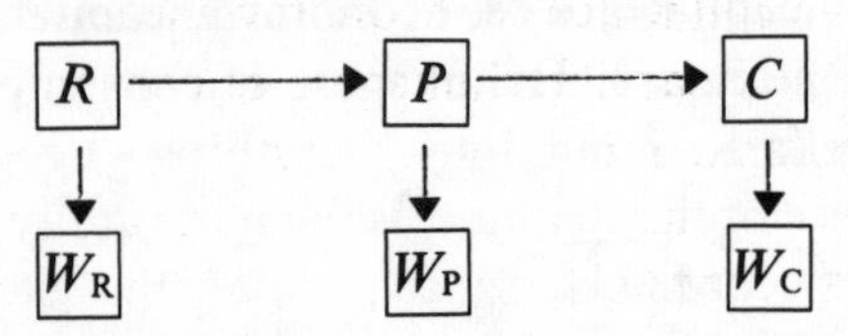

Now, as it happens there is an interesting relationship between R and the sum of the waste flows generated in any period of time. If we forget for the moment about production going to create capital stock, then *the amount of waste in any period is equal to the amount of natural resources used up*. That is

$$R = W = W_R + W_P + W_C$$

The reason for this equivalence is the *First Law of Thermodynamics*. This law essentially states that we cannot create or destroy energy and matter. Whatever we use up by way of resources must end up somewhere in the environmental system. It cannot be destroyed. It can be converted and dissipated. For example, coal consumption in any year must be equal to the amount of waste gases and solids produced by coal combustion. Some of it will appear as slag, some as carbon dioxide and so on. This equivalence is not a hard and fast one once we consider capital formation, for then some of the resource flows become 'embodied' in capital equipment. But, at the same time, capital equipment constructed in past periods will be wearing out, so it will appear as a waste flow. In any given period, then, we shall have a more complicated relationship between R and W.

The relevance of the First Law of Thermodynamics was given prominence in one of the most celebrated and justly famous essays of the twentieth century. 'The economics of the coming spaceship Earth' was written in 1966 by Kenneth Boulding. Boulding's conception was of planet Earth as a 'spaceship'. If we think of a spaceship going on a long journey it will have only one external source of energy – solar energy. It will have a stock of resources depending on whatever was put aboard before take-off. But as that stock is reduced, so the expected lives of the spacemen are reduced unless, of course, they can find ways to recycle water and materials and generate their own food. The spaceship is, of course, Earth and Boulding's essay was pointing to the need to contemplate Earth as a closed economic system: one in which the economy and environment are not characterised by linear interlinkages, but by a *circular* relationship. Everything is an input into everything else. Simply saying that the end purpose of the economy is to create utility, and to organise the economy accordingly, is to ignore the fact that, ultimately anyway, a closed system sets limits, or boundaries, to what can be done by way of achieving that utility.

The linear system can now be converted to a circular system in light of Boulding's contribution. We now have:

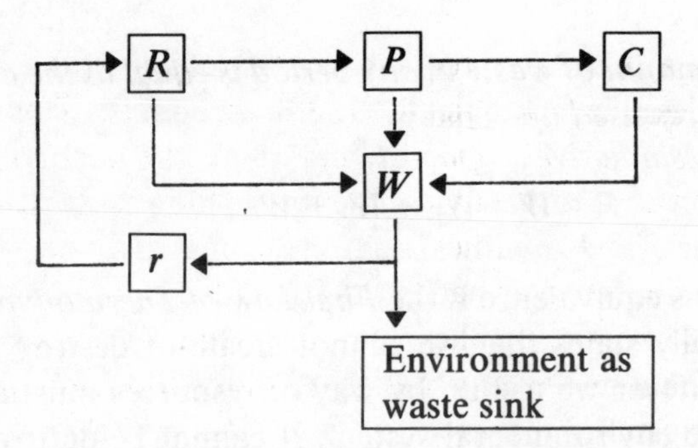

The box *r* is recycling. We can take some of the waste, *W*, and convert it back to resources. We are all familiar with bottle banks for recycling glass bottles. The lead in junked car batteries is generally recycled. Many other metals are recycled. Some waste paper returns to be pulped for making further paper, and so on. But a great deal of waste, indeed the majority of it, is not recycled. As the diagram shows, it goes into the environment.

Why is not all waste recycled? It is here that the *Second Law of Thermodynamics* becomes relevant. Boulding drew attention to the second law, but another economist, Nicholas Georgescu-Roegen, has been the most prolific and forceful advocate of the second law's relevance to economics. In terms of the circular flow diagram above there is a basic reason for the lack of recycling, apart, that is, from missed opportunities. The materials that get used in the economy tend to be used *entropically* – they get dissipated within the economic system. Of the many hundreds of components in a car it is possible to recycle only a few of them – maybe the aluminium in some parts, the steel in the car body, lead from the batteries. The wood and plastics are generally impossible to extract without the expenditure of such large sums of money that it would not make any sense. In other cases it is not technically feasible to recycle. Think of the lead in leaded gasoline. It cannot be captured from the car exhaust and returned to the economic system. Moreover there is a whole category of resources that cannot be recycled – energy resources. Even if we capture the carbon dioxide from burning fossil fuels, it does not create another fuel. We can capture some of the sulphur oxides and recycle the sulphur, but we cannot recycle energy. Entropy therefore places a *physical* obstacle, another 'boundary', in the way of re-designing the economy as a closed *and sustainable* system.

Now consider what happens to that proportion of the waste flow that we cannot recycle. It goes into the environment. The environment has a capability to take wastes and to convert them back into harmless or ecologically useful products. This is the environment's *assimilative capacity* and it is the second major economic function of natural environments. So long as we dispose of waste in quantities (and qualities) that are commensurate with the environment's assimilative capacity, the circular economic system will function just like a natural system, although, of course, it will still draw down the stocks of any natural resources that do not renew themselves ('exhaustible' resources). The system will therefore still have a finite life determined by the availability of the exhaustible natural resources and other considerations we shall shortly introduce. But if we dispose of wastes in such a way that we damage the *capability* of the natural environment to absorb waste, then the economic function of the environment as waste sink will be impaired. Essentially, we will have converted what could have been a renewable resource into an exhaustible one. The assimilative capacity of the environment is thus a resource which is finite. So long as we keep within its bounds, the environment will assimilate waste and essentially return the waste to the economic system.

The resources box, R, in the diagram can be expanded to account for two types of natural resource. *Exhaustible resources (ER)* cannot renew themselves and include such resources as coal, oil, and minerals. *Renewable resources (RR)* have the capacity to renew themselves. A forest produces a 'sustainable yield', so that if we cut X cubic metres of timber in any year, the stock of trees will stay the same as long as the trees have grown by X cubic metres. The same is true of fish. Some resources are mixes of renewable and exhaustibles – soil would be one example. Some renewable resources are very slow-growing, some are fast-growing. Clearly, if we harvest a renewable resource at a rate faster than the rate at which it grows, the stock will be reduced. In this way a renewable resource can be 'mined', treated like an exhaustible resource. If we wish to sustain renewable resources we must be careful to harvest them at a rate no greater than their natural regenerative capacity. The resource sub-sector now appears as:

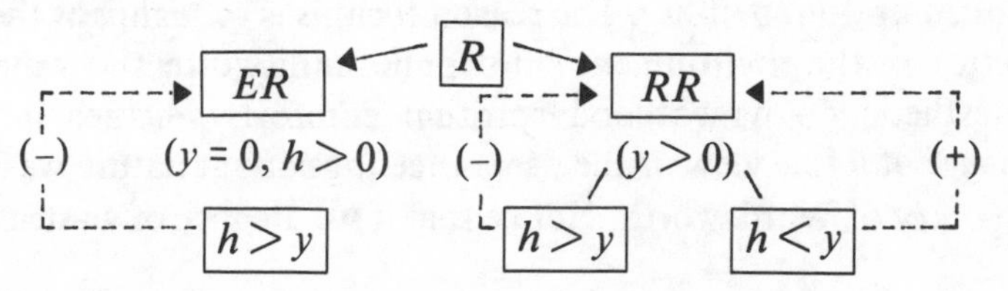

where y refers to the yield of the resource, and h to the rate at which it is harvested (extracted, exploited). The plus sign tells us that if $h < y$ the resource stock grows, and if $h > y$ the stock falls (the minus sign).

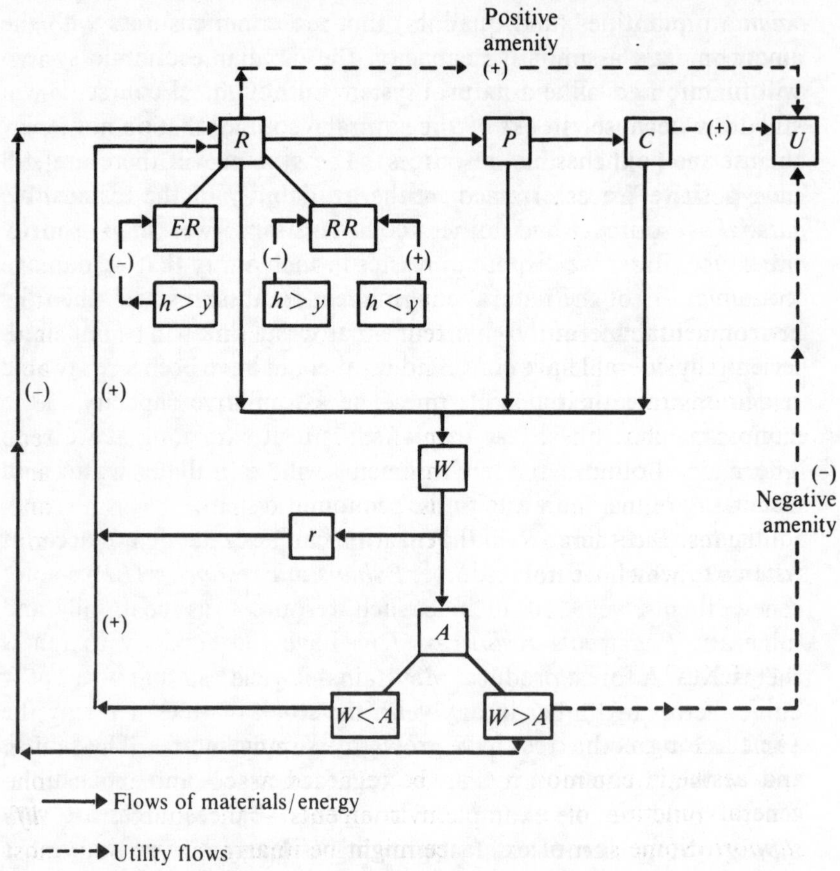

Figure 2.4 The circular economy.

We are now in a position to complete our picture of the circular economy. Instead of being an open, linear system, it is closed and circular. The laws of thermodynamics ensure that this must be so. In Figure 2.4 we show the full picture. We have added back in the flow of consumption to utility. The reason for this is to highlight the third function of the environment – it supplies utility directly in the form of aesthetic enjoyment and spiritual comfort, whether it is the pleasure of a fine view or the deeper feelings about nature we find in the poetry of Wordsworth. Notice that if we dispose of wastes, W, in excess of the assimilative capacity, A, of the environment, we shall damage this third function. Polluted rivers detract from this economic function.

By looking at this circular flow, sometimes called a *materials balance* model, we have been able to identify clearly three *economic* functions of the environment – as resource supplier, as waste assimilator, and as a direct source of utility. They are economic functions because they all have a positive economic value: if we bought and sold these functions in the market-place they would all have positive prices. *The dangers arise from the mistreatment of natural environments because we do not recognise the positive prices for these economic functions.* This is not the fault of economics or economists (although it is often made out to be in the environmental literature). Indeed, environmental economists have been at considerable pain to point out these economic functions and to demonstrate their positive price. Nor is it intrinsic to modern economics that these economic functions should be ignored. Ignorance of economic functions lies elsewhere in the personal and social aims of individuals, groups, communities, pressure groups and politicians. But there is a problem with the perception of economic systems to which we now turn.

2.4 EXISTENCE THEOREMS

The three economic functions, resource supply, waste assimilation and aesthetic commodity, can be regarded as components of one general function of natural environments – the function of *life support*. Some sort of existence might be imaginable without most natural resources, though not without all of them. But for the foreseeable future we need to survive and, more so, we need them to fulfil human values. The problem we face is that the design of economies – whether free market, planned, or mixed – offers us no guarantee that the life support functions of natural environments will persist. Modern economics spends quite a lot of time trying to determine whether equilibria *within* the economic system exist – for example, whether we can have equilibria between supply and demand in money markets, goods markets, and labour markets and whether there is some set of market-clearing prices which will secure all these equilibria.

But we seem to have no comparable analysis that demonstrates whether any particular economy is consistent with the natural environments which are necessarily linked to that economy. They are consistent in one sense – economies exist and natural environments exist. What we do not know is what needs to occur for them to *co-exist* in equilibrium. We do not have an *existence theorem* that relates the scale and configuration of an economy to the set of environment–economy interrelationships underlying that economy. Because we have no such theorem, our planning of the workings of economic systems – and 'planning' here includes letting the economy operate with free markets – risks the running down, the depreciation of the natural environment's functions. Economies may survive, and may survive for long periods of time in such states of disequilibrium. But if we are interested in *sustaining* an economy, it becomes important to establish some conditions for the compatibility of economies and their environments. This is an issue that we consider in Chapter 3.

3 · THE SUSTAINABLE ECONOMY

3.1 RULES FOR SUSTAINING CLOSED ECONOMIES

Chapter 2 showed how the 'open' linear economy construct of modern economics textbooks needs to be revised to allow for the economic functions of natural environments and the thermodynamic equivalences between resource extraction and waste discharges. The development of this 'closed' model of the economy-with-environment immediately raised a broad question about the capability of the natural environment to sustain the economy. To sustain something means making it last, to keep it in being and make it endure. Sustaining an economy does not just mean keeping it in existence: it may be comparatively simple to have an enduring economy in which the standard of living declines over time. Few people would disagree with the idea that the economy needs to change over time in order to improve that standard of living. How we define 'standard of living' is a moot point. It clearly cannot be something as single-valued as real income per capita. Equally, we cannot deny the important role that real income plays in improving the happiness of people. We might therefore think of the standard of living as a set or 'vector' of components – the utility from real incomes, education, health status, spiritual well-being, and so on. Some would stress one component more than the other, but precisely how our standard of living objective is interpreted does not matter for the issues in question in this chapter. The issue is, then, *how should we treat natural environments in order that they can play their part in sustaining the economy as a source of improved standard of living*?

Chapter 2 has already suggested some of the guidelines we might use. We observed that the first two functions of the natural environment – resource supplier and waste assimilator – implied certain rules of resource and environmental management if we wished to think of those functions being maintained over lengthy periods of time. These rules were:

1. Always use renewable resources in such a way that the harvest rate (the rate of use) is not greater than the natural regeneration rate.
2. Always keep waste flows to the environment at or below the assimilative capacity of the environment.

Symbolically, the rules are:

(1) $h < y$
(2) $W < A$

If we observe rules (1) and (2) we know that the stock of renewable resources and the stock of assimilative capacity will not fall. Those stocks are therefore available in any future period to sustain the economy still further. Implicit in the rules we have used, therefore, is the idea that *the resource stock should be held constant over time.*

We investigate this idea of holding the resource stock constant in more detail in Section 3.3. For the moment, note some of the many caveats that need to be made even at this stage. First, we have ignored exhaustible resources. In physical terms their stock cannot be held constant unless we use none of them! Second, y and A are not static. We can manage natural resources so as to improve the sustained yield and the waste assimilative capacity: river flow can be augmented, forests can be managed to improve timber yield, pasture and grazing land can be fertilised and seeded, and so on. Third, the rules of environmental management seem to elevate the role of natural resources and the environment's assimilative capacity. The implication is that we somehow cannot manage without them or, at least, we can only manage for some limited (non-sustainable) period of time. It is indeed the case that environmental economics places emphasis on the economic functions of natural environments – the subject would hardly exist if this were not so. But are natural resources essential? This question is considered below.

3.2 COMPLEMENTARITY AND TRADE-OFFS

The rules of environmental management outlined above tend to imply that we should not let the stock of renewable resources and amount of waste assimilative capacity decline. It is a little more convenient to think of assimilative capacity as one more renewable resource, the resource of waste degradation capability. So, the rules reduce to a basic statement that the stock of renewable resources should not decline over time. Since exhaustible resources must, by definition, be exhausted one day, we need to consider how the management rules can be modified to allow for them. Two ways in which they can be integrated into the rules are as follows:

1. To ensure that as exhaustible resources are depleted, their reduced stock is compensated for by increases in renewable resources.
2. To allow for the fact that a given standard of living can be secured from a *reducing* stock of resources.

The first modification allows for *substitutability* between exhaustible and renewable resources. An example might be the substitution of fossil fuel energy by solar, wind, tidal and wave energy sources. The second modification allows for increased *efficiency* of resource use. It is indeed the case that most advanced economies now use less energy to produce one unit of Gross Domestic Product than they did a hundred years ago.

Clearly, our simple management rules are already becoming more complex. The idea of holding the stock of renewable resources at least constant over time in order to ensure sustainability needs modification to allow for offsetting influences: (a) the need to expand renewable resources to compensate for declining exhaustible resource stocks, and (b) the reduced need for *all* resources to sustain a given standard of living (since the efficiency consideration is likely to apply to renewable resources as well). We have no real way of telling which influence is the more important without detailed empirical investigation.

But there is a further factor which has a major influence on the equation: population growth. A given standard of living may be supportable with less resource inputs over time, but if population grows rapidly the effect of the increased demand for resources can quickly 'swamp' such efficiency gains. Since the world's population is growing very fast, and since there is little prospect of slowing it down by deliberate management, resource depletion is more, rather than less, likely. Notice that we do not need to embrace ideas of 'catastrophe' or 'doomsday' when looking at the consequences of resource depletion. Our interest is in whether improvements in the

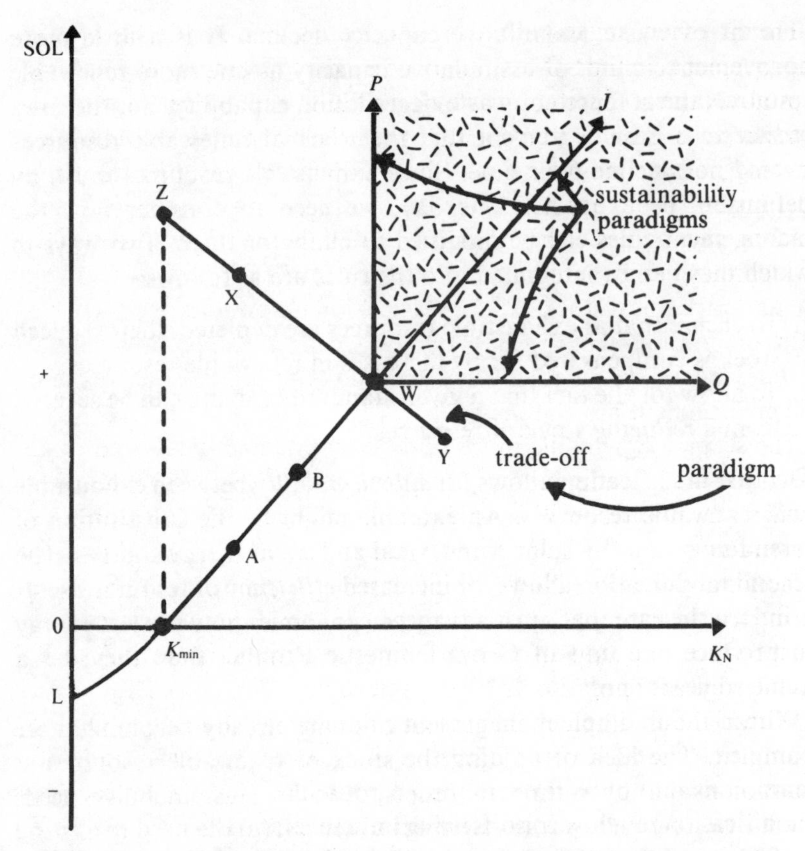

Figure 3.1 Paradigms of environment–standard of living relationships.

average standard of living can be achieved if the stock of natural resources declines.

Figure 3.1 shows how the issue might be characterised. The vertical axis shows the standard of living (SOL) while the horizontal axis shows the stock of environmental resources, or 'natural capital' (K_N). The origin 0 is best interpreted as some positive subsistence standard of living, so that reductions to the negative part of the vertical axis imply serious reductions in livelihoods, nutritional decline, extreme poverty and perhaps starvation at point L; K_{min} corresponds to a minimum level of the natural capital stock necessary to meet the subsistence standard of living. Two extreme views about the relationship between natural capital stock and the SOL can now be illustrated.

The first view suggests that for economies with low levels of K_N, improvements in the SOL can only be achieved by increasing natural capital. Natural capital growth and SOL are to be viewed as *complements*. An illustrative path by which such poor economies can develop, then, is shown by $K_{min}WJ$ which we refer to as a 'sustainability paradigm'. Once the economy has 'taken off' and reaches, say, point W then it might be possible to improve the SOL by operating anywhere in the shaded section *PWQ*, i.e. expanding the natural capital stock or at least keeping it constant. But, while a path of development (raising the SOL) in which natural capital *declines* might be feasible it can only be temporary. Development and environment are complements.

The second view is the more traditional one and is referred to as the 'trade-off' paradigm. What happens here is that the economy is always at a point like W, and development can *only* be secured by surrendering some K_N for improvements in the SOL. The path is something like XWY. If we want more environment, the SOL as defined here must decline. If we want more SOL, natural capital must be reduced. On this view, the same situation occurs at A or B, i.e. the trade-off situation is always relevant.

Within these extremes there is any number of variations. On the assumption that all would agree on the K_{min} concept, the trade-off situation might not be relevant until a point like W has been reached. Then, SOL increases can be achieved by travelling from W to X to Z at which point minimum natural capital stocks are reached again. On this variant, then, environment and development are complements only in the early stages of development. Once take-off is achieved, they become substitutes but with the caveat that they can be traded off against each other only up to a certain limit, and only for certain environmental functions. Thus it is easy to see how environmental *amenity* trades off against development, but the life-support and waste assimilation functions of natural environments are not substitutable. This composite view might be regarded as relevant to modern-day development issues. In the Sahel, for example, it is difficult to envisage development without natural resource augmentation. In the richer nations of the West, development and some environmental services may trade off against each other. The relevance of K_{min}, however, could be significant, as with the effect of trace gases on the ozone layer, the effects of fossil fuel burning on atmospheric carbon dioxide (the 'greenhouse effect'), and so on.

Figure 3.1 is meant to be illustrative only. But is raises the question of how a path like WXZ, in which natural capital declines, is feasible if our interest is in a sustainable economy. We have already suggested one way in which such a path could be consistent with sustainability – increases in the efficiency with which resources are used. The primary source of such changes in efficiency is technological progress. But we should not think of technological change as a 'free good' – it too brings side effects. Fossil fuel burning was a major technological advance, but it has also brought us problems of pollution. We also noted that population growth could adversely affect the chances of moving on to a path like WXZ in Figure 3.1.

One other type of substitution needs to be considered in this context. Economists speak of substituting man-made capital (K_M) – machines, factories, roads – for natural capital. Indeed, traditional economic growth has proceeded on this basis: machines have substituted for animal power, electricity for fuel wood, artificial fertilisers for organic manures, and so on. If this is true natural capital may be inessential for raising the SOL. In terms of Figure 3.1, the path WXZ can be achieved by surrendering K_N for K_M. If the two types of capital were equally productive of increments in SOL we would be indifferent between them, or we might favour K_N for other attributes – its aesthetic qualities for instance – but if K_M can be demonstrated to be more productive, then the choice may favour K_M.

3.3 MAINTAINING THE NATURAL CAPITAL STOCK

The discussion above suggested two reasons why the idea of maintaining the natural capital stock need not, after all, be essential to a sustainable economy: technological change which improves the efficiency of resource use, and the substitution of more productive man-made capital for natural capital. These issues need to be considered further in the context of a more general rationale for conserving and augmenting natural capital.

K_M and K_N substitution

One immediate problem with the distinction between K_M and K_N is that man-made capital is not independent of natural capital. The latter is often needed to make the former. Recall the First Law of Thermodynamics from Chapter 2: this reminded us that to produce anything we must consume some natural resources. The idea of substitution might be rescued if we can demonstrate that the extra productivity in K_M outweighs the extra natural resources that get used up in the production of K_M. At this stage all we can say is that this is not obvious.

The second caveat about K_M and K_N substitution is that natural capital fulfils other economic functions. The natural capital we are talking about includes the world's tropical forests, ocean habitats, wetlands and fisheries, atmosphere and stratospheres, and so on. In all cases there are life support functions which are not served by man-made capital. These include climate regulation, watershed protection and the maintenance of the stock of biological resources. To say that K_M is more 'productive' than K_N thus begs the question to some extent, for it is important to deal with the multifunctionality of natural resources, a feature not shared by man-made capital. To this factor we have to add differences in the pollution profile of the two types of capital: using fossil-based electricity is more polluting than using solar energy. Even the once dreamed-of clean and cheap nuclear energy is no longer regarded in the same light.

The third caveat is that substitutability may not be relevant to all natural resources. Neoclassical economics tends to work with the idea of fairly smooth substitution between inputs. It is because of this substitution that it is possible, analytically anyway, to obtain results which reduce the emphasis we might wish to place on natural resources. But natural resources are not like other resources in that their many functions include their role in, for example, the maintenance of biogeochemical cycles in the environment and on which mankind depends. Only if we can substitute wholesale for these functions can we sustain the idea of trading off between K_M and K_N.

Technological progress

Even if the substitutability between types of capital is brought into question, we are left with technological progress as a way of reducing the natural resource input to SOL generation. There can be no question of the importance of this source of increased efficiency. Past visions of the future in which communication does not require resource-intensive travel by automobile, aeroplane or train are now real as information technology advances. The caveats are two-fold. First, new technology is not necessarily less polluting. Second, will technological progress continue for ever, or at least for a very long time? Mankind's inventiveness shows little sign of abating. If anything it has increased in the twentieth century. But the most optimistic view of the role which technology plays in freeing us from dependence on natural resources depends on some almost indefinitely renewable resource which eventually takes over when exhaustible resources have gone. In the literature this is often called the 'backstop technology'. Several technologies have often been cited as backstop technologies: energy from fast-breeder reactors, energy from fusion reactors, energy from converting shale oil. One observation is that fast-breeder technology has been run down in at least one country (the United Kingdom) because of expense and lack of promise. Fusion reactors appear to be no nearer a backstop technology than they were, and grave doubts have been expressed about their environmental cost. The point is not a categorical one. There may indeed be backstop technologies that will free us from natural resources, but they cannot be brought into existence simply by assuming that they are there.

Sustainability, uncertainty and irreversibility

One of the problems of reaching very definite conclusions about the role which natural environments play in supporting and sustaining economic systems is that we face considerable scientific uncertainty about that role. We do not understand fully how trace gases function in the atmosphere and stratosphere; the chemistry of acid rain is still being developed; the role of ocean currents in climate determination is open to debate, and the ways in which natural forest stands protect soils, rivers and microclimates still need more research. If we could be sure of the benefits of substituting man-made capital for natural capital then the trade-off between them would not be a serious one. But we are not sure of the ways in which environments function, either internally or in terms of their interactions with the economy. Moreover, if we do decide to surrender natural capital there is often a sting in the tail: irreversibility. If we make a mistake, very often we cannot correct it afterwards. Tropical forests cannot be created, feasibly anyway. Desertified land is very difficult to reclaim. Once a species is lost it has gone forever.

The presence of uncertainty and irreversibility together should make us more circumspect about giving up natural capital. As information and understanding increase so the trade-off decision might be made with more certainty about the consequences. Until then, caution should be the order of the day. In terms of Figure 3.1, the trade-off curve ZXWY is flatter: reductions in K_N may achieve only limited sustainable increases in SOL. This issue of valuation of environmental services will occur again several times in this book.

Resilience

Considerable attention has been paid in recent years to the problems faced by the poorest countries in the world. Invariably they rely on natural resources far more directly than those of us in advanced economies. Fuel usually means fuel wood, water comes directly from surface and ground water sources without treatment, shelter requires wood, food supplies depend on subsistence agriculture and hence on soil quality. The sustainability of these societies depends on the maintenance of the stocks of these natural resources. Yet a stock can be maintained and the society can still be non-sustainable because the margin of flexibility is so low. It may require just a few years of drought, one war, one dramatic crisis, for the society to be set back many years in terms of its development prospects. If resource stocks were bigger there would be a greater margin of flexibility to adjust to these external 'shocks'. Since man-made capital is frequently not available in these societies we cannot argue that man-made capital would ensure as much if not more resilience. It might, but the option is not there. In these circumstances, more natural capital can mean more resilience to shocks and hence a more sustainable society.

Intergenerational equity

Another reason for maintaining the resource stock is to ensure broadly equal access to it by different generations. Intergenerational equity relates to the idea of fairness or justice between different generations. If it is accepted as a social goal – an issue that is discussed in Chapter 14 – then maintaining K_N has added force. Once again, it must be recalled that we can create and destroy K_M, but the possibilities of increasing K_N are less. We can grow more trees, set land aside to revert back to wilderness, restock the oceans. But in many cases the environmental losses incurred are irreversible. Any irreversibility now means the removal of an option for future generations – they cannot secure access to the resource if it has been made extinct.

Rights in nature

No one doubts that humans have rights. But do other sentient beings also have rights? The animal rights movement may not appeal to everyone, but it does advocate rights for sentient beings using arguments that cannot be regarded as silly. But if we accept that animals have rights, one of those rights must be to existence in order to exercise other rights. When we destroy natural capital we are invariably destroying habitat, the environments that wild animals require for their existence. Reducing K_N is thus likely to conflict with animal rights, and this deserves consideration as a further argument for protecting K_N. Chapter 14 explores the issue further.

3.4 THE MEANING OF CONSTANT CAPITAL STOCK

Our discussion so far has suggested that sustainability can be analysed in terms of a requirement to maintain the natural capital stock. This requirement ensures that we observe the 'bounds' set by the functioning of the natural environment in its role of support system for the economy. How far it is possible to relax this requirement depends on what we believe about the degree of substitutability between renewable and exhaustible resources, and between man-made capital and natural capital. It also depends on the behaviour of technological progress in reducing the resource input to a unit gain in the standard of living, and on the effects of population growth in dissipating the capital stock. There can be no hard and fast conclusion here, but the issues must be raised if we are to understand better the thrust of some of the modern environmentalist movement. Moreover, we need to express these arguments about sustainable development in terms of the underlying economic concepts.

We have talked about the requirement that the natural capital stock be constant. We have not explained what this might mean.

There are several interpretations. First, we could say that the capital stock is constant if its physical quantity does not change. But we have no way of adding up the different physical quantities (tonnes of coal, cubic metres of wood, litres of water, etc.). The standard economic approach would be to value each type of resource in money terms and compute the overall aggregate money value. If this could be done, in the same way as we make estimates of the 'national wealth' – i.e. the stock of man-made capital – then we could rephrase the K_N requirement in terms of a constant real value of the stock of natural assets.

Second, we could think in terms of the unit value of the services of K_N. That is, we could look at the prices of natural resources and aim to keep these constant in real terms. Provided we are satisfied that prices reflect absolute scarcity – an issue discussed in Chapter 16 – constant real prices will imply a constant natural capital stock in this modified sense. One obvious problem here is that many resources do not have observable prices. We would need to find implicit or 'shadow' prices in some way.

Third, we could think of a constant value of the resource flows from the natural capital stock. This is different from constant prices because we would allow quantity to decline but the price to rise, keeping value constant.

3.5 EXISTING AND OPTIMAL CAPITAL STOCKS

Conserving the natural capital stock is consistent with several situations. The stock in question might be that which exists at the point of time that decisions are being taken – the existing stock – or it might be the stock that should exist. The latter is clearly correct in terms of the application of neoclassical economics principles to resource issues. Economics would argue that there are costs and benefits of changing the natural capital stock. If it is reduced it will be for some purpose. For example, much tropical forest clearance takes place for agricultural purposes. Wetlands are similarly drained to gain the fertile soil for crop growing. Natural habitats are reduced for housing development, and so on. Thus, each destructive act has benefits in terms of the gains from the use to which the land is put. In the same way, using the atmosphere or the oceans as 'waste sinks' has benefits in that alternative means of disposal are often more

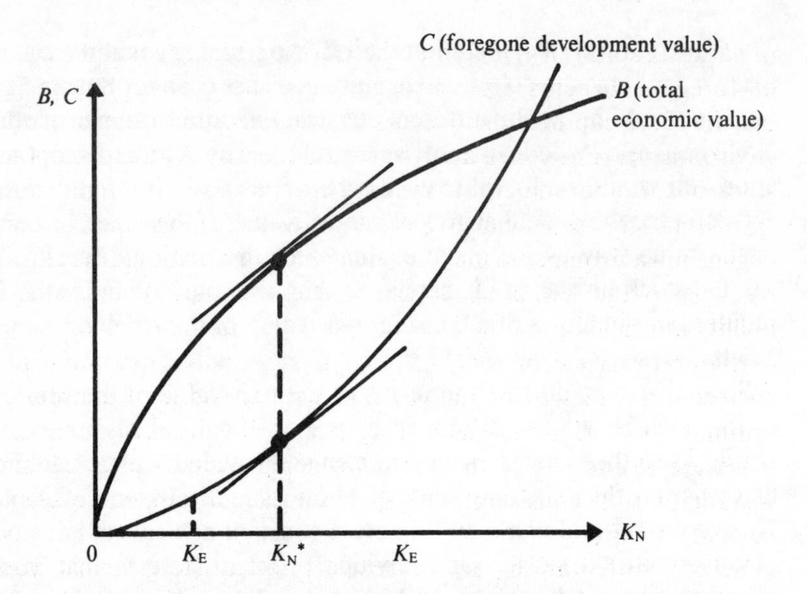

Figure 3.2 The costs and benefits of environmental change. K_N is the natural capital stock. *B* shows the benefits from increasing it, benefits that accrue as use and non-use values. *C* is the cost of increasing the natural capital stock and these stocks are the foregone benefits from using the natural assets for some other purpose. K_N^* is the optimal stock.

expensive. The environment as a waste sink thus reduces production and consumption costs compared to what they would have been. Environmental destruction also has costs since a great many people use natural environments (for wildlife observation, recreation, scientific study, hunting and so on). These 'use benefits' are lost (i.e. there are costs of destruction) if the land is converted for some other purpose. Similarly, one of the benefits of keeping the atmosphere unpolluted is that we avoid the damage that is done by pollution, e.g. better health, and globally, the avoidance of impacts such as global warming through trace-gas emissions. Natural environments do not just have 'use values'. Many people like to think of environments being preserved for their own sake, an 'existence value'. These 'non-use' values need to be added to the use values to get the *total economic value* of the conserved resource or environment (see Chapter 9).

Figure 3.2 depicts the cost–benefit comparison. The stock of natural assets is shown on the horizontal axis and costs and benefits are shown on the vertical axis. The cost curve shows that as the stock of natural capital (K_N) increases there are increasing costs in the form of foregone benefits from *not* conserving the environment. The benefit curve captures the benefits to users and non-users of natural environments. Economic analysis would identify K_N^* as the optimal stock of the environment. If the existing stock is to the right of K_N^* then it will be beneficial in net terms to reduce the stock, i.e. to engage in environmental degradation and destruction. If the existing stock is to the left of K_N^* then improvements in environmental quality are called for.

If our overview of the meaning of sustainable development is correct it appears to be inconsistent with the idea of maintaining optimal stocks of natural assets, or, at least, it will only be consistent if we are to the left of the optimum depicted in Figure 3.2 (since sustainability is consistent with increasing environmental assets) or coincident with it.

Several observations are in order. First, existing stocks would generally be regarded as being below optimal stocks in many developing countries. For some Sahelian countries they are significantly below the optimum in that desertification and deforestation actually threaten livelihoods. Nor is there evidence that the further reduction of soil quality, tree cover or water supplies will result in some form of surplus which can be re-invested in other, man-made capital assets. To some extent, therefore, deliberations about what precisely constitutes an optimum are redundant in the contexts of these countries.

The second observation relates to the identification of the 'optimum' in Figure 3.2. To say that capital stocks 'should' be optimal is tautologous. The interesting feature of optimality is how the benefits of augmenting natural capital are calculated. The critical factor here is that the multifunctionality of natural resources needs to be recognised, including the role as integrated life-support systems. Thus, a cost-benefit analysis that compares the 'value' of, say, afforestation with the opportunity cost of land in terms of foregone development values needs more careful execution than might otherwise appear to be the case. How far life-support functions such as contributions to geochemical cycles can be captured by cost-benefit is open to question. In the face of uncertainty and irreversibility, conserving what there is could be a sound risk-averse strategy. Put another way, even in countries where it might appear that we can afford to reduce natural capital stocks further, there are

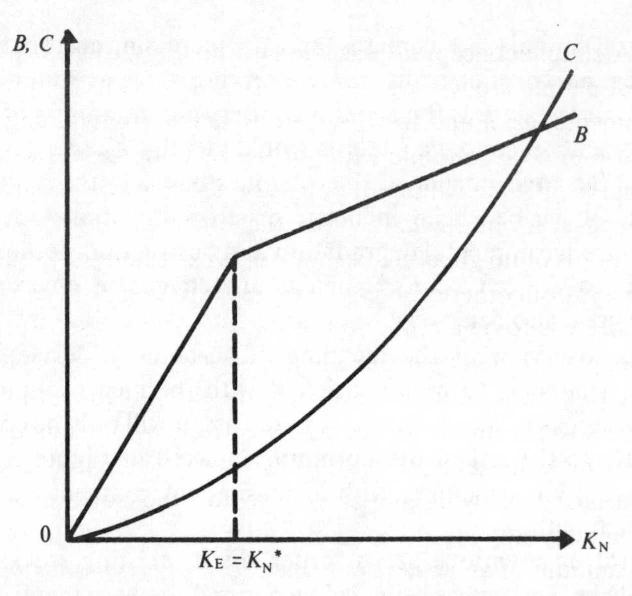

Figure 3.3 Costs and benefits of conservation when the valuation function is kinked. The benefit function of Figure 3.2 is now kinked at the existing stock of natural capital, making the existing and optimal stocks probably coincident.

risks from so doing because of (a) our imperfect understanding of the life-support functions of natural environments, (b) our capability to substitute for those functions even if their loss is reversible in theory, and (c) the fact that losses are often irreversible. There is therefore a rationale in terms of *uncertainty* and *irreversibility* for conserving the existing stock, at least until we have a clearer understanding of what the optimal stock is and how it might be identified.

A third observation is that optimality tends to be defined in terms of economic efficiency, whereas conservation of the natural capital stock serves other social goals. That is, Figure 3.2 is helpful as far as it goes, but it does not embrace the 'non-efficiency' benefits of natural capital stocks. These include serving certain distributional goals, both within current generations and between current and future generations. Of course, we have to be sure that these non-efficiency goals cannot be served better by converting natural capital into man-made capital, an issue to which we return.

A fourth reason for supposing that existing stocks are important arises from recent research on the use of willingness-to-pay and willingness-to-accept measures of benefit (see Knetsch and Sinden,

1984). A simple conceptual basis for estimating a benefit is to find out what people are willing to pay to secure it. Thus, if we have an environmental asset and there is the possibility of increasing its size, a measure of the economic value of the increase in size will be the sums that people are willing to pay to ensure that the necessary land or other asset is obtained. Whether there is an actual market in the asset or not is of no great relevance. We can still find out what people would pay if only there was a market (see Chapter 10). In the same way, if there is to be a reduction in the size of the asset, we can ask what people are willing to accept to give it up. Economic theory predicts that the difference between these willingness-to-pay and willingness-to-accept measures (the 'equivalent and compensating variation' measures of welfare gain) will not differ significantly. That is, a measure of willingness to pay for a small gain will be approximately equal to the requirement for compensation to give up a small amount of an asset. Empirical work suggests otherwise, with very large discrepancies between willingness to pay and willingness to accept being recorded. Prospect theory offers a rationale for compensation requirements being very much larger. Essentially, what exists is seen as a reference point and attitudes to surrendering some of what is already owned or experienced are quite different to those that come into play when there is the prospect of a gain. Put another way, the valuation function B in Figure 3.2 is 'kinked' at the existing stock of assets. The result of modifying Figure 3.2 is shown in Figure 3.3. The existence of the kink means that the optimal level of K_N is likely to be at the point of the kink: existing and optimal natural capital stocks coincide. In terms of the 'constant capital' idea in sustainable development, it implies that a high valuation should be placed on reductions in the existing capital stock, thus supporting the view that conservation of existing stocks itself has a high priority.

Overall, while there is a powerful case in analytical economics for thinking in terms of maintaining optimal rather than existing natural capital stocks as the basic condition for sustainability, there are also sound reasons for conserving at least the existing capital stock. For poor countries dependent upon the natural resource base, optimal stocks will in any event be above the existing stock. In other cases there is a rationale in terms of incomplete information about the benefits of conservation (the failure to appreciate and measure multifunctionality), uncertainty and irreversibility for conserving the existing stock. Additionally, resource conservation serves non-efficiency objectives whereas optimality tends to be defined in terms of efficiency only. Finally, even in terms of efficiency, the existence of a valuation function which is kinked at the existing endowment of natural resources adds emphasis to the conservation of existing stocks.

PART II

THE ECONOMICS OF POLLUTION

4 · THE OPTIMAL LEVEL OF POLLUTION

4.1 POLLUTION AS EXTERNALITY

The economic definition of pollution is dependent upon *both* some *physical* effect of waste on the environment *and* a human reaction to that physical effect. The physical effect can be biological (e.g. species change, ill-health), chemical (e.g. the effect of acid rain on building surfaces), or auditory (noise). The human reaction shows up as an expression of distaste, unpleasantness, distress, concern, anxiety. We summarise the human reaction as a *loss of welfare*. As Chapter 2 indicated, terms such as 'utility' or 'satisfaction' are, for our purposes, synonymous with welfare.

We now need to distinguish two possibilities for the economic meaning of pollution. Consider an upstream industry, which discharges waste to a river, causing some loss of dissolved oxygen in the water. In turn, suppose the oxygen reduction causes a loss of fish stock in the river, incurring financial and/or recreational losses to anglers downstream. If the anglers are not compensated for their loss of welfare, the upstream industry will continue its activities as if the damage done downstream was irrelevant to them. They are said to create an *external cost*. An external cost is also known as a *negative externality*, and an *external diseconomy*. If we were considering a situation where one agent generates a positive level of welfare for a third party, we would have an instance of an *external benefit* (*positive externality*, or *external economy*).

An external cost exists when the following *two* conditions prevail:

1. An activity by one agent causes a *loss of welfare* to another agent.
2. The loss of welfare is *uncompensated*.

Note that *both* conditions are essential for an external cost to exist. For example, if the loss of welfare is accompanied by compensation by the agent causing the externality, the effect is said to be *internalised.* This distinction will be made clearer shortly.

4.2 OPTIMAL EXTERNALITY

The first fundamental feature of the different definitions of externality has already been noted: the physical presence of pollution does not mean that 'economic' pollution exists. The next observation is equally important, but much less easy to understand – *even if 'economic' pollution exists it is unlikely to be the case that it should be eliminated.* This proposition can be demonstrated using Figure 4.1.

In Figure 4.1, the level of the polluter's activity, *Q*, is shown on the horizontal axis. Costs and benefits in money terms are shown on the vertical axis. MNPB is 'marginal net private benefits'. A formal derivation of MNPB, in the context where the polluter is a firm, is given in Appendix 4.1. But an intuitive explanation is also possible. The polluter will incur costs in undertaking the activity that happens to give rise to the pollution, and will receive benefits in the form of revenue. The difference between revenue and cost is *private net benefit.* MNPB is then the marginal version of this net benefit, i.e. the extra net benefit from changing the level of activity by one unit. MEC is the 'marginal external cost', i.e. the value of the extra damage done by pollution arising from the activity measured by *Q*. It is shown here as rising with output *Q*. We consider other possible shapes for MEC in Appendix 5.2.

We are now in a position to identify the *optimal level of externality.* It is where the two curves intersect, i.e. where MNPB = MEC. Why is this? We first offer an intuitive explanation. Since the two curves are marginal curves, the areas under them are 'total' magnitudes. The area under MNPB is the polluter's total net private benefit, and the area under MEC is total external cost. On the assumption that the polluter and sufferer are equally deserving – i.e. we do not wish to weight the gains or losses of one party more than another's – *the aim of society could be stated as one of maximising the sum of benefits minus the sum of costs.* If so, we can see that triangle OXY is *the largest area of net benefit obtainable.* Hence, *Q**

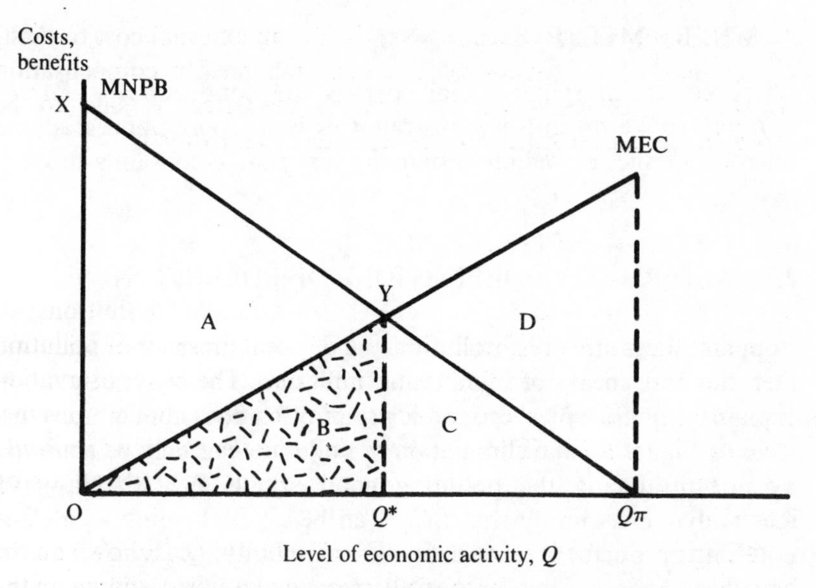

Figure 4.1 Economic definition of optimal pollution.

is the optimal level of activity. It follows that the level of physical pollution corresponding to this level of activity is the optimal level of pollution. Finally, the optimal amount of economic damage corresponding to the optimal level of pollution Q^* is area OYQ^* – area B in Figure 4.1. Area OYQ^* is known as *the optimal level of externality*.

This result can also be derived formally. At Q^*

$$\text{MNPB} = \text{MEC} \tag{4.1}$$

but (from Appendix 4.1)

$$\text{MNPB} = \text{P} - \text{MC} \tag{4.2}$$

where MC is the marginal cost of producing the polluting product. Hence

$$\text{P} - \text{MC} = \text{MEC} \tag{4.3}$$

or

$$\text{P} = \text{MC} + \text{MEC} \tag{4.4}$$

Now, MC + MEC is the sum of the marginal costs of the activity generating the externality. It is *marginal social cost* (MSC). Hence, when

$$\text{MNPB} = \text{MEC}, \text{P} = \text{MSC} \tag{4.5}$$

'Price equals marginal social cost' is the condition for *Pareto optimality*. We do not demonstrate this here – any undergraduate microeconomics or welfare economics text should contain a proof.

4.3 ALTERNATIVE DEFINITIONS OF POLLUTION

Popular literature on pollution, and sometimes the scientific literature too, speaks of 'eliminating' pollution. The above discussion explains why the typical economic prescription does not embrace this idea. In Figure 4.1 the elimination of pollution can only be achieved by not producing the polluting good at all. But, the laws of thermodynamics imply that there can be no such thing as a non-polluting product. Hence to achieve zero pollution we would have to have zero economic activity. Calls for 'no pollution' thus appear illogical.

The situation is not quite as extreme as this, however. We need to modify Figure 4.1 in an important respect if we are to try to make compatible the economist's and the scientist's presciptions about desirable levels of pollution. In Chapter 2 we saw that the natural environments which receive waste products can be characterised as having a certain 'assimilative capacity' – they can receive a certain level of waste, degrade it and convert it into harmless or even beneficial products. If the level of waste, W, is less than this assimilative capacity, A, then some externality will still occur as the process of degradation and conversion takes place. But if W exceeds A a further process of degradation will also occur, for A itself will be impaired. Disposing of waste to environments that cannot handle it simply reduces the capacity of that environment to deal with more waste.

To some extent we can capture this idea of assimilative capacity by observing that the MEC curve in Figure 4.1 should really have its origin at some positive level of economic activity Q_A. Below this level, the only kind of externality will be 'temporary' – the environment will eventually return to normal once the waste degradation process has taken place. On the assumption that we can ignore this temporary externality for the moment, the MEC curve appears as in Figure 4.2. (Note that MEC begins at Q_A only if people

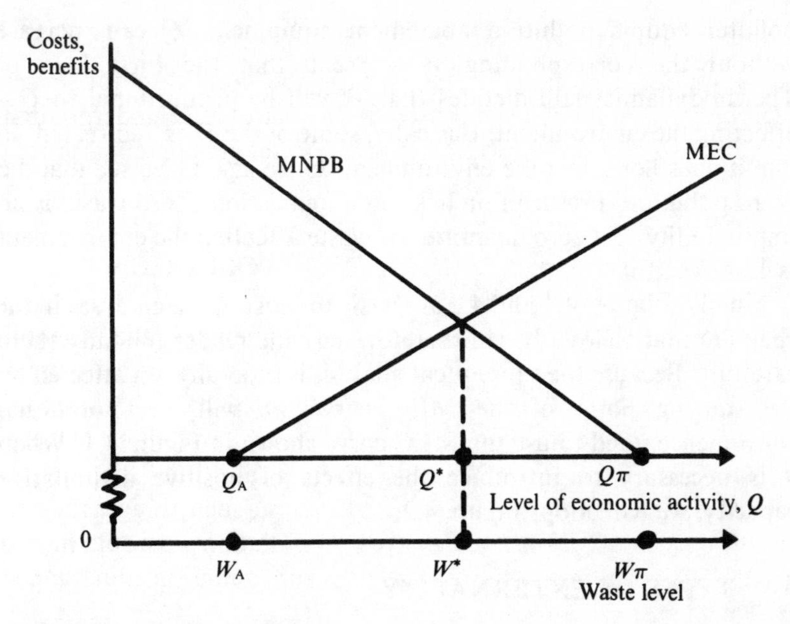

Figure 4.2 Optimal pollution levels with positive assimilative capacity.

notice the physical effects then. Otherwise it can begin even further to the right along the horizontal axis. In the extreme, if people do not care about the physical effects of the waste flows there is no MEC curve at all.)

Figure 4.2 does not alter any of the analysis about the economically optimal level of externality. The findings of the previous section stand. But we can now see that the idea of 'zero pollution' is not, after all, quite so silly as if first appeared. Zero pollution is still non-optimal, as Figure 4.2 shows, but it does not entail zero economic activity. In a static world the difference between the economist's optimum and the scientist's prescription is likely to be significant. As we shall see later in this text, once dynamic considerations are introduced the difference is not so marked, and may not exist at all.

Figure 4.2 also shows how the level of economic activity relates to the level of waste emitted. Assuming waste is directly proportional to the level of activity we can simply translate any amount of Q into some corresponding level of W. Just as Q^* is the optimal level of economic activity, so W^* is the optimal level of waste-producing pollution. Later we shall have occasion to modify this picture: if the polluter adopts pollution abatement equipment, Q can increase without the corresponding W – recall that the First Law of Thermodynamics still dictates that W will be proportional to Q – affecting the environment. Basically, some of the W is 'redirected' so that it does not affect the environment. Once again, we see that the 'zero pollution' prescription has some foundation. Zero waste is an impossibility, but zero quantities of waste affecting the environment is less fanciful.

Finally, Figures 4.1 and 4.2 are basic to most of the analyses in the chapters that follow. It will therefore pay the reader to study them carefully. Because the subsequent analysis is generally not affected by the starting point of the MEC curve we will, for notational convenience, tend to use the MEC curve shown in Figure 4.1. When it is necessary to introduce the effects of positive assimilative capacity, we will adopt Figure 4.2.

4.4 TYPES OF EXTERNALITY

We are now in a position to define some further terms. In terms of Figure 4.1,

Area B	= the optimal level of externality
Area A + B	= the optimal level of net *private* benefits for the polluter
Area A	= the optimal level of net *social* benefits
Area C + D	= the level of *non-optimal* externality which needs to be removed by regulation of some sort
Area C	= the level of net private benefits that are socially unwarranted
Q^*	= the optimal level of economic activity
$Q\pi$	= the level of economic activity that generates maximum *private* benefits

Figure 4.1 thus demonstrates a very important proposition: in the *presence of externality there is a divergence between private and social cost*. If that divergence is not corrected the polluter will continue to operate at a point like $Q\pi$ in Figure 4.1. At $Q\pi$, private benefit is maximised at A + B + C, but external cost is B + C + D. So, net social benefit = A + B + C – B – C – D = A – D, which is clearly less than A, the net social benefits when the polluter's activity is regulated to Q^*.

Externality level C + D is said to be *Pareto relevant* because its removal leads to a 'Pareto improvement', i.e. a net gain in social benefits. Externality level B is *Pareto irrelevant* because there is no need to remove it.

4.5 WHO ARE THE POLLUTERS?

We have deliberately refrained from classifying polluters. The typical 'image' is that polluters are firms. But it is also the case that polluters are individual people – car drivers create noise and cause accidents, people who play radios in and out of doors cause noise nuisance, and so on. Indeed, the general combinations are as follows:

Externality Generator	*Externality Sufferer*
Firm	Firm
Firm	Individuals
Individuals	Firm
Individuals	Individuals
Government	Firm
Government	Individuals

The inclusion of government as a creator of externality acknowledges that governments often generate external effects through poor legislation and rules.

4.6 CONCLUSIONS

1. Scientists tend to define pollution differently to economists.
2. For the economist, pollution is an *external cost* and occurs only when one or more individuals suffer a *loss of welfare.*
3. Even then, economists do not typically recommend the *elimination* of externality because they argue that the *optimal externality* is not zero.
4. The idea of 'zero pollution' is not, however, absurd. At least two considerations make it more reasonable than it appears at first sight. These are (a) the fact that the environment tends to have a positive assimilative capacity, and (b) the fact that it is possible, to some extent, to divorce economic activity from waste flows

affecting the environment by introducing pollution abatement.

5. It is wrong to think of 'polluters' only as firms: individuals pollute. So do governments.
6. Caveat – the analysis in this chapter has assumed *perfect competition.* As we shall see, some of the conclusions do not hold when we relax this assumption.

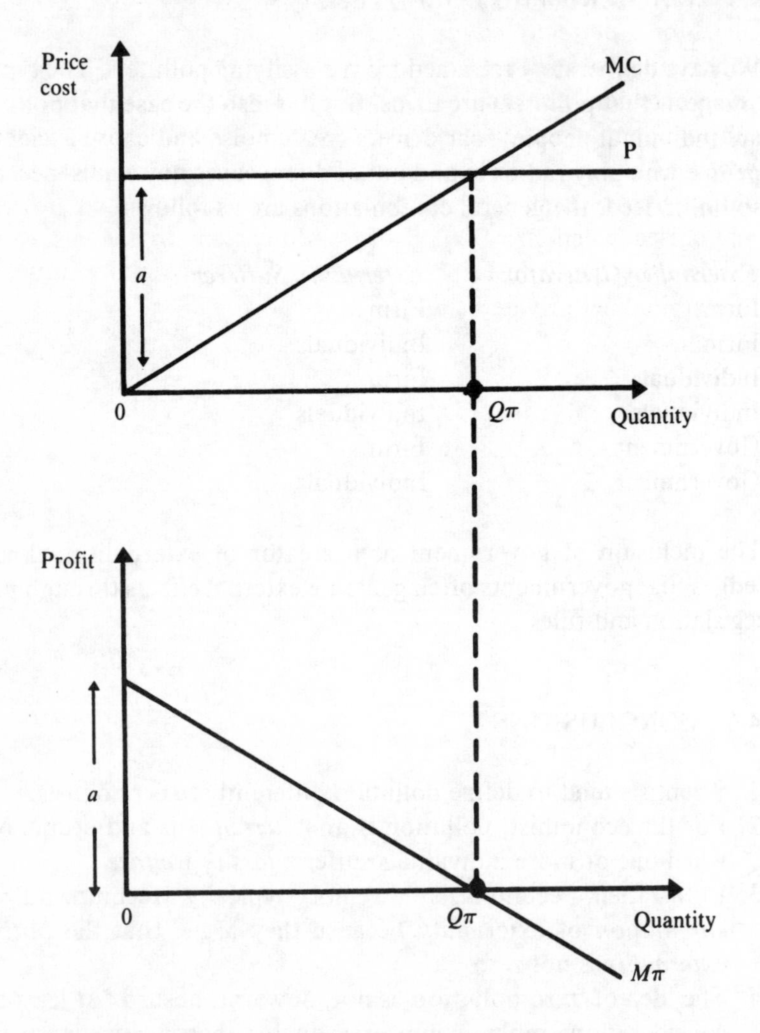

Figure A4.1 Deriving the MNPB curve

APPENDIX 4.1: DERIVING A MARGINAL NET PRIVATE BENEFIT CURVE

Chapter 4 introduced MNPB in a general way. To give it more formal meaning we can look at how it is derived in the context of the theory of the firm. Figure A4.1 shows a demand and marginal cost curve for a *perfectly competitive firm*. (The type of competition is important – we shall see later in the text that the definition of MNPB given here does *not* hold for imperfectly competitive conditions.) By subtracting marginal cost (MC) from price (P), we derive a *marginal profit* curve ($M\pi$). $M\pi$ shows the extra profit made by expanding output by one unit. Clearly, total profits, the area under $M\pi$, are maximised when $M\pi = 0$. Profit is equivalent to the *net benefit* obtained by the firm. Hence, marginal profit is formally equivalent to marginal net private benefits.

5 · THE MARKET ACHIEVEMENT OF OPTIMAL POLLUTION

5.1 PROPERTY RIGHTS

Chapter 4 demonstrated that a socially optimal level of economic activity does not coincide with the private optimum if there are external costs present. The issue arises therefore of how to reach the social optimum. Some form of intervention by government would seem to be necessary. Before looking at the various forms of regulation that might be applied, it is important to probe a little further to be sure that markets will not 'naturally' achieve the optimal level of externality.

It is the contention of one school of thought that even if markets may not secure the optimum amount of externality, they can be very gently 'nudged' in that direction without the necessity for full-scale regulatory activity involving taxes or standard-setting. This basic idea was first propounded in a paper by Ronald Coase (1960). To understand the argument we have first to establish the concept of 'property rights'.

Despite the apparent meaning of the phrase, a property right relates to the right to *use* a resource. This might mean the right to cultivate crops on land that is owned, the right to use one's own house, and the right to use the natural environment in a particular way. Such rights are rarely, if ever, absolute: they are circumscribed in some way by the generally accepted rules of society. The right to cultivate land does not usually carry with it the right to grow opium poppies or even giant hogweed (which is capable of causing quite severe skin irritation). The rights are said to be 'attenuated'. Note that 'property' has a much wider meaning than in everyday language, it can refer to any good or resource. Similarly, the environment is a resource and hence 'property'.

Rights can be *private*, i.e. owned by readily identifiable individuals, or *communal* where the use of the property in question is shared with others. The latter kind of property is known as *common property*. Before the enclosures of land in England, grazing land was often common property: many individuals could graze their animals on the land. In a great many developing countries, land is owned communally. We consider in Chapters 16 and 17 whether the way in which property rights are held helps to explain the process of natural resource degradation, but for the moment we are interested in the general concept of property rights.

5.2 THE POTENTIAL FOR MARKET BARGAINS IN EXTERNALITY

Figure 5.1 repeats the basic optimal externality diagram in Chapter 4. Recall that, left unregulated, the polluter will try to operate at $Q\pi$ where his profits are maximised. But the social optimum is at Q^*. The workings of the market and the goal of a social optimum appear to be incompatible.

Now consider a situation in which *the sufferer has the property rights*. What this means is that the sufferer has the right *not* to be polluted and the polluter does not have the right to pollute. In that case the starting point is surely the origin in Figure 5.1. The sufferer will prefer that no pollution at all takes place and, since he has the property rights, his view will hold the biggest sway. But now consider whether the two parties – polluter and sufferer – might 'bargain' over the level of externality. Suppose the issue is whether to move to point d or not. If they moved to d, the polluter would gain Oabd in total profit, but the sufferer would lose Ocd. But since Oabd is greater than Ocd, there is potential for a bargain. Very simply, the polluter could offer to *compensate* the sufferer by some amount *greater* than Ocd, and less than Oabd. The polluter will still have a net profit. Moreover, the sufferer would be better off: although he would lose Ocd, he would gain more than that in compensation. If such a bargain could be struck, the move to d would be seen to be an improvement for both parties (such a move is known as a 'Pareto improvement' since at least one party is better off and no party is any worse off). But if the move from O to d is a social improvement so is the move to e (simply repeat the argument). Indeed, so is a further

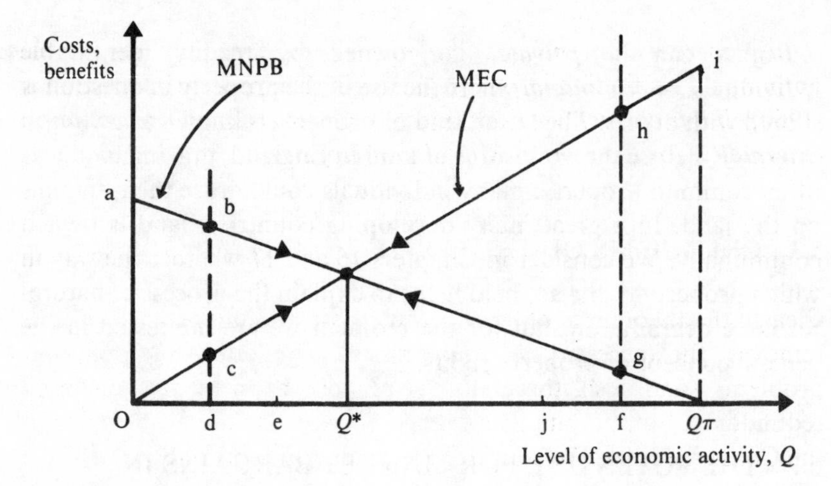

Figure 5.1 Optimal pollution by bargaining.

move to Q^*. But any move to the right of Q^* is not feasible because the polluter's net gains then become less than the sufferer's losses – hence the polluter cannot compensate the sufferer to move beyond Q^*. Thus, if we start at O and the property rights belong with the sufferer, there is a 'natural' tendency to move to Q^*, the social optimum.

Now imagine that the property rights are vested in the polluter. The starting point is $Q\pi$ because that is the point to which the polluter will go given that he has every right to use the environment for his waste products. But it is now possible for the two parties to come together again and consider the move from $Q\pi$ *back* to f. But this time the sufferer can compensate the polluter to give up a certain amount of activity. Since the sufferer would have to tolerate a loss of fhi$Q\pi$ if the move to f does not take place, he will be willing to offer any amount less than this to make the move. The polluter will be willing to accept any amount greater than fg$Q\pi$, the profits he will have to surrender. The potential for a bargain exists again and the move to f will take place. But if the move to f is a social improvement, so is the move from f to j and from j to Q^*. Hence Q^* is once again the level of activity to which the system will gravitate.

So long as we can establish a bargain between polluter and polluted, the market will, on the above argument, take us to Q^* which is the social optimum. The potential importance of the argument can now be seen, for *regardless of who holds the property* *rights, there is an automatic tendency to approach the social optimum.* This finding is known as the 'Coase theorem', after Coase (1960). *If it is correct, we have no need for government regulation of externality, for the market will take care of itself.*

5.3 CRITICISMS OF THE COASE THEOREM

Clearly the theorem is of considerable potential importance since it removes the necessity of government regulation of pollution problems (and also threatens to render the next few chapters redundant!) But, despite its elegance, there are many problems with the Coase theorem. We consider the main criticisms only.

The state of competition

Chapter 4 was careful to point out that the analysis of optimal externality assumed perfect competition. It was on this basis that we saw that

$$MNPB = P - MC$$

and, hence,

$$(MNPB = MEC) \text{ entails } (P = MSC)$$

In terms of the bargaining approach, what is being assumed is that MNPB is the polluter's *bargaining curve*. It is this to which he refers when deciding how much to pay, or how much to accept, in compensation. But suppose that perfect competition does not prevail. Then P – MC is no longer the bargaining curve because it will not be equal to MNPB. If the polluter is a firm, it should be fairly evident that his bargaining curve is his marginal profit curve (see Appendix 4.1) and, under imperfect competition, this is equal to *marginal revenue* minus marginal cost, i.e.

$$MNPB = MR - MC$$

Under imperfect competition, MR is not equal to P because the demand curve is above the marginal revenue curve. It follows that the bargaining solution does not apply under imperfect competition.

How serious this is as a criticism depends on two things. First it depends on how different we think the real world is from perfect competition. While some economists would argue that the amount of competitive 'imperfection' (or monopoly) is not very great, our view is that perfect competition is a convenient fiction for constructing economic models, but it is remote from describing the real world. Thus, the existence of imperfect competition provides the basis for a serious criticism of the Coase theorem. The second point is more complicated and is dealt with more formally in Appendix 5.1. The possibility exists that the bargaining curve of the polluter can be defined as one relating jointly to the interests of polluters and consumers. They need then to bargain with the sufferers of the pollution. While the approach is technically correct, it requires a rather fanciful involvement of producers (polluters), consumers and sufferers all in one bargain. It does not therefore seem at all realistic.

The absence of bargains and the existence of transaction

The second criticism of the Coase theorem is that we are probably all rather hard-pressed to think of real-world examples of such bargains taking place. It is true that some electricity-generating authorities 'bargain' with the local population to accept nuclear power stations or waste disposal facilities, perhaps offering cash compensation or a contribution to local facilities. There are also examples of international bargains between countries that suffer pollution and countries that create it, but they typically involve common property resources, and we deal with that issue later. But Chapter 2 indicated that externality is likely to be pervasive because of the materials balance principle. We should therefore be able to point to many such bargains rather than to isolated examples. The fact that we do not observe many examples of the bargains taking place suggests that there are either obstacles to them, or that the Coase theorem is not rooted in real-world economics.

The response of those who believe in the market bargain approach is that there are indeed obstacles to bargaining in the form of *transactions costs*. Such costs include those of bringing the parties together, organising often widely distributed and difficult-to-identify sufferers, the actual bargain itself and so on. If the transactions costs are so large that any *one* party's share of them outweighs the expected benefits of the bargain, that party will withdraw from the bargain, or not even commence it. Moreover, it seems likely that transactions costs will fall on the party that does not have the property rights. But transactions costs are real costs – we have no reason for treating them differently to other costs in the economy. Thus, if transactions costs are very high all we appear to be saying is that the costs of the bargain outweigh any benefits. In that case it is *optimal* that no bargain occurs.

Carried to this level the argument quickly becomes redundant, for what it says is that bargains will either take place or they will not. If they do, then the amount of externality emerging will be optimal (by the Coase theorem). If they do not take place, it is also optimal for it simply means that transactions costs exceed expected net benefits from the bargain. We have an unfalsifiable theory about optimal externality. It says that all the externality we observe is optimal externality and hence there is no need to do anything about it. But the proof involves non-falsifiable statements and hence the argument is non-falsifiable.

Nonetheless, the transactions costs argument serves to remind us of some important caveats in any recommendation about regulation of externality:

1. Simply because we observe externality it does not mean that something should be done on grounds of economic efficiency – we might be observing Pareto-irrelevant externality (Chapter 4). This kind of mistake is in fact very common, as with statements to the effect that 'all' pollution should be eliminated, or tobacco smoking should be prohibited and so on.
2. The existence of high transactions costs might explain why government intervention occurs. For high transactions costs do not entail that the externality is optimal at all – instead it may simply be that government intervention is cheaper and can achieve optimality.

Letting T = transactions costs, B = the gain from the bargain for the party bearing the transactions costs, and G = the cost of government intervention, we might summarise the possibilities as follows:

- If $T < B$, a bargain might take place (see below for reasons why they might not occur in this context).
- If $T > B$, a bargain will not occur, but some other regulatory approach might occur.
- If $T > G < B$, government regulation is likely to occur, and it will be efficient.

Finally, note that while transactions costs may leave some of the bargaining theory intact, their existence means that the optimal level of activity is no longer invariant with the allocation of property rights. It will matter who bears the transactions costs.

Identifying the bargaining parties

Even if transactions costs are less than the benefits to be obtained from a bargain, no bargain may take place. Many pollutants are long-lived – they stay in the environment for long periods of time and may affect people years, decades or even hundreds of years from now. If so, the people who are going to be affected by the pollution may not yet exist, and it is then not possible to speak of the two parties coming together to bargain. Toxic chemicals, radioactive waste, ozone layer depletion and global carbon dioxide pollution all fit this category, among many others. At best, some groups in the present generation would have to bargain *on behalf* of future generations. The idea of future generations having such representatives is of course not fanciful – many regulations reflect that kind of interest – and typically we expect governments to take on this role. But the contexts involved are usually common property ones and the outcome is usually some attenuation of the rights of polluters.

A further problem of identifying the polluters and the sufferers arises in cases of *open access resources*. An open access resource is one owned by nobody (common property resources are owned by an identifiable *group*). In such cases it is not clear who would bargain with whom since no one individual has an incentive to reduce his or her access to the resource.

Lastly, even in conventional pollution contexts it is often difficult to say who the polluters and sufferers are. Sufferers may be unaware of the source of pollution from which they suffer, or even unaware that damage is being done. This is often the case for air pollutants and water pollutants. Indeed, this situation seems likely to characterise the majority of pollution situations. The costs of generating the information for the sufferers need to be added to the costs of transacting any bargain. The likelihood of bargains being socially efficient even if they occurred is also remote given the need to identify damage done and its distribution among sufferers. Of course, this kind of problem will arise for regulatory solutions as well. Governments have to find information on damage.

Common property contexts

We noted earlier that property rights can be private or communal. In the communal case a kind of mutual bargain among users of the property can occur. Each user agrees to restrict his usage of the resource in the interest of its longer-term sustainable use for the community as a whole, and for later generations. This is called a *cooperative* solution to a problem of *assurance*. Each individual needs assurance that others will also behave in a cooperative fashion, otherwise there will be a temptation to 'break ranks' and seek the maximum private gain. Despite a voluminous theoretical and empirical literature on such 'game theoretic' situations, it is not easy to say why some common property contexts are subject to cooperative solutions and others break down. But from the bargaining theory point of view the important point to note is that each user of the common property is the polluter (or resource user) *and* each individual user is also the beneficiary. In terms of the previous diagrams, MNPB and MEC 'belong' to the same people. Rational cooperative individuals will therefore net out the costs and benefits to arrive at their own personal Q^* so that the sum of the individual positions will be the social optimum. Nonetheless it can pay an individual to move beyond Q^* if he or she judges they can 'get away with it' and make fairly large short-term gains at the expense of the other users now and in the future.

Threat-making

One other problem with the bargaining solution is that it offers potential for making an economic activity out of threat-making. If a sufferer compensates a polluter because the polluter has the property rights, it is open to other 'polluters' to enter the situation and to demand compensation. Threat-making is hardly a rational use of scarce economic resources. Possibly the situation can be corrected by carefully defining who is entitled to property rights, e.g. by denying them to potential threat-makers, but it has to be acknowledged that compensation schemes for potential polluters have suffered this difficulty. In some countries it is possible to receive government cash for *not* engaging in cultivation, the idea being to protect environmentally valuable land and reduce agricultural surpluses. It seems likely that some farmers could say that they are going to farm an area of wetland even if they never intended to, gaining 'compensation' in the process.

The Coase theorem is important in forcing advocates of environmental regulation to define their terms and justify their case more carefully than they might otherwise have done. But there are many reasons why bargains do not, and cannot, occur. An investigation of those reasons may help to explain why government regulation is the norm in pollution contexts.

APPENDIX 5.1: RESURRECTING THE COASE THEOREM UNDER IMPERFECT COMPETITION

Buchanan (1969) has suggested a way in which the Coase theorem might be resurrected under imperfect competition. Figure A5.1 shows the imperfectly competitive firm together with the profit maximising position, $Q\pi$, the bargaining outcome if marginal profit is the bargaining curve, and the bargaining outcome if the curve P − MC is used as the polluter's bargaining curve. We see that

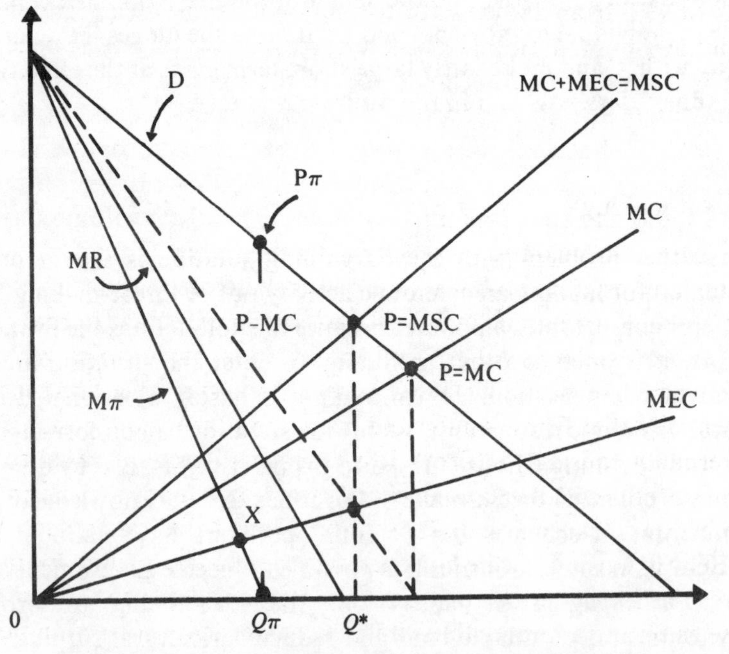

Figure A5.1 Coasian bargains and imperfect competition.

P – MC = MEC does secure an optimum. But P – MC is not equal to marginal profit, so we need to re-interpret P – MC. It is in fact a 'marginal surplus' curve, the marginal change in combined producer and consumer surplus. If this is set equal to MEC and the two curves are bargaining curves, then an optimal outcome occurs. The implication is that the bargain now needs to take place between the polluter, the consumer of the polluter's product, *and* the sufferers. Such 'tripartite' bargaining restores the Coase theorem. The problem, of course, is exactly what this means in practice since it is difficult to envisage such tripartite bargaining taking place.

APPENDIX 5.2: NON-CONVEXITY AND THE MARKET BARGAIN THEOREM

Several writers have pointed out that normal presentations of externality contexts assume 'well-behaved' marginal external cost and marginal profit functions such that a unique, stable equilibrium is secured. Figure A5.2 shows some possible results of assuming 'non-convexity'. In (*a*) we show a decreasing MEC function which cuts MNPB from above. In this situation it can be seen that point E is not an optimum (total external costs exceed total private benefits) nor is it a stable equilibrium since, to the right of E, polluters can compensate sufferers to accept pollution increases, and to the left of E, sufferers can compensate polluters back to zero output. In (*b*) MEC slopes downwards but cuts MB from below. In this case E is both stable and an optimum. This situation in (*a*) causes difficulty for the bargaining solution, although we may note that, if property rights are vested in polluters, and $Q\pi$ is therefore the starting point, the absence of a bargain will be Pareto optimal if total external costs at $Q\pi$ are less than total private benefits. More to the point, we must ask whether a declining MEC is at all realistic. One argument is essentially that firms cannot lose more than their fixed costs. If the externality causing the firm's loss reaches an amount equal to the firm's profits calculated as an excess over *variable* costs, the firm will close down, causing a discontinuity in the MEC curve such that MEC = 0. It is not clear, however, whether this particular argument gives rise to any serious problem. It is perhaps better to think of this case as setting a limit within which any externality correction policy can take place. Nor does it mean that the MEC curve has to slope downwards over its whole length, merely that we have an eventual discontinuity. Baumol and Bradford (1972) have, however, provided a more general point about non-convexity. Again, the analysis is in terms of inter-firm externalities but the results appear to hold for all categories of externality. The argument is that the very existence of external effects may be sufficient to induce a non-convexity situation. Consider two industries, P which produces some commodity with pollution as a side effect, and S which produces a commodity that does not pollute but suffers the pollution from P. In

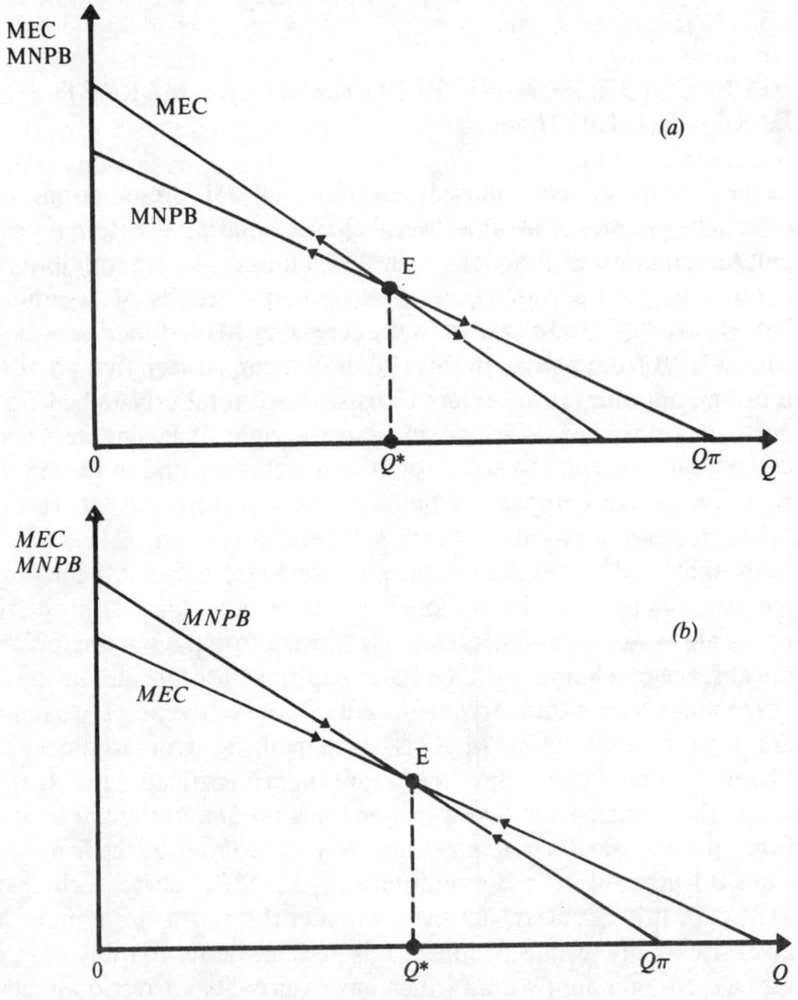

Figure A5.2 Non-convex MEC curves.

Figure A5.3(*a*) line OA shows the output possibilities of industry P. For 4 units of work (i.e. 4 units of leisure) 10 units of output are secured. For 8 units of work, 20 units of output are secured. In Figure A5.3(*b*) the output possibilities of industry S are shown. Line OB shows input/output combinations when industry P does not produce at all. Line OH shows reduced outputs per unit input when industry P does produce, such that S suffers the externality. Now consider the combined production possibilities. We can choose to produce at A in Figure A5.2(*a*) in which case we shall have the combination (–8L, 20P, OS), where L is leisure. Or we can choose B in Figure A5.2(*b*) in which case we have (–8L, OP, 400S). But if we choose to put *some* labour into producing some of both commodities, we shall face the effects of the externality. Thus, if we put 4 units of labour in each we shall have 10 units of P, but only 100 units of S (not 200) because output line OH is now operative. Hence, we shall have the combination (–8L, 10P, 100S). If we plot the output combinations to produce a production possibility curve we obtain Figure A5.4.

Line PP_1 shows the production possibility curve if there are *no* externalities; it is a constant returns curve as we would expect. Line PP_2 shows the production possibility curve if the externality exists. *The externality itself has generated non-convexity in the production possibility curve.*

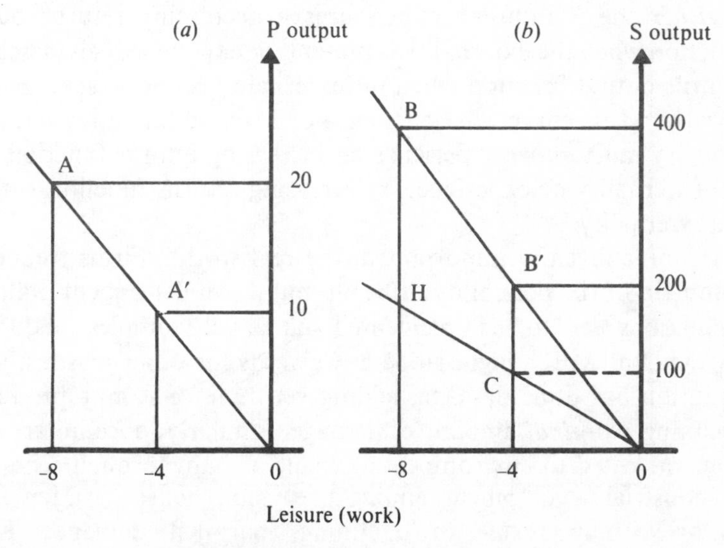

Figure A5.3 Polluting and polluted firms.

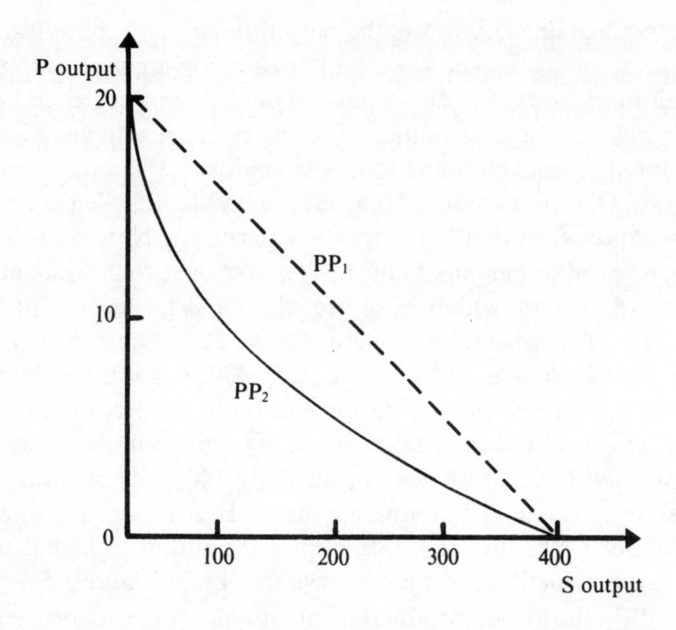

Figure A5.4 Production possibilities with externality.

In Figure A5.3 the total loss of the S industry increases at a constant rate. This means that the MEC curve is horizontal. The result that the production possibility curve is concave is *strengthened* if *either* the S industry experiences a decreasing returns output function when the externality is present, *or* experiences an increasing returns output function when the externality is *not* present. In each case the MEC curve will be rising. For MEC to fall, however, the S industry must *either* experience an increasing returns function with the externality *or* experience a decreasing returns function without the externality.

Is non-convexity important in the real world? Firms frequently complain of the possibility of being shut down if stringent pollution regulations were to be implemented, but actual examples are difficult to find. But MEC might slope downwards for other reasons. Once pollution has done damage, adding yet more of it may produce a declining *marginal* amount of damage. Similarly, once an area has been subjected to environmental blight, perhaps through a housing or industrial development, adding a few more houses or a few more factories may result in declining marginal damage. These observations are sufficient to show that MEC need not slope upwards or be horizontal, but for all practical purposes non-convexity does not seem to be a significant detraction from the analytics of pollution control pursued in the remainder of this volume.

6 · TAXATION AND OPTIMAL POLLUTION

6.1 INTRODUCTION

Recall that the aim of pollution regulation is assumed to be one of finding ways of reaching Q^*, the socially optimal level of pollution. Chapter 5 asked whether we needed to look for any government-initiated 'economic instruments' – taxes, regulations, etc. – at all. We concluded that 'markets in externality' were feasible in a limited number of cases, but that, generally, some form of intervention would be required.

Many economists advocate a particular type of intervention – a tax on the polluter based on the estimated damage done. Damage is another word for external cost. Such a tax is known as a *Pigovian tax*, after Arthur C. Pigou (1877–1959) who was Professor of Political Economy at Cambridge University from 1908 to 1944. In his *Economics of Welfare* (first published in 1920) he proposed a tax as a suitable means of equating private and social cost. Pigovian taxes tend to be known today as *pollution charges*, and some examples of charges which approximate Pigovian taxes do exist.

In this chapter we look at the theoretically 'ideal', or 'optimal' Pigovian tax. It is as well to remember, however, that no real-world charge could come close to the theoretically correct Pigovian tax. Instead of 'optimal' levels of pollution and optimal taxes, we tend to speak of 'acceptable' levels of pollution. It so happens that pollution charges in general are not very common. The main form of regulatory instrument used throughout the world is the *standard*. We will offer some explanations in this chapter for the general neglect of tax/charge solutions.

6.2 THE OPTIMAL PIGOVIAN TAX

Look at Figure 6.1 which repeats the pollution diagram introduced in Chapter 4. If we imposed a tax on each unit of the level of activity giving rise to pollution, and made the tax equal to t^*, we can see that such a tax would have the effect of shifting MNPB left towards (MNPB – t^*). Very simply, t^* has to be paid on each unit of activity, so that the marginal net benefit is reduced by t^*. The polluter will now aim to maximise private net benefits, subject to the tax, and this occurs at Q^*. The tax t^* is thus an optimal tax (because it achieves the social optimum at Q^*). How is t^* determined? It is equal to MEC at the optimum. *This defines an optimal Pigovian tax – it is equal to the marginal external cost (i.e. marginal pollution damage) at the optimal level of pollution.* A *damage function* tells us how pollution damage varies with the level of pollution emitted, *and* what the monetary value of that damage is. (It should then be possible to relate it back to the level of activity of the polluter.) Indeed, there are quite a few steps involved in finding such damage functions. The sequence is:

Economic activity of the polluter → Pollution emissions → Pollution concentration in the environment → Pollution exposure → Physical damage function → Monetary value of damage

Appendix 6.1 shows this sequence in more detail for power station emissions. The need to find the whole damage *function* (or a good part of it) arises because we want to find the optimal level of pollution – i.e. we need at least some part of MEC in Figure 6.1. A single point is no good to us if we are designing pollution taxes. We review the techniques for finding damage functions in Chapter 10.

But not only do we need a good part of the MEC function, we also need to know MNPB. If the polluter is a firm this may be very difficult because of commercial confidentiality of information. Indeed, many economists consider that the government, as the taxing authority, is in a poor position to extract this information. This *asymmetry of information* between the polluter and the regulator is often regarded as an objection to any form of government intervention.

In practice, these informational difficulties may not be overwhelming. We may only be concerned to get the right direction

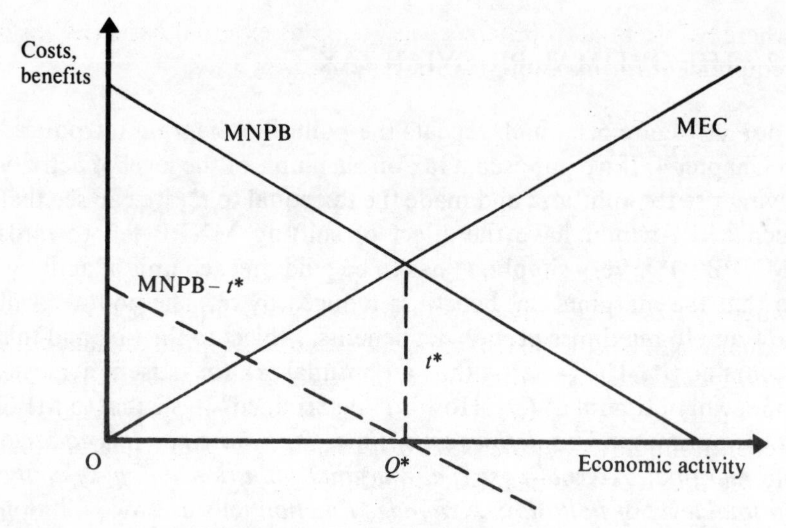

Figure 6.1 The optimal pollution tax.

of change in pollution levels, rather than achieve a theoretical optimum. If so, charges are surely a proper weapon in the regulatory armoury.

6.3 ILLUSTRATING THE OPTIMAL PIGOVIAN TAX MATHEMATICALLY

Net social benefits (NSB) are made up of the gross benefits of the polluting activity *minus* private costs *C*, *minus* external costs, EC: i.e.

$$\text{NSB} = PQ - C(Q) - \text{EC}(Q) \tag{6.1}$$

where P is price, Q is output (polluting activity) and P is parametric (i.e. P does not depend on Q as it would under imperfect competition). Then,

$$\frac{\partial \text{NSB}}{\partial Q} = P - \frac{\partial C}{\partial Q} - \frac{\partial \text{EC}}{\partial Q} = 0 \tag{6.2}$$

is a first-order condition for maximising NSB. Hence

$$P = \frac{\partial C}{\partial Q} + \frac{\partial \text{EC}}{\partial Q} = \frac{\partial \text{SC}}{\partial Q} \tag{6.3}$$

where SC is equal to private costs (C) plus external costs (EC), is a requirement for maximum NSB. Alternatively,

$$P - \frac{\partial C}{\partial Q} = \frac{\partial \mathrm{EC}}{\partial Q}$$

or (6.4)

$$\frac{\partial \mathrm{NPB}}{\partial Q} = \frac{\partial \mathrm{EC}}{\partial Q}$$

where NPB is net private benefits, i.e. price minus private costs. Equation (6.3) is the rule that price of the polluting product must equal *marginal social* cost. Equation (6.4) rearranges equation (6.3) to give the optimisation rule we have been using, i.e. marginal net private benefits should equal marginal external costs. Using equation (6.3) we see that it can be met if we impose a tax, t^*, where

$$t^* = \frac{\partial \mathrm{EC}}{\partial Q^*} \tag{6.5}$$

where Q^* is the level of activity, solving equation (6.3). Then,

$$P = \frac{\partial C}{\partial Q^*} + t^* \tag{6.6}$$

6.4 POLLUTION CHARGES AND PROPERTY RIGHTS

There is a further 'problem' with the pollution charge. Figure 6.2 repeats Figure 6.1, but this time we have shaded in the amounts of tax charged. Thus, if the polluter continued to produce at $Q\pi$ he would be liable for a total pollution tax bill of ObdQ^* + Q^*de$Q\pi$ (the reader should confirm that these are equal to areas acdQ^* and Q^*d$Q\pi$f, respectively). Now, Q^*de$Q\pi$ – the dotted area – will not be paid because the tax bill exceeds the net private benefits of output $Q^*Q\pi$. Instead the polluter will move back to Q^* to avoid the tax, just as the theory requires. So far there are no surprises. But once at Q^* the polluter still pays ObdQ^* *despite the fact that he is now emitting the optimal amount of pollution.* The polluter appears to be being penalised twice – once by losing profits (assume the polluter is a firm) to get back to Q^* in order to avoid the tax, and again when he is operating at the optimal level of pollution.

Is this socially justified? The answer to this is that it depends on

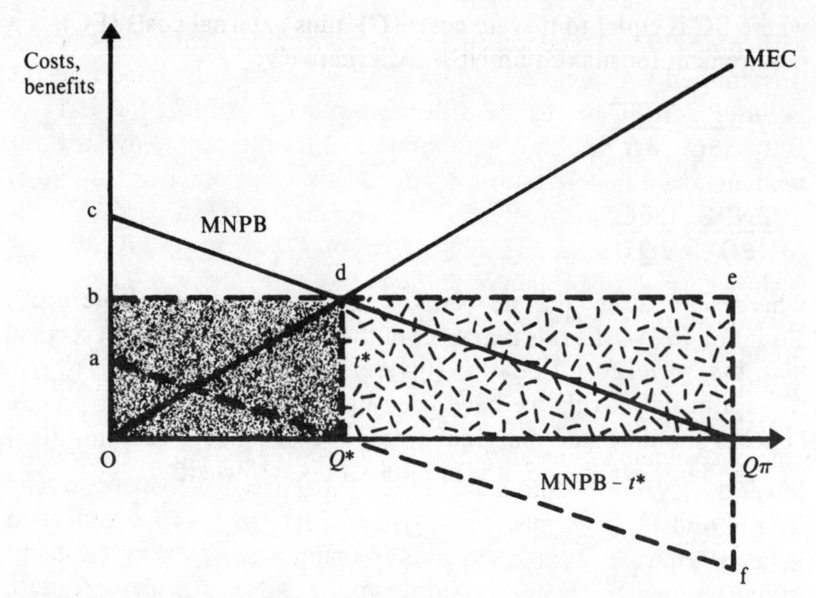

Figure 6.2 Pollution taxation and property rights.

our view of *property rights*. If the firm has *no* right to use the environment for emitting wastes, then the pollution charge ObdQ^* is a charge for using property belonging to others (the state, say). If the firm has *every* right to use the environment as it sees fit, then not only is the charge on optimal pollution wrong, but so is the charge that would apply between Q^* and $Q\pi$ in Figure 6.2, i.e. the charge concept is wrong altogether. Lastly, we might say that the firm has *no* right to pollute above Q^*, but *every* right to emit the optimal level of pollution (associated with OQ^*).

It is evident, then, that the design of pollution tax depends on what view is taken of the polluter's rights to use the environment as a 'waste sink'. Those rights may be enshrined in law, but are often a mix of legal interpretation and traditional practice. Appendix 6.2 raises a fourth issue about the design of Pigovian taxes.

6.5 POLLUTION CHARGES AND ABATEMENT COSTS

A feature of pollution charges is that they should encourage the installation of pollution abatement (or 'control') equipment. Thus, it is possible to remove particulate matter and sulphur from chimneys with 'precipitators' and 'scrubbing' equipment, to treat sewage before it is emitted to water, and so on. So far we have assumed that the polluter adjusts to the pollution charge by altering the level of activity giving rise to the pollution. In order to allow for the abatement equipment option we introduce a new diagram. In Figure 6.3 we see the familiar MEC curve but we have dispensed with the MNPB curve. Instead, MAC is a *marginal abatement cost* curve. (It is shown as a straight line for convenience: in reality it is likely to be curvilinear or 'stepped'.) The horizontal axis now shows the level of pollution. MAC shows the extra costs of reducing the level of pollution by expenditures on abatement. For example, the marginal cost of reducing pollution just below level W_1 is MAC_1. The marginal cost of reducing pollution below W_2, however, is MAC_2. That is, the lower the level of pollution the higher is the marginal cost of reducing it still further. This may seem odd at first sight, but it reflects a general empirical observation. It is comparatively cheap to 'clean up' initial amounts of heavy pollution, but once we get to very little pollution reducing it further requires advanced forms of treatment,

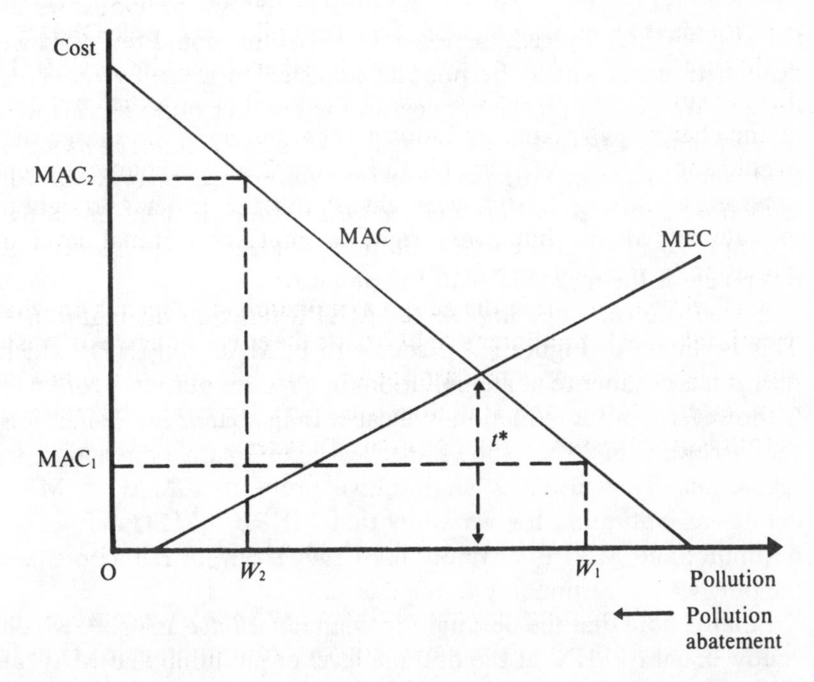

Figure 6.3 Optimal pollution: the abatement cost–external cost approach.

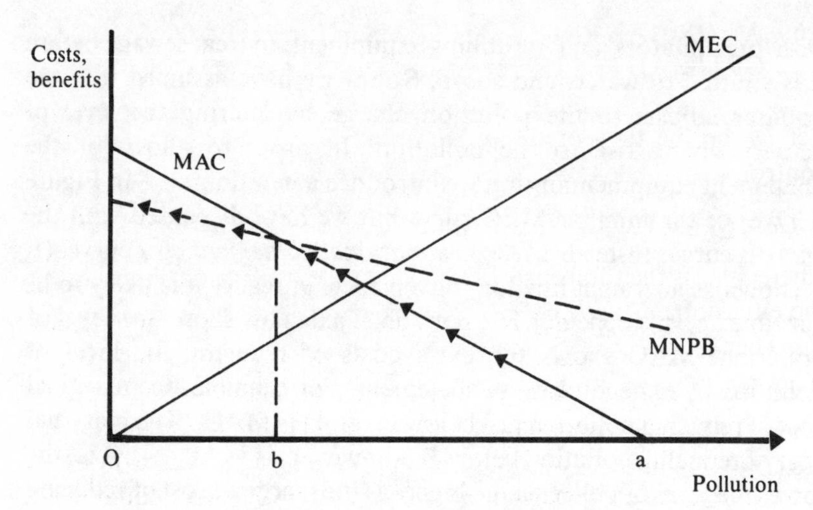

Figure 6.4 The abatement cost–net benefit relationship.

using chemicals, special filtering equipment, and so on. Hence the general shape of MAC.

Now the optimal level of pollution in Figure 6.3 is where MAC = MEC. This looks very similar to our previous result (MNPB = MEC). Indeed, there is a formal connection. Previously we dealt with cases where the polluter adjusted to a tax by *reducing output*. We noted that the net cost to the polluter of doing this was the *foregone profit* (net private benefit). *So,* MNPB *could be thought of as an abatement cost curve in the context where only output reductions can be used to reduce pollution.* MAC is then simply the analogue of this cost curve, but in a context where abatement equipment is the means of reducing pollution.

Indeed, we can superimpose the MNPB function on Figure 6.3. This is shown in Figure 6.4. From a to b, MAC < MNPB which means it is cheaper to abate pollution than reduce output. From b to O, however, output reduction is cheaper than abatement. Hence it is the 'arrowed line' that shows the 'least cost' path of reaction to regulation. This provides an intuitive proof that MAC = MEC defines an optimum, for we know that MNPB = MEC defines an optimum, and MNPB is simply MAC when output reductions are the only way of responding to regulation.

Finally, note that the optimal Pigovian tax is once again t^*, which is now equal to MEC at the optimal level of pollution and MAC at the same pollution level.

6.6 A FORMAL PROOF THAT MAC = MEC PRODUCES OPTIMAL POLLUTION

Let Q_C be the flow of economic output produced *with* pollution control, and Q_N be the flow *without* control. Then,

$$Q_C = Q_N - \text{TAC} \tag{6.7}$$

where TAC is the total costs of abatement. Let the value of services of the environment *with* pollution control be E_C, and *without* control E_N. Then

$$E_C = E_N - \text{TEC} \tag{6.8}$$

where TEC is the total external (damage) cost. Total social benefits are $(Q_C + E_C)$ in the economy, so

$$\begin{aligned} \text{TSB} + Q_C + E_C &= Q_N - \text{TAC} + E_N - \text{TEC} \\ &= Q_N + E_N - [\text{TAC} + \text{TEC}] \end{aligned} \tag{6.9}$$

Now, pollution, W, affects TSB, TAC and TEC, so

$$\frac{\partial \text{TSB}}{\partial W} = -[\frac{\partial \text{TAC}}{\partial W} + \frac{\partial \text{TEC}}{\partial W}] = 0 \tag{6.10}$$

is a condition for maximising TSB. Or

$$(-)\text{MAC} = \text{MEC} \tag{6.11}$$

(The minus sign simply indicates that we 'read' MAC from right to left.)

Note that equation (6.9) also tells us that maximising TSB is the same as *minimising* (TAC + TEC), i.e. minimising the *sum* of abatement and damage costs. This result is used in some textbook presentations.

6.7 PIGOVIAN TAXES AND IMPERFECT COMPETITION

The main difficulty with Pigovian taxes highlighted so far is the need to know both the MNPB (or MAC) and MEC functions. But, just as we discovered with the Coase theorem, relaxing the assumption of perfect competition causes problems.

Figure 6.5 shows the imperfectly competitive firm with private marginal cost, MC, and marginal social cost curve, MSC. MEC is

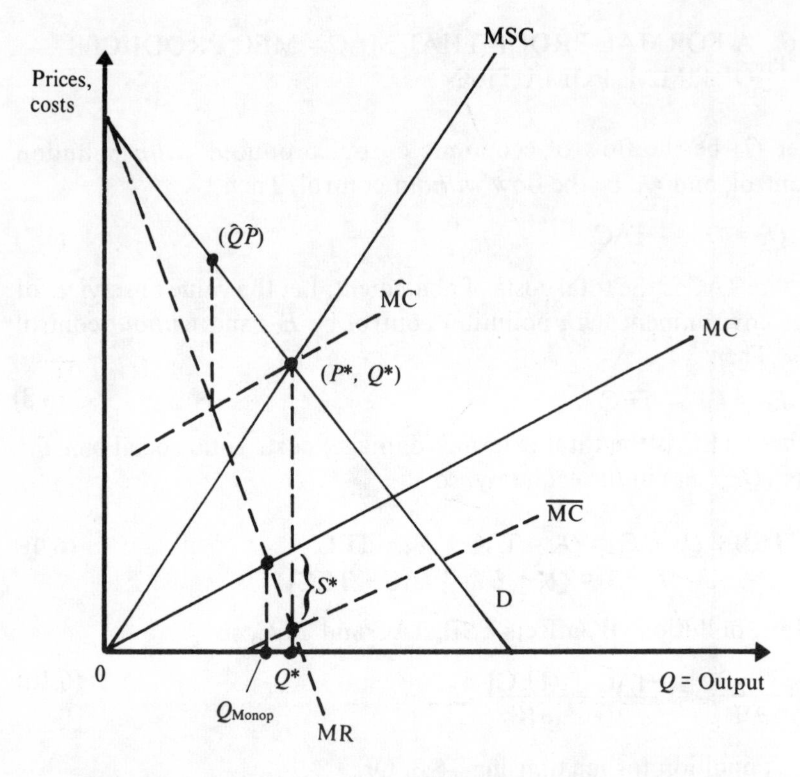

Figure 6.5 Pigovian taxes/subsidies and imperfect competition.

then the vertical distance between MSC and MC. Optimal output is given by Q^* where P = MSC. But if we place a tax equal to MEC at Q^*, the apparent requirement for an optimal tax, the effect is to shift MC upwards to $\hat{MC}$. But the firm equates marginal revenue and marginal cost, so its after-tax profits are maximised where MR = $\hat{MC}$. This produces output $\hat{Q}$ and price $\hat{P}$, and this is decidedly *non*-optimal.

How do we get to Q^*? From Figure 6.5 it is evident that we need to move MC *down* to $\overline{MC}$ so that MC = MR gives Q^* and P^*. But this implies a *subsidy* of S^*, not a tax. *Moreover, this subsidy is equal to the distance between (the old) MC and the MR at the optimum.*

We can in fact repeat the exercise and come up with a positive tax. Figure 6.6 shows this result. Note that MSC is drawn so as to be significantly different to MC. To find the tax we need to construct

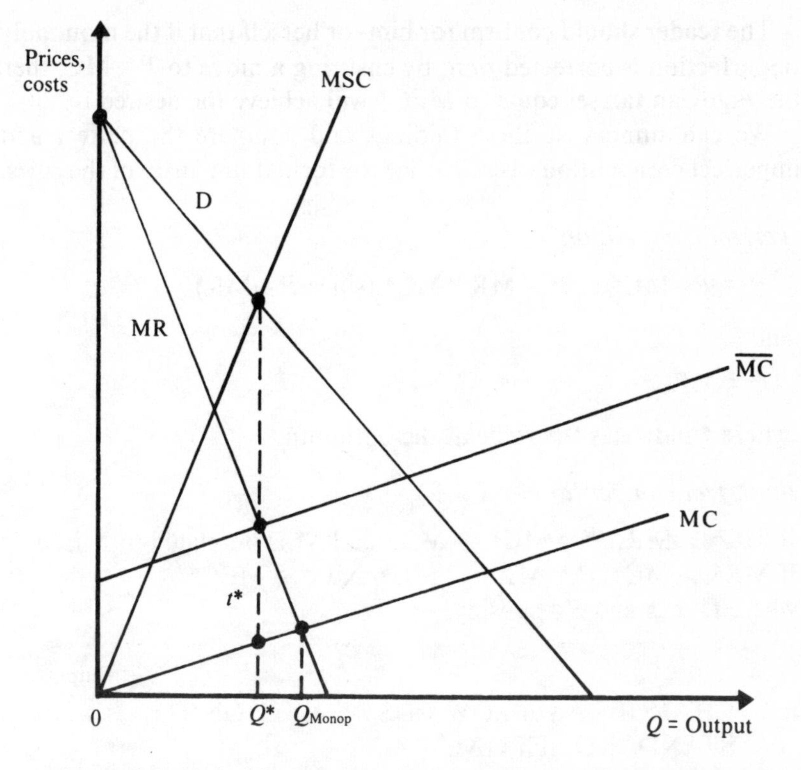

Figure 6.6 A positive Pigovian tax under imperfect competition.

$\overline{MC}$ to intersect MR to give P^*, Q^*. This time we see that a positive tax emerges.

The difference between the two cases is that the positive tax emerges when MR > MC at Q^*, and the subsidy (negative tax) emerges when MC > MR at Q^*. Which of these cases applies is determined by the size of MEC: the bigger MEC is, the more likely it is that MR > MC and a positive tax will be appropriate. This suggests that pollution taxes are still appropriate under imperfect competition if the externality is large relative to private costs. But in neither tax nor subsidy case does the optimal tax equal MEC*. Why not? What is happening is that we are asking the pollution tax to solve two imperfections in the market, the externality *and* the existence of monopoly (downward-sloping demand curves). It is hardly surprising that the result is less precise than under perfect competition when only one imperfection, the externality, is present.

The reader should confirm for him- or herself that if the monopoly imperfection is corrected *first*, by ensuring a move to P = MC, then the Pigovian tax set equal to MEC* will achieve the desired result.

We can summarise these findings and compare the perfect and imperfect competition cases. Notice the formal similarity of the rules.

Perfect competition

$t^* = P - \text{MC}^*$ or $t^* = \text{MR} - \text{MC}^*$ (since $P = \text{MR}$)

and

$t^* = \text{MEC}^*$

where * indicates the value at the optimum.

Imperfect competition

If $\text{MC}^* > \text{MR}$, $S^* = \text{MC}^* - \text{MR}^*$ and S^* is unrelated to MEC*
If $\text{MR}^* > \text{MC}^*$, $t^* = \text{MR}^* - \text{MC}^*$ and $t^* = \text{MEC}^*$
where t = tax and S = subsidy.

6.8 CHARGES AS A LOW-COST SOLUTION TO STANDARD SETTING

There is one other feature of the tax/charge solution to externality which is of some importance. *It is that, compared to standards set without taxes, charges will tend to be a lower-cost method of achieving a given standard.* This result is due to Baumol and Oates (1971). We offer a diagrammatic proof here.

Figure 6.7 shows *pollution reduction* (i.e. level of abatement) on the horizontal axis, and money values on the vertical axis. MAC_1, MAC_2 and MAC_3 are then marginal abatement cost curves for three *different* firms producing the same product. Note that the MAC curves slope upwards from the left to the right, and not from right to left as in Figure 6.3. This is simply because we have put pollution *reduction* on the horizontal axis, as opposed to pollution levels. MAC for each firm differs simply because of different technologies in use in the firms: we have no reason to suppose they will be the same. We observe that, for any given level of abatement, firm 1 has the highest costs of abatement, firm 2 the next highest and firm 3 the lowest. For simplicity, we assume $S_1S_2 = S_2S_3$ and $S_1 + S_2 + S_3 = 3S_2$.

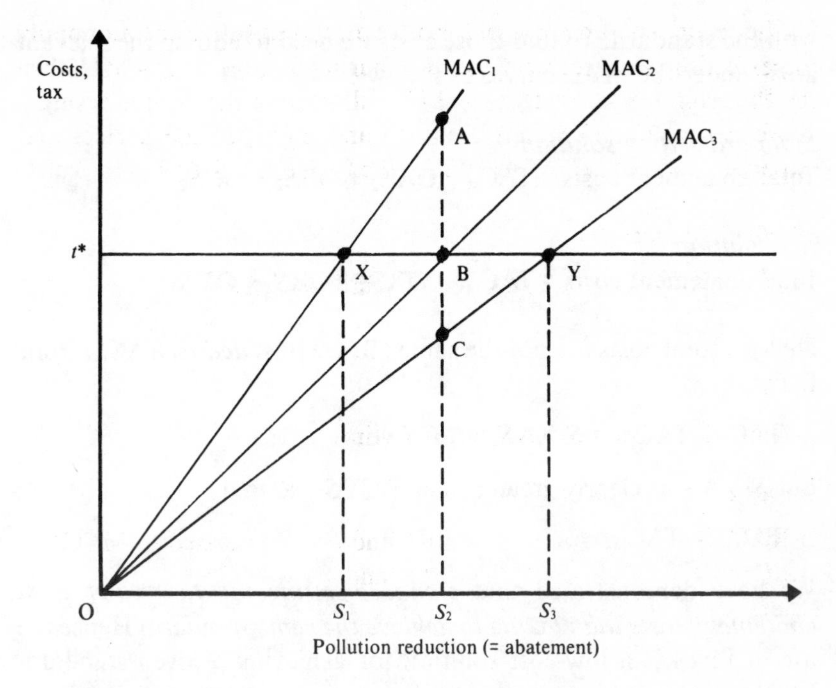

Figure 6.7 Taxes as a low-cost method of achieving a standard.

Now suppose we set a *standard* which says that we have to achieve S_2 of pollution abatement. One way of doing this is simply to tell each firm to abate pollution by an amount OS_2. This means firm 1 will go to point A, firm 2 to point B, and firm 3 to point C. We achieve the *overall* standard of $3S_2$.

An alternative is to set a tax t^*. The tax is set so that firm 1 goes to point X, firm 2 to point B, and firm 3 to point Y. To see why, consider firm 1. Up to S_1 it is *cheaper* for the firm to abate pollution than it is to pay the tax (t^* lies above MAC_1). After S_1, however, the firm will pay the tax rather than abate (t^* lies below MAC_1). A similar analysis applies to the other firms. Note that the *overall* standard $3S_2$ is achieved, but that firm 1, with the highest abatement costs, has abated pollution *less* than S_2, while firm 3, with the lowest abatement costs, has abated *more* than S_2.

There seems nothing to choose between the standard-setting approach and the taxation approach. Both achieve the overall standard of $3S_2$. But there is a difference in the *costs* of compliance with the standard. To find those costs we need to add up the relevant areas under the MAC curves as follows:

Standard-setting solution
Total abatement costs = $TAC_{st} = OAS_2 + OBS_2 + OCS_2$

Tax solution
Total abatement costs = $TAC_{tax} = OXS_1 + OBS_2 + OYS_3$

The two total costs are not the same. To see this, *deduct* TAC_{tax} from TAC_{st},

$$TAC_{st} - TAC_{tax} = S_1XAS_2 - S_2CYS_3$$

But S_1XAS_2 is clearly greater than S_2CYS_3, so that

$$TAC_{st} > TAC_{tax}$$

We have demonstrated that *standard-setting incurs greater total abatement costs than taxing to achieve the same standard.* Hence the use of taxes is a low-cost solution for achieving a given standard. Whether it is the *least*-cost solution to standard-setting depends on what *other* mechanisms we have for achieving a standard – e.g. marketable permits (see Chapter 8). Typically, we can only find out by 'simulating' pollution control – i.e. by devising computer simulations which 'mimic' the actual situation and then assessing the response to each method of securing a standard.

Notice that we have said nothing about the standard being optimal. To find an optimal standard we need information on the damage function. What is being demonstrated is that, *even where we impose 'accessible' standards, a tax has an important role to play.*

6.9 WHY ARE POLLUTION TAXES NOT WIDESPREAD?

Pollution taxes have many virtues. They make use of market mechanisms by charging a price for hitherto unpriced but valuable services provided by the natural environment. To *some* extent, they 'mimic' the market since the tax could be varied to reflect increasing scarcity of these services. They have optimality properties if both damage costs and abatement costs are known, and, even if they are not known, they have least-cost (i.e. 'cost effectiveness') properties. Yet in the real world, pollution taxes are the exception, not the rule. Not only are the charges limited in extent, their formulation tends to owe little to the theory outlined in this chapter. Why is this? Pezzey (1988) offers a number of interesting suggestions to explain the limited role of taxes, and these are discussed below.

Uncertainty about the justice of Pigovian taxes
Industry will always understandably resist new taxes. But this is not adequate to explain opposition if the situation is that some form of regulation will be introduced. One fear, however, is that the tax will go 'beyond' taxing Pareto-relevant (non-optimal) pollution, to taxing for optimal pollution, and even for physical pollution in a $W<A$ context. Industry *might* tolerate the former (which Pezzey calls the 'standard' *polluter pays principle*) but not the latter (the 'extended' *polluter pays principle).*

Lack of knowledge of the damage function
A strict Pigovian tax requires that we know at least part of the MEC curve, which is the marginal interpretation of the overall total external cost function, or 'damage function'. The ways in which damage functions might be measured are reviewed in Chapter 10. The judgement of many economists and perhaps even more pollution control agents is that damage functions are very difficult to estimate in practice. Moreover, they argue, even if we secure some estimates it is not difficult to find other experts who will argue for different damages, opening the way for disputes about the legal basis for a tax or charge. This objection has some validity, and the charge that the damage figures can be 'massaged' could be serious in countries where it is possible to dispute the basis of taxation in the courts. But the idea that an 'optimal' Pigovian tax can be calculated is unrealistic. The point of damage estimates is to obtain some overall 'feel' for the levels of damage, not to find accurate numbers (even if they could be found). The kind of information needed would tell us whether we are very wide of the mark in taking a particular pollutant or whether we are in the right 'ballpark'. Moreover, the use of taxes to regulate consumption and production is not unusual in modern economies. For example, few would dispute that tobacco and alcohol taxes have a social cost 'component'. In the same way, taxes on pollution should bear *some* relationship to social cost estimates.

The status quo

Pollution regulation has, by and large, grown from earlier public health laws. These were formulated mainly in the nineteenth century when the only real mechanism for controlling pollution was direct regulation based on standards and backed up by inspection and penalties for transgression. Taxes are thus a 'new' idea in the context of pollution control. Newness is not generally welcome in regulatory circles, not least because the regulator wants to know why the existing system is inadequate. It is not just a matter of pointing to the desirable characteristics of taxes: it is necessary also to show that alternative systems, and especially the one already in place, are worse than the proposed one. There are indeed benefits in 'sticking with what we've got'. Particular concerns will be whether regulatory taxes are compatible with the existing legal system, and what the transitional costs are.

APPENDIX 6.1: THE EMISSIONS–DAMAGE SEQUENCE FOR AN ELECTRICITY POWER STATION

Figure A6.1 shows how pollution from power stations impacts on the economy and individuals. Depending on the choice of abatement-control technology, a coal-fired power station discharges residual gases and matter into the area, and solid wastes to ground and water. This is 'transferred' to the atmosphere according to local meteorological conditions and topography. These ambient concentrations relate to human exposure according to various behavioural conditions, e.g. time spent at work, at home and travelling. Exposure and ambient concentration tend to be the same thing for non-human 'receptors' of pollution. The 'dose-response' relationship then links damage done to emissions or exposure, giving rise to various impacts which are localised and some that may travel across national boundaries. To find society's valuation of the damage done we need to adopt monetary valuation techniques as discussed in Chapter 10.

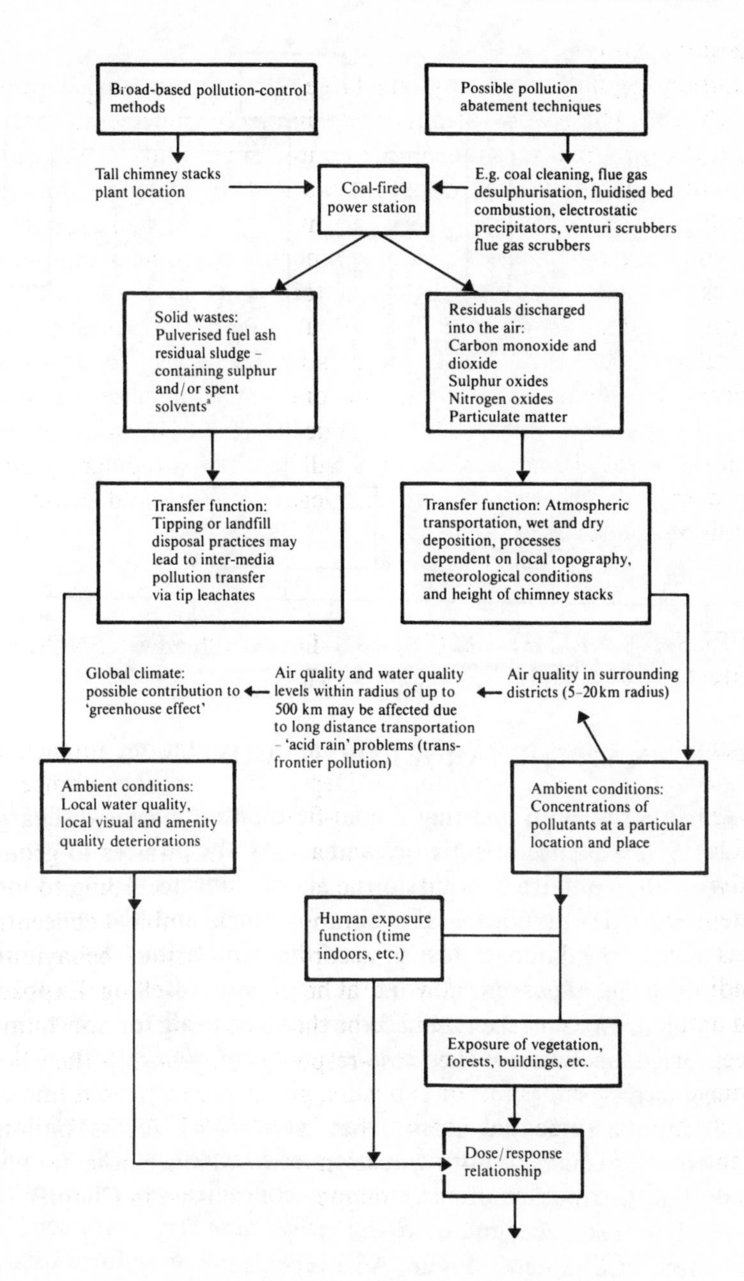

[a]Depending on which, if any, of the abatement techniques are operative.

Figure A6.1 Links in the environmental chain associated with power station pollution and damage impacts. (*Continued overleaf.*)

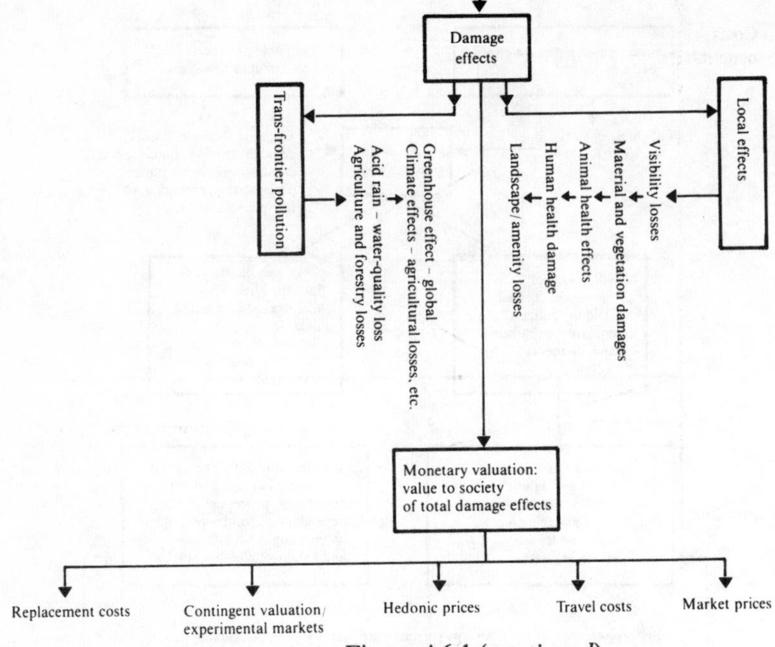

Figure A6.1 (*continued*)

APPENDIX 6.2: THE TARGET OF A PIGOVIAN TAX

We noted in the main text that damage estimation took place in the context of a sequence of activities and events: the pollution-creating activity, the pollutant emissions, their concentration in the environment, the exposure, the damage done, and the monetary evaluation of the damage. The 'proper' Pigovian tax is related to the *monetary value of damage done* at the optimum. But we saw this was complicated. Taxes, if they are to be introduced at all, are more likely to be levied on *emissions or ambient concentrations* measured in physical terms.

We cannot assume that emissions and concentrations are related in a one-to-one fashion. Ignoring atmospheric factors, which will alter the spatial distribution of the emission–concentration relationship, we need to take account of *assimilative capacity*, a concept we introduced in Chapter 2. Figure A6.2 repeats the basic form used in the text, but the level of emissions is also shown. Up to emission level e_1, the environment assimilates the waste. Thereafter, emissions cause physical changes which, we assume, cause immediate economic damage. The effect, then, is that MEC only begins to rise at A, the

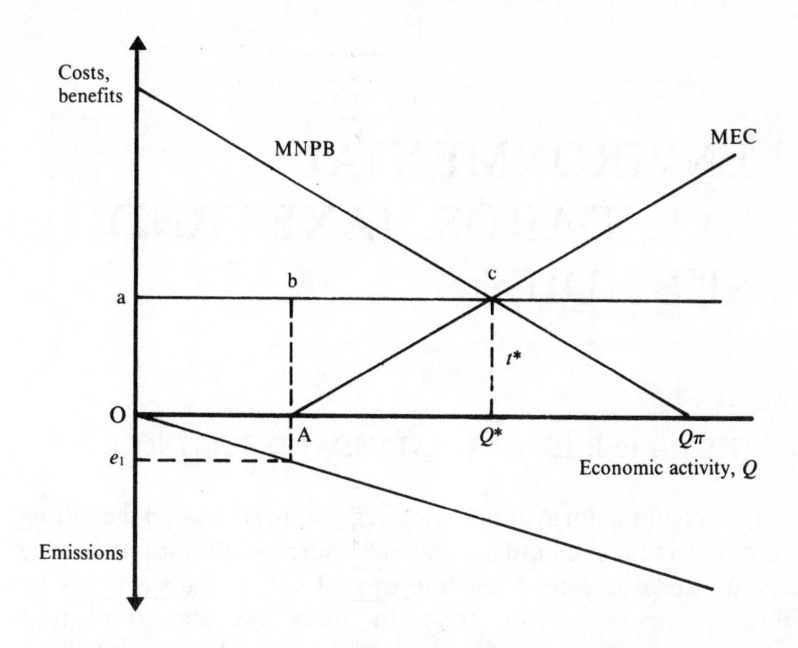

Figure A6.2 Three incidences of a pollution tax.

level of activity corresponding to the assimilative capacity of the environment.

Now suppose a tax is imposed such that the polluter pays OacQ^* in tax. We now see that, not only is he paying in the context of optimal externality, he is also paying a tax of OabA which corresponds to his 'use' of the environment even though no physical pollution is occurring (on other than a temporary basis). This seems to add to the 'unfairness' of a blanket tax on all emissions, regardless of whether they create any damage at all (associated with OA), optimal damage (associated with AQ^*), or non-optimal damage (associated with $Q^*Q\pi$).

Once again, however, it depends what we think the polluter is paying for: all damage done, non-optimal damage only, or the use of the environment. In the last case, the polluter is paying for the assimilative functions of the environment and it is then proper to pay some form of tax associated with OA. Nonetheless, the tax in Figure A6.2 is based on damage, so that it seems odd to use a tax based on one criterion – damage done – to reflect another criterion, namely use of a scarce resource.

7 · ENVIRONMENTAL STANDARDS, TAXES AND SUBSIDIES

7.1 THE INEFFICIENCY OF STANDARD-SETTING

The most common form of pollution regulation is through the setting of environmental standards. Chapter 6 indicated reasons as to why taxes are not widespread and are treated with some suspicion by polluters. Standard-setting tends to imply the establishment of particular levels of environmental concentration for the pollutant, for example *X* micrograms per cubic metre, or a percentage of dissolved oxygen in water or a level of decibels that are not to be exceeded. Standards are most likely to be set with reference to some health-related criterion, for example a level of contaminants that must not be exceeded in order that water is safe for drinking, concentrations of sulphur dioxide and particulate matter that are consistent with the avoidance of respiratory illness, and so on.

The problem with standard-setting is that it is virtually only by accident that it will produce an economically efficient solution, i.e. it is unlikely to secure the optimal level of externality. To see this consider Figure 7.1 which repeats the familiar pollution diagram. A standard *S* is set and this corresponds to pollution level W_s and economic activity level Q_s. Setting standards also entails having some monitoring agency which oversees polluters' activity and which has the power to impose some penalty. If it has no powers of punishment the only incentive the polluter has to stay within the standard is some form of social conscience. Typically, then, standards are associated with penalties – polluters can be prosecuted or at least threatened with prosecution. In many countries actual legal cases against polluters are rare because the pollution inspectorate uses its powers to alter the polluter's behaviour before the case comes to court.

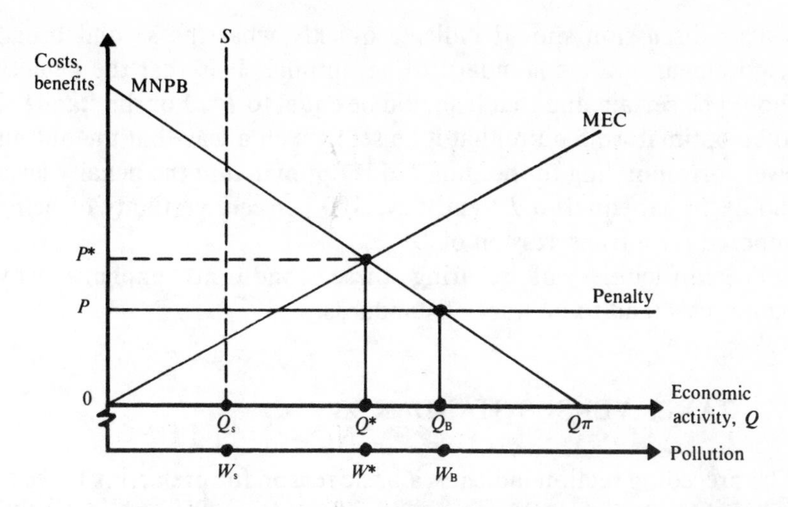

Figure 7.1 The inefficiency of standards.

Suppose the penalty in question is set at P in Figure 7.1. For the standard to work, then, the polluter must only pollute up to the maximum permitted level Q_S. It will be evident that Q_S is not optimal since it is less than Q^*. Indeed, unless the standard is set at Q^* it will not be optimal. The standard could coincide with the optimum provided the optimum was identifiable, a problem that is common to the Pigovian tax solution as well. So far, then, there is not much to choose between standards and taxes – both seem to require detailed information on the MNPB and MEC functons for an optimum to emerge.

But the penalty P also happens to be inefficient in this case. The polluter has an incentive to pollute up to Q_B. Why? He will do so because the total penalty up to Q_B is less than the net private benefits from polluting. He will not go beyond Q_B because further pollution attracts a penalty in excess of marginal net benefits. Strictly, we need to rephrase this finding in terms of the probability of the penalty being suffered. Remember, the polluter has to be caught by the pollution inspector and that is often difficult where, for example, there are many polluters in the area, each contributing a comparatively small amount to the total level of pollution. The calculation that the polluter does, therefore, is to compare the penalty *multiplied* by the probability of facing the penalty, with the net benefit of polluting. Even if the penalty is certain in Figure 7.1, it still pays to pollute up to Q_B.

This discussion should indicate quickly what the second broad requirement is for a standard to be optimal. It is that the penalty should be certain and that it should be equal to P^*. For the standard to be optimal we require that it be set in such a way that the output level corresponding to the standard is optimal, *and* the penalty level should be set equal to P^* and have 100 per cent certainty of being imposed for a transgression of Q^*.

The difficulties of securing these conditions explains why economists tend to be wary of standards.

7.2 TAXES VERSUS STANDARDS

The preceding section indicates a basic reason for preferring taxes to standards. Other considerations are also relevant and are discussed below.

Taxes as least-cost solutions

In Chapter 6 it has already been demonstrated that if a standard is to be adopted, a tax is the best way of achieving it. Clearly, this is not an issue of the superiority of taxes over standards, but a demonstration that a 'mix' of standards and taxes will, generally, be preferable to the adoption of standards alone.

Uncertainty and the benefit function

Figure 7.2 shows the basic pollution diagram but it is assumed that there is some uncertainty about the precise location of the benefit function. MNPB(true) shows the actual one and MNPB(false) the wrong one. The decision-maker assumes that MNPB(false) is the correct curve. Is the cost of his mistake bigger under a standard or a tax? So long as MEC and MNPB have the same (but opposite signed) slopes, the costs of being wrong are the same and there is no reason to prefer a tax to a standard. Thus, the tax t is set on the basis of trying to secure the optimal level of pollution assuming MNPB(false) is the correct curve. But MNPB(true) is the correct curve and hence the polluter, knowing this, goes to the point where MNPB(true) equals t. The effect is too much pollution (Q' instead of Q^*). The loss associated with the excess pollution is the area under MEC between Q^*Q' *minus* the area under MNPB(true) between Q^*Q'. This is shown as the triangle bde.

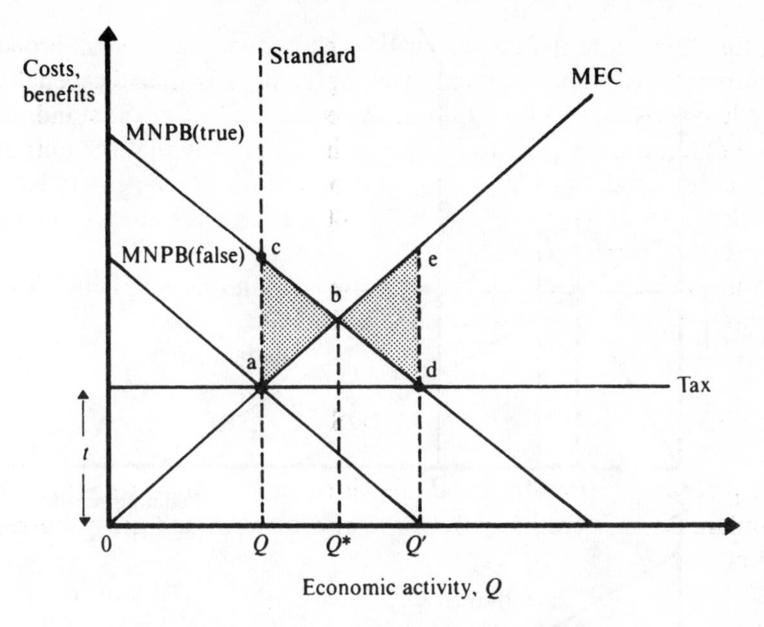

Figure 7.2 Equivalence of tax and standard.

Now assume the regulatory authority decides to set a standard, still believing in MNPB(false). The standard is set at Q. Provided the standards is rigidly enforced (but see Section 7.1), the level of activity is at Q, below the optimum Q^*, and with a loss of abc. It will be seen that the two shaded triangles are of equal size and hence there is nothing to choose between a tax and a rigidly enforced standard.

Figure 7.3 repeats the analysis but this time the two curves have different slopes. In case (*a*) the MEC curve is steeper than MNPB, and in case (*b*) it is less steep. Observation will show that in case (*a*) the tax solution produces a very much larger loss of welfare, i.e. the standard is to be preferred. In case (*b*) the standard produces the bigger loss – the tax is to be preferred. Notice that all these results hold just the same if it was the MEC function about which we are uncertain.

Clearly, the information requirements for making a rational choice between taxes and standards are quite formidable. Essentially, if the regulator does not know the *location* of MNPB but knows the *relationship between the slopes of MNPB and MEC* then he can make the right decision. But the regulator is very unlikely to know the relative slopes of the functions if he does not know even the scale of one of them.

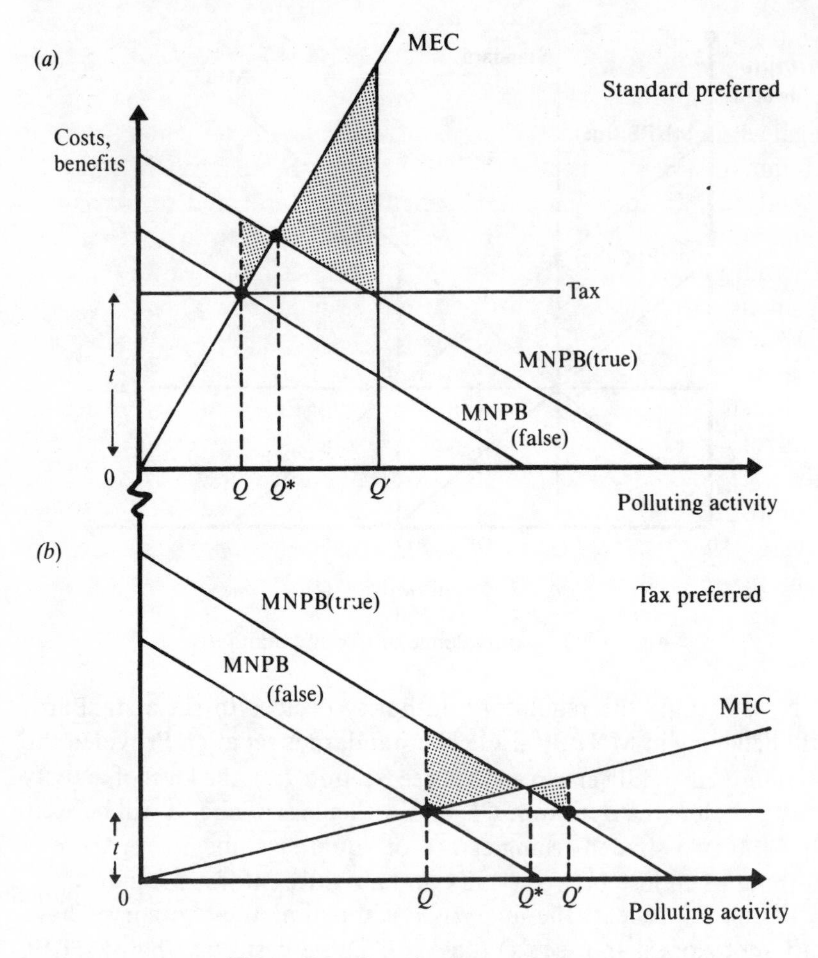

Figure 7.3 Standards versus taxes.

Dynamic efficiency

Taxes are superior to standards in one other respect. Inspection of Figure 7.1 shows that up to Q_s the polluter has no incentive to abate pollution. He faces no penalty for wastes emitted up to that point. But it may be socially desirable to encourage polluters to search continually for lower cost technologies for reducing pollution. Under the standard-setting approach this incentive does not exist. With a tax, however, the polluter still pays the tax on the optimal amount of pollution – recall the discussion in Chapter 6 – and hence has a continuing incentive to reduce pollution.

Administrative costs

The tax solution is certainly costly to implement. It is also open to legal wrangling if the tax is based on a measure of the economic value of damage which is disputed by the polluter. Since industry typically spends significant sums on challenging standards and regulation in general, it is not clear that this is a real criticism of the tax solution. The administrative costs of imposing the tax may also differ little from those involved in ensuring that standards are kept. In both cases monitoring is required. Standard-setting implies that a penalty system be in place and implementable. Taxes require that fees be collected. Some economists have argued that technology-specific controls are cheapest to administer, i.e. regulations of the form that a given technology must be used. Again, however, there must be monitoring and a penalty system for disobeying the requirement. Overall, it is far from clear that standards are cheaper to administer than taxes – only individual case studies will decide the issue.

Outright prohibition

There is one circumstance in which a tax is self-evidently inferior to a standard. This is where the pollutant is so damaging that an outright ban on its use is called for. In such circumstances we are effectively saying that the MEC curve is vertical – there are infinite marginal damage costs associated with the use of the pollutant. Alternatively, there is such uncertainty that we decide it is too risky to use the pollutant. This situation fits a number of ecotoxins and food additives. Clearly, there is no point in having a tax in these circumstances since the revenues would never be collectable.

7.3 POLLUTION REDUCTION SUBSIDIES

We have concentrated on regulatory mechanisms that use the 'stick' – a tax or a penalty for exceeding a standard. But why not approach the issue differently and encourage polluters to install abatement equipment by having a subsidy on the amount of pollution reduced? Like standards, subsidies are not popular with economists. It is important to understand the nature of a subsidy in this context. The idea is to give payments to firms who pollute below a certain prescribed level. Let the subsidy be S per unit of pollution, the

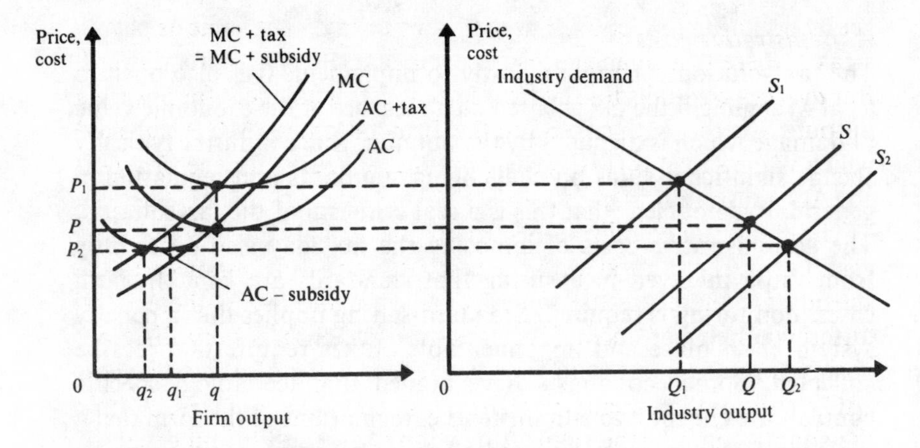

Figure 7.4 Taxes versus subsidies for individual firms (*left*) and the industry (*right*)

prescribed level be W and the actual level achieved by a polluter be M; M is below W. The subsidy payment is then

$$\text{Subsidy} = S\,(W - M)$$

Figure 7.4 illustrates what happens. The diagram shows the position of each individual *firm* on the left and the *industry* on the right. The distinction turns out to be important. The initial points are P,q for the firm, with price being equal to the lowest point on the average cost curve AC, and P,Q for the industry with aggregate supply curve S. Note that the P = AC condition means that we are considering an industry in which there is free exit and entry. First consider the effect of a tax. This will shift AC and MC upwards for the firm, bringing about a new *short-run* equilibrium where the ruling price, P, equals the new marginal cost at q_1 for the firm. But the ruling price is now below the new average cost so firms will exit the industry, shifting the industry supply curve to the left. A new long-run equilibrium is therefore P_1,Q_1 for the industry and P_1,q for the firm. This is fairly straightforward and as we would expect.

The effect of the subsidy is a little more difficult to analyse. This *raises* the firm's MC curve. If the subsidy is the same amount as the tax, the curve will shift to (MC + subsidy) which is the same as (MC + tax). This seems odd – surely subsidies will *lower* the MC curve? In this case this is not so and the formulation of the subsidy explains why. As the firm expands output, it *foregoes* a subsidy which it could get by pollution reduction. Foregoing a subsidy is the same as paying a tax – there is a financial loss in each case. So MC shifts upward. But *average* cost falls for the firm since it gets a payment for lowering output. So, the MC curve for the firm becomes (MC + subsidy) which is the same as (MC + tax), but the AC curve for the firm falls to (AC – subsidy).

The short-run equilibrium is where price equals the new marginal cost, i.e. q_1, the same as with the tax. The short-run responses to the subsidy are therefore the same as those for the tax – there is no difference between them. The long-run response is very different however. In the short run, price now exceeds the new *average cost* (AC – subsidy) and hence new firms will enter the industry, shifting the supply curve to the right. A new long-run equilibrium occurs at P_2,Q_2, and P_2,q_2 for the individual firm.

What happens to pollution? The relevant comparison is what happens in the long run. Under the tax, industry output falls and hence pollution falls. Under the subsidy, however, industry output expands the pollution expands. Even though pollution per firm has fallen in Figure 7.4, the number of firms has increased. A subsidy, then, runs the risk of altering the exit and entry conditions into the polluting industry in such a way that, instead of reducing pollution, it may actually increase it.

8 · MARKETABLE POLLUTION PERMITS

8.1 THEORY OF MARKETABLE PERMITS

The idea of pollution permits was introduced by J.H. Dales (1968). As with standard-setting, the regulating authority allows only a certain level of pollutant emissions, and issues permits (also known as pollution 'consents' or certificates) for this amount. However, whereas standard-setting ends there, the pollution permits are tradeable – they can be bought and sold on a permit market.

Figure 8.1 illustrates the basic elements of marketable permits. MAC is the marginal abatement cost curve which, as Chapter 6 showed, can also be construed as the MNPB function if the only way of abating pollution is to reduce output. The horizontal axis shows the level of emissions and the number of permits: the easiest assumption to make is that one permit is needed for each unit of emission of pollution. The optimal number of permits is OQ^* and their optimal price is OP^*. That is, the authorities, if they seek a Pareto optimum, should issue OQ^* permits. S^* shows the supply curve of the permits: their issue is regulated and is assumed not to be responsive to price.

The MAC curve is in fact the demand curve for permits. At permit price P_1, for example, the polluter will buy OQ_1 permits. He does this because, in terms of control strategies, it is cheaper to abate pollution from Q_2 back to Q_1 than to buy permits. To the left of Q_1, however, it is cheaper to buy permits than to abate pollution. MAC is thus the demand curve for permits.

8.2 THE ADVANTAGES OF MARKETABLE PERMITS

Why do the permits have to be marketable? There are six main attractions of marketability.

1. Cost minimisation

Figure 8.2 repeats Figure 8.1, but omits the MEC curve. It also shows the overall MAC curve as being the sum of the individual polluter's MAC curves. We assume just two polluters for simplicity. This aggregation is legitimate because it was shown above that the MAC curve is the demand curve for permits: adding the curves up is therefore the same as aggregating any set of demand curves. By reference to the individual MAC curves of the two polluters we can see how many permits are purchased. Polluter 1 buys OQ_1 permits, and polluter 2 buys OQ_2 permits at price P^*. Note that the higher

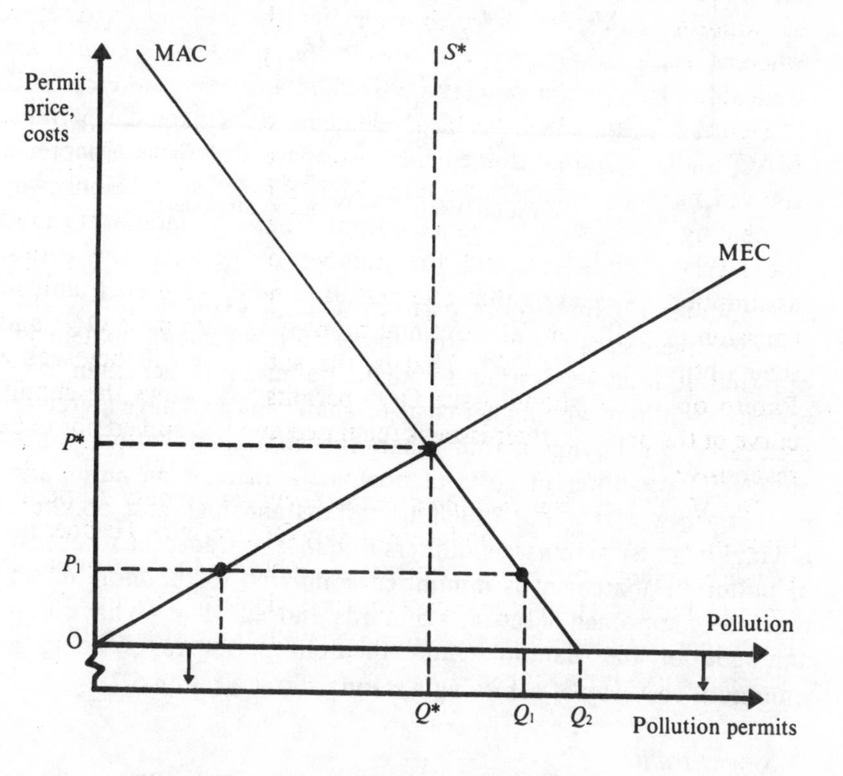

Figure 8.1 The basic analytics of marketable permits.

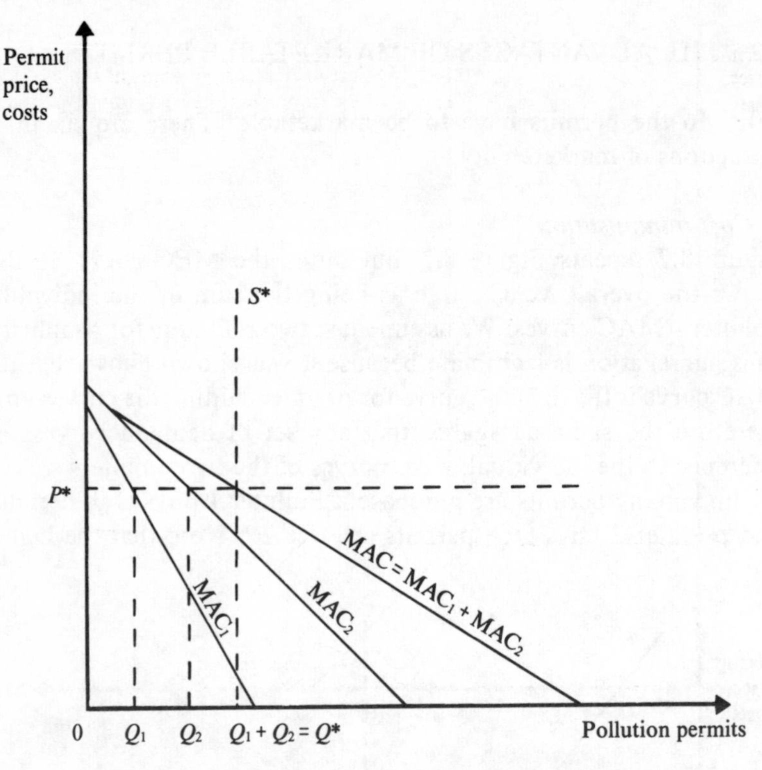

Figure 8.2 Cost minimisation with marketable permits.

cost polluter (2) buys more permits. This gives us a clue to the cost-effectiveness of permits. Polluters with low costs of abatement will find it relatively easier to abate pollution rather than buy permits. Polluters with higher costs of abatement will have a greater preference for buying permits than for abating pollution. Since polluters have different costs of abatement there is an automatic market – low-cost polluters selling permits and high-cost polluters buying them. By giving the polluters a chance to trade, the total cost of pollution abatement is minimised compared to the more direct regulatory approach of setting standards. Indeed, what we have is an analogue of the Baumol–Oates theorem about taxes being a minimum-cost way of achieving a standard (see Section 6.7).

2. New entrants

Suppose new polluters enter the industry. The effect will be to shift

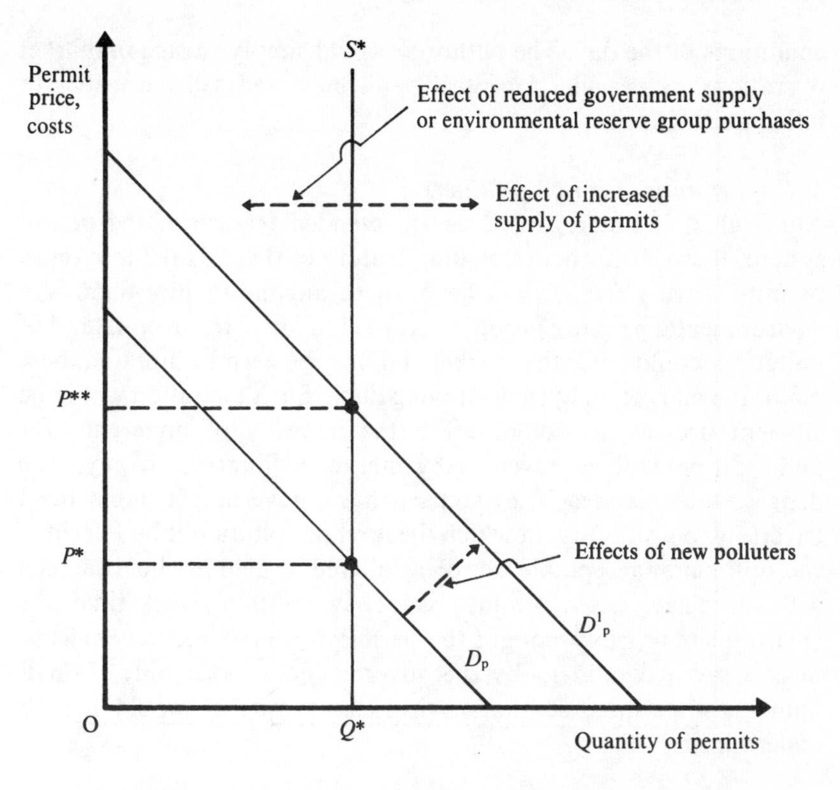

Figure 8.3 Changing the supply and demand for permits.

the aggregate pollution permit demand curve to the right, as in Figure 8.3. As long as the authorities wish to maintain the same level of pollution overall, they will keep supply at S^* and the permit price will rise to P^{**}. The new entrants will buy permits if they are high abatement cost industries, otherwise they will tend to invest in pollution control equipment. Once again, the overall cost minimisation properties of the permit system are maintained. But suppose the authorities felt that the increased demand for permits should result in some relaxation in the level of pollution control. Then they could simply issue some new permits, pushing the supply curve S^* to the right. Alternatively, if they felt that the old standard needed tightening they could enter the market themselves and buy some of the permits up, holding them out of the market. The supply curve would shift to the left. In short, the permit system opens up the possibility of varying standards with comparative ease to reflect the conditions of the day. The authority would simply engage in market operations, rather like a central bank buys and sells securities to influence their price.

3. *Opportunities for non-polluters*

Although it is not regarded as an intended feature of the permit system, there is another intriguing feature of them. If the market in permits is truly free, it will be open to anyone to buy them. An environmental pressure group, concerned to lower the overall level of pollution, could enter the market and buy the permits, holding them out of the market, or even destroying them. Such a solution would be efficient because it would reflect the intensity of preference for pollution control, as revealed by market willingness to pay. The danger with this idea is, of course, that a government might react adversely to a situation in which the level of pollution it had decided was optimal or acceptable was being altered by people who disagreed with it. They might simply issue new permits each time the environmental group bought the permits. In practice, the environmental group would lobby the government to issue only a small number of permits, so that environmental quality would not be undermined.

4. *Inflation and adjustment costs*

Permits are attractive because they avoid some of the problems of pollution taxes. As we saw in Chapter 6, even where a standard is set and taxes are used to achieve it, there are risks that the tax will be mis-estimated. With permits it is not necessary to find both the desirable standard and the relevant tax rate; it is necessary only to define the standard and find a mechanism for issuing permits. Moreover, if there is inflation in the economy, the real value of pollution taxes will change, possibly eroding their effectiveness. Because permits respond to supply and demand, inflation is already taken care of. Taxes also require adjustment because of entry to, and exit from, the industry. Permits, as we have seen, adjust readily to such changes, whereas taxes would require adjustment.

5. *The spatial dimension*

We have tended to assume that there are just a few polluters and that the points at which the pollution is received (the 'receptor points') are also few in number. In practice we are likely to have many emission sources and many receptor points. If we are to set taxes with at least a broad relationship to damage done, it will be necessary to vary the taxes by source since different receptor points will have different assimilative capacities for pollution. Additionally, there are likely to be *synergistic* effects. That is, several pollutants may combine to produce aggregate damages larger than the sum of the damages from single pollutants. This raises the spectre of a highly complex and administratively burdensome system. To a considerable extent permits avoid this spatial problem. To investigate this further we need to look briefly at different types of permit systems.

6. *Technological 'lock-in'*

Permits are also argued to have an advantage over charges systems with respect to 'technological lock-in'. Abatement expenditures tend to be 'lumpy'; to increase the level of effluent removal, for example, it is frequently necessary to invest in an additional type of abatement process. Adjustments to changes in charges are therefore unlikely to be efficient unless the changes in the charge can be announced well in advance and can be backed by some assurance that a given charge level will be fairly stable over the short and medium term. The charge approach also risks underestimating abatement costs. For example, if the aim is to achieve a given standard, then, together with the regulating authority's assessment of abatement costs, this will determine the relevant charge. If the authority is wrong about the abatement costs, however, the charge could be set too low in the sense that polluters will prefer to pay it than to invest in abatement equipment, thus sacrificing the desired standard. This reluctance of polluters to invest in equipment will be strengthened by the previously discussed 'lumpiness' factor. A permit system generally avoids this problem of lumpy investment, the authority's uncertainty about abatement costs, and polluters' distrust of charges. This is so because the permits themselves are issued in *quantities* equal to the required standard, and it is prices that adjust. The consequences of an underestimate of abatement costs in the presence of permits is simply that the price of permits is forced up (since the demand for them is determined by abatement costs, as we saw), whereas the environmental standard is maintained (Rose-Ackerman, 1977).

8.3 TYPES OF PERMIT SYSTEMS

The literature has tended to classify three types of permit system. The *ambient permit system* (APS) works on the basis of permits defined according to exposure at the receptor point. Quality standards might vary according to the receptor point: there is no need for each receptor point to have the same ambient quality standard. Under an APS, then, permits have to be obtained from the market in permits at the receptor point. This means that the trade in permits will not be on a one-for-one basis; it will be necessary to trade on the basis of the number of permits required to allow a given amount of pollution concentration at the receptor point. Each polluter, then, may face quite complex markets – different permit markets according to different receptor points, and hence different prices.

The *emissions permit system* (EPS) is much simpler. It simply issues permits on the basis of source emissions and ignores what effects those emissions have on the receptor points. Within a given region or zone, then, the polluter would have only one market to deal with and one price, the price of a permit to emit pollutants in that area. Trade in permits is on a one-for-one basis.

The APS has obvious complications for the polluters and may well be an administrative nightmare for the regulators as well. The EPS is simpler but has other problems. By not discriminating according to receptor points it is unlikely to discriminate between sources on the basis of the damage done. It will therefore be inefficient. Put more formally, the price of permits will not approximate the marginal external cost. Second, any one area is likely to experience some concentration of pollution in specific small areas – so-called 'hot spots' – where actual concentrations exceed the standard. Because the EPS is emission-based across a wider area, it will not take account of this failure to observe the standard at all points. The simple technique of re-defining the area so that the hot spot is contained within a narrower zone to which the standard applies really amounts to turning the EPS into an APS, and we are back to the complexities of many markets and prices. The EPS also works on the basis of a one-for-one trade within the defined zone – there is no trade outside the zone. With the APS, however, all receptor points are taken into account. EPS could thus result in damage outside the zone being ignored.

To overcome these difficulties a third system has been proposed.

This is the *pollution offset* (PO) system. Under the PO system, the permits are defined in terms of emissions, trade takes place within a defined zone, but trade is not on a one-for-one basis. Moreover, the standard has to be met at all receptor points. The exchange value of the permits is then determined by the effects of the pollutants at the receptor points. The PO system thus combines characteristics of the EPS (permits are defined in terms of emissions, and there is no trade outside the defined area) and the APS (the rate of exchange between permits is defined by the ambient effects).

Which is the best system? Tietenberg (1985) has reviewed much of the evidence. His review suggests that EPS is more expensive than APS in terms of the total abatement costs likely to be involved. But the APS is also judged to be a largely unworkable system because of its complexity. How then does EPS fare in comparison to the more traditional standard-setting, or 'command-and-control' systems? The evidence is varied and is not easy to compare as the two systems might have different amounts of emission control because of difficulties in the spatial configuration of the requirements to meet the standard. The PO system was not evaluated.

8.4 PERMIT TRADING IN PRACTICE

There is some experience of pollution permit trading in the United States. The Clean Air Act (1970) established National Ambient Air Quality Standards (NAAQSs) which were to be implemented by the individual states under State Implementation Plans (SIPs). The Act marked the introduction of federal control, through the Environmental Protection Agency (EPA), over what had previously been a state responsibility alone. The SIP for each state had to indicate to EPA how the state would implement the ambient standards for all pollutants other than 'new sources' which were controlled directly by standard-setting by EPA.

In 1977 the Clean Air Act was amended to allow for the fact that many states were not meeting the ambient standards. Areas not meeting the standards were declared to be *non-attainment* regions. Stringent regulations were applied to these regions. All 'reasonably available control technologies' (RACTs) had to be applied to existing plant, and there had to be 'reasonable further progress' in achieving annual reductions so that the standard could be achieved. New sources were subject to construction permits which were conditional on the use of the 'lowest achievable emission rate' (LAER), the lowest emission rate demonstrated to have been achieved elsewhere. In the area where standards had been met, the focus switched to prevention of significant deterioration (PSD), i.e. to ensuring that the areas did not deteriorate.

The other main change in 1977 was the introduction of an emissions trading programme. Basically this operates through an *emission reduction credit*. Suppose a source controls emissions more than it is required to do under the standard set. Then it can secure a credit for the 'excess' reduction. The credit could then be traded in several ways. The first way is through a policy of *offsets*. These can be used in non-attainment areas, allowing new sources to be established, and which thus add to emissions, provided there is a credit somewhere else in the region. The new source effectively buys the credits from existing sources, the overall pollution level is not increased, and new industry is not unduly deterred from setting up in non-attainment regions that would otherwise suffer a loss of income and employment.

The second way is through a *bubble policy*. A 'bubble' is best thought of as an imaginary glass dome covering several different sources of pollution, either several points within one plant, or several different plants. The aim is not to let the overall emissions from the imaginary bubble exceed the level required by the standard-setting procedure. If any one point exceeds the RACT standard, for example, it can be compensated for by securing emission reduction credits elsewhere within the bubble.

The third procedure utilises *netting*. This is similar to the bubble, but relates to sources undergoing modification and which wish to avoid the rigours of being classified as a new source and subjected to the stricter standard (LAERs). Again, so long as plant-wide emissions do not increase, the modified source can increase emissions if there are emission reduction credits to offset the increase.

Lastly there is *banking* whereby sources can store up emission reduction credits for use later in a netting, bubble or offset context.

These components have a clear affinity with the permit trading systems discussed previously. The actual progress of these legislative features of the US policy is complex and varied. An overall evaluation of the policy is difficult, but several general observations stand out. First, trading has tended to result in better air quality, although there are exceptions. Second, there appear to have been significant cost savings. Third, the offset policy probably has assisted regions which would otherwise have suffered economically because of firms being unable to set up in non-attainment regions. Fourth, administrative costs have been high. Fifth, it is probable that abatement technology introduction has been stimulated by the policy. By 1986 the total number of bubbles in existence was thought to be about 250; 3,000 offset transactions were reported. The amount of netting appears not to be known and banking has had a very limited impact.

9 · MEASURING ENVIRONMENTAL DAMAGE I: TOTAL ECONOMIC VALUE

9.1 THE MEANING OF ENVIRONMENTAL VALUATION

The preceding chapters have discussed alternative ways of correcting excessive pollution levels – letting a market in externality develop, taxes, standards and marketable permits. It was shown that some form of regulatory approach will generally be required – it is very unlikely that markets in externality will develop. The remaining instruments of regulation can be used in two sets of circumstances:

1. Situations where no attempt is made to identify the economically optimal level of pollution.
2. Situations where efforts are made to determine the optimum and then achieve it.

In the first case there is no requirement to measure the external cost curve (MEC). We determine a standard, perhaps on health-related criteria, and find the best way of achieving that standard. We saw that taxes and marketable permits had attractive characteristics in this respect. In the second case we have first to identify the optimum, or approximate it, and then set the standard or tax accordingly. As we saw, to do this we need also to know the private benefit function of the polluter (MNPB).

This provides the first justification for trying to *measure* environmental damage, i.e. to identify the MEC curve. It is important to recognise that the measurement in question is in *money* terms. If it was in any other units we could not identify the optimum because the MNPB curve (or the abatement cost curve, MAC) is measured in these units. For the purposes of this chapter, therefore, 'valuation' means money valuation.

The idea of putting a money value on damage done to the environment strikes many as illicit, even immoral. The justification for monetary valuation lies in the way in which money is used as a *measuring rod* to indicate gains and losses in utility or welfare. That is, money is the means of measurement. It must not be confused with more popular concepts about making money as an *objective* – crude greed, profit at the expense of others, the pusuit of Mammon. The reason money is used as the measuring rod is that all of us express our preferences every day in terms of these units – when buying goods we indicate our 'willingness to pay' (WTP) by exchanging money for the goods, and, in turn, our WTP must reflect our preferences. We might use any other units provided they can be applied meaningfully to both the benefit and cost sides of the pollution picture, and provided both reflect the preferences of individuals. Some attempts have been made to find other units – notably energy units – but, even if they can be applied to both sides of the picture, they have no meaning in terms of *preference* revelation. Accordingly, money units remain the best indicator we have. Environmental economists simply have to bear the burden of trying to explain what the use of money measures means, and what it does not mean. Misunderstanding is something we can reduce, but probably not eliminate.

Because money valuation relates back to individual preferences, it does however follow that any rejection of preference as the proper basis for decisions about the environment will entail rejection of the use of money values, or economic values as we shall call them. This is important. Many commentators on environmental economics observe that there is a multiplicity of values – we cannot subsume duty, obligation, keeping promises, love, and natural justice under economic values. What is more, each type of value has a different *moral standing* according to the viewpoint of the individual. Some see duty as the dominant moral rule; others see consistency (doing unto others only that which you would wish to see done to you, for example); still others see natural justice as the important rule. Chapter 15 discusses these profound issues in more detail. In this chapter we begin with the assumption that it is economic value that counts, although, as we shall see, the detection and measurement of those values seems to raise many of the concerns that the critics express about economic values.

Table 9.1 Pollution damage (in billions) in the Netherlands.

Pollution	Cumulative damage to 1985		Annual damage 1986	
	Dfl	US$	Dfl	US$
Air pollution	4.0–11.4	1.2–3.0	1.7–2.8	0.5–0.8
Water pollution	n.a.	n.a.	0.3–0.9	0.1–0.3
Noise nuisance	1.7	0.5	0.1	0.0
Total	5.7–13.0	1.7–3.5	2.1–3.8	0.6–1.1

Sources: (1) Netherlands Ministry of Public Housing, Physical Planning and Environmental Management, *Environmental Program of the Netherlands 1986–1990*, The Hague, 1985. (2) J. B. Opschoor, 'A Review of Monetary Estimates of Benefits of Environmental Improvements in the Netherlands', OECD Workshop on the Benefits of Environmental Policy and Decision-Making, Avignon, France, October 1986.

9.2 THE USES OF ECONOMIC VALUE

We have already identified a major use to which economic value measurements can be put: they should enable us to identify, or at least approximate, the optimum. We may wish to do this *ex ante*, i.e. before deciding on a type of environmental regulation. We may wish to do it *ex post*, i.e. after a regulation has been imposed, to see if the regulation has got us nearer to the optimum.

A separate use for economic value measurements is to demonstrate the importance of environmental policy. Many of the gains from environmental policy do not show up in the form of immediate monetary gain: the benefits are to be found more in the quality of life than in any increment to a nation's economic output. But it is essentially a historical accident that some gains in human welfare are recorded in monetary terms in the national accounts and others are not. By and large, this is explained by the fact that the accounts measure gains to economic sectors in which property rights – whether private or public – have been well defined. The third party effects of economic activity – noise, air pollution, water pollution, etc. – do not show up in the accounts either because the ill-defined or absent rights to clean air, peace and quiet and pure water mean that no monetary transfer takes place between polluter and polluted, or because such transfers as do take place (e.g. through court action) are not part of the national accounting conventions. Thus environmental benefits tend to be less 'concrete', more 'soft' than market-place benefits. The temptation is to downgrade them by comparison.

We can view the widespread support for environmental policy as a reflection of the inappropriateness of this downgrading process. In reality, the environment is valued highly and one task in environmental policy is to record and measure these environmental values in whatever ways possible.

It is possible to illustrate the way in which benefit estimation techniques have been used to measure the importance of damage to the environment and, conversely, the benefits of environmental policy.

Table 9.1 shows estimates for the costs of environmental damage in the Netherlands. Note that these are damage estimates arising from pollution. A good many types of damage did not prove capable of 'monetisation', so that, if the monetised figures are accepted, actual damage exceeds the estimates shown. Various techniques were

Table 9.2 Pollution damage in the Federal Republic of Germany (1983–85)

Pollution	DM billion	US$ billion
Air pollution		
Health (respiratory disease)	2.3–5.8	0.8–1.9
Materials damage	2.3	0.8
Agriculture	0.2	0.1
Forestry losses	2.3–2.9	0.8–1.0
Forestry recreation	2.9–5.4	1.0–1.8
Forestry (other)	0.3–0.5	0.1–0.2
Disamenity	48.0	15.7
Water pollution		
Freshwater fishing	0.3	0.1
Ground water damage	9.0	2.9
Recreation	n.a.	n.a.
Noise		
Workplace noise	3.4	1.1
House price depreciation	30.0	9.8
Other	2.0	0.7
Total	103.0	33.9

Source: Adapted from data given in W. Schulz, 'A Survey on the Status of Research Concerning the Evaluation of Benefits of Environmental Policy in the Federal Republic of Germany', OECD Workshop on the Benefits of Environmental Policy and Decision Making, avignon, France, 1986.

used to derive the figures and considerable caution should be exercised in quoting or using them. They are, at best, 'ball park' numbers. Nonetheless, they show that even measured damage is a significant cost to the economy – the totals shown are 0.5–0.9 per cent of the Netherlands' GNP.

Table 9.2 presents similar estimates for the Federal Republic of Germany. Again, many items have not been valued and differing techniques are used to derive the estimates. The figures shown total to over 100 billion Deutschmarks annual damage (about US $34 billion), the major part of which is accounted for by the disamenity effects of air pollution (which is likely to include some of the separately listed air pollution costs), and the effects of noise nuisance on house values. The important point is that, if the estimates can be accepted as being broadly in the area of the true costs, pollution damage was costing an amount equal to 6 per cent of the Federal Republic of Germany's GNP in 1985.

Table 9.3 shows estimates for the USA for the year 1978. However, in this case the figures are for *damage avoided* by environmental policy. That is, taking the total of $26.5 billion, the argument is that, in the absence of environmental policy, pollution damage would

Table 9.3 The benefits of pollution control in the USA (1978).

Pollution	US$ billion
Air pollution	
Health	17.0
Soiling and cleaning	3.0
Vegetation	0.3
Materials	0.7
Property Values[a]	0.7
Water pollution[b]	
Recreational fishing	1.0
Boating	0.8
Swimming	0.5
Waterfowl hunting	0.1
Non-user benefits	0.6
Commercial fishing	0.4
Diversionary uses	1.4
Total	26.5

Source: A.M. Freeman, *Air and Water Pollution Control: A Benefit-Cost Assessment*, Wiley, New York, 1982.
[a] Net of property value changes thought to be included in other items.
[b] At one half the values estimated for 1985.

have been $26.5 billion higher in 1978 than it actually was. The total shown in Table 9.3 would be 1.25 per cent of GNP in 1978. The marked divergence between this figure and the percentage suggested for Germany is partly explained by the absence of estimates for noise nuisance, and by the very low figure for property value changes.

9.3 COSTS, BENEFITS, WILLINGNESS TO PAY AND WILLINGNESS TO ACCEPT

We have seen that an underlying purpose in attempting a monetary measure of the environment is to provide a check on the economic rationality of investing in environmental improvement. The cost of such improvements is measured in money terms and the monetary sum involved should approximate the value to society of the resource used up. Since resources are scarce it is important to establish that the gain from the policy exceeds the resource cost, and this can only be done by measuring the benefit in the same units as the costs. In fact expenditures should be undertaken until the extra *benefits* are just equal to the extra *costs*. In formal terms, *marginal benefit* should equal the *marginal cost* of providing that benefit. In turn, this equivalence meets the requirement that the scarce resources in the economy be used in their most efficient way, i.e. given a certain level of resources, the 'marginal benefit equals marginal cost' rule maximises the total net benefit that can be achieved with these resources.

As noted previously, it is important to understand that the concept of benefit is interpreted in a particular way. The basic idea is that 'what people want' – individuals' preferences – should be the basis of benefit measurement. The easiest way to identify these preferences is to see how people behave when presented with choices between goods and services. We can reasonably assume that a positive preference for something will show up in the form of a *willingness to pay* for it. In turn, each individual's willingness to pay will differ. Since we are interested in what is socially desirable, we can aggregate the individual willingness to pay to secure a total willingness to pay. The willingness-to-pay (WTP) concept thus gives an automatic monetary indicator of preferences. While we can safely assume that people will not be willing to pay for something they do not want, we cannot be sure that WTP as measured by market prices accurately

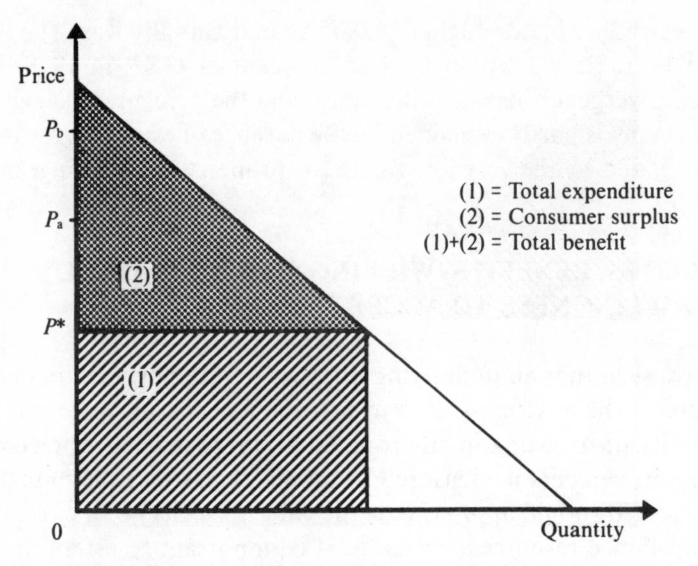

Figure 9.1 A demand curve for environmental goods.

measures the whole benefit to either individuals or society. The reason for this is that there may be individuals who are willing to pay *more* than the market price. If so, their benefit received is larger than market price indicates. The 'excess' that they obtain is known as *consumer surplus*.

Accordingly we can write the following fundamental rule:

Gross WTP = Market price + Consumer surplus

The idea can be illustrated with the aid of a diagram showing a demand curve. Figure 9.1 shows that the market price, determined by forces of supply and demand in this case, is P^*. Since it is not possible to charge a different price to each and every individual buying the good, P^* becomes the market price for everyone. But individual A can be seen to be willing to pay a higher price: P_a. Similarly, individual B is willing to pay a price P_b. The total amount of benefit obtained is in fact the entire area under the demand curve shown by the two shaded areas. The shaded rectangle is the total expenditure by individuals on this particular good, and the shaded triangle is the consumer surplus. The two areas together then measure total benefit.

The intuitive basis to monetary benefit measurement is thus rather simple. People reveal their preferences for things they desire by

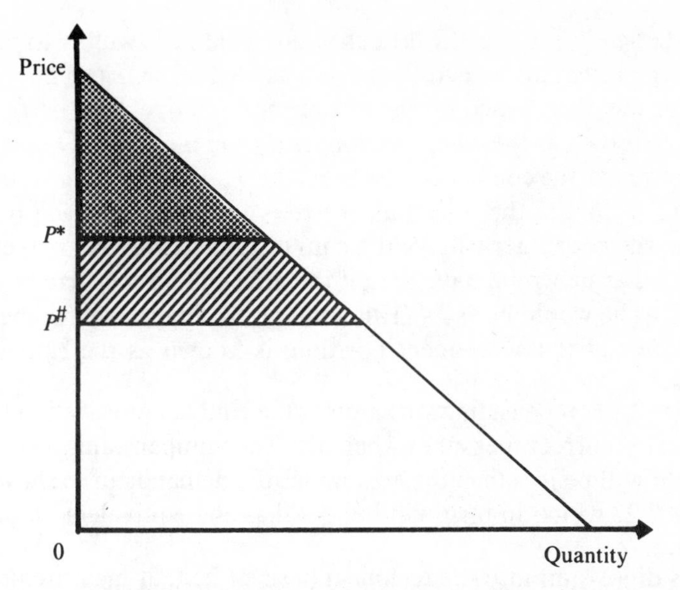

Figure 9.2 Changes in price and welfare gains.

showing their willingness to pay for them. Market price is our initial guide to what people are willing to pay and hence total expenditure on the good is our first approximation of benefit received. But since there will be people willing to pay more than the market price, and hence who secure a surplus of benefit over expenditure, *gross* WTP will exceed total expenditure. What we seek in benefit measurement, then, is a measure of areas under demand curves.

As it happens, the strict requirements for areas under demand curves to measure benefits is more complicated than this. Demand curves of the kind shown in Figure 9.1 have the same income level as we move up or down the demand curve. Along such demand curves, known as *Marshallian demand curves*, income is held constant. We require that individuals' welfare, well-being or 'utility' be held constant, which somehow means correcting the demand curve for the fact that utility varies as we move up and down the demand curve. Such adjustments have been worked out in the economics literature. Figure 9.2 shows the same demand curve as Figure 9.1 but this time P^* falls to $P^{\#}$ because of some change in the market. It will be evident that the price falls make the consumer 'better off' because the total shaded area (consumer surplus) has actually increased. The gain from the price fall is shown by the heavy shaded area.

Hypothetically, we can ask the consumer what he is willing to pay to secure the price fall so as to leave him as well off at *P#* as he was at *P**. This measure, based on the income and relative price pertaining to *P**, is known as the *compensating variation* measure of benefit. If instead we ask the consumer how much he would be willing to accept in order to forego the price fall, the relevant base point will be *P#*. That is, the consumer will want a sum of money that will make him as well off as he would have been if the price fall has occurred, i.e. as well off as he would be at *P#*. This sum, pertaining to the income and price levels at the subsequent position, is known as the *equivalent variation.*

Either the compensating variation or the equivalent variation is the technically correct measure of benefit. The compensating variation measure will be less than the area under the demand curve shown in Figure 9.2, which in turn will be less than the equivalent variation measure.

This digression into the technical basis of benefit measurement is important because it reveals that we have two basic concepts of benefit: one based on willingness to pay (WTP) and another based on willingness to accept (WTA). The theory of economics tells us that these ought not to differ very much but, as we shall see, some empirical studies suggest that there may be marked differences between the two. To obtain some idea of why this appears to happen consider the intuitive basis of the two measures. WTP has already been explained: individuals reveal their preferences for an environmental gain by their willingness to pay for it in the market place (we consider in a moment the fact that most environmental goods and services have no markets). But we are often faced with the problem of how we value an environmental *loss*. In that case we can ask how much people are willing to pay to prevent the loss or how much they are willing to accept in the way of compensation to put up with the loss. In short, there will be two measures of benefit gained from an environmental improvement and two measures of loss, or 'damage', from an environmental deterioration. The measures are:

1. WTP to secure a benefit.
2. WTA to forego a benefit.
3. WTP to prevent a loss.
4. WTA to tolerate a loss.

Why should these measures differ? Individuals appear to view losses differently to gains, a phenomenon that psychologists refer to as 'cognitive dissonance'. Given an initial position, they see an extra benefit as being worth so much, but a removal of some part of what they already have is seen differently, perhaps as containing some infringement against what they regard as being theirs 'by right'. Certainly, the phenomenon of asymmetry in the valuation of gains and losses in relation to some initial position is known to psychologists. They differentiate the benefit case from the loss case, referring to the former as having a 'purchase structure' and the latter as having a 'compensation structure'. How the values differ in the two contexts depends very much on what is considered by the individual as being the 'normal' state.

If WTA and WTP do differ significantly, then we have a problem for the measurement of environmental benefits, for many cases will involve the prevention of a loss rather than securing a benefit. It is likely then that the 'compensation structure' will be more important in these cases than the 'purchase structure'. A policy of preventing the loss may not be justifiable if the measure of benefit is based on WTP to prevent the loss, but justifiable if the benefit is measured as WTA compensation to tolerate the loss. It seems fair to say that this issue is not resolved in the environmental economics literature. Psychologists express little surprise that WTP and WTA are not the same; some economists find that they differ in many studies; others find that they may converge if the study is formulated in a particular way, and economic theorists tend to dispute that WTP and WTA can differ so much simply because the theory says that they ought not to differ (and hence there must be something wrong with the empirical studies).

9.4 TOTAL ECONOMIC VALUE

We are now in a position to explore the nature of the economic values embodied in the demand curve of Figure 9.1. While the terminology is still not agreed, environmental economists have gone some considerable way towards a taxonomy of economic values as they relate to natural environments. Interestingly, this taxonomy embraces some of the concerns of the environmentalist. It begins by distinguishing user values from 'intrinsic' values. User values, or user benefits, derive from the actual use of the environment. An angler, wildfowl hunter, fell walker, ornithologist, all use the natural environment and derive benefit from it. Those who like to view the countryside, directly or through other media such as photograph and film also 'use' the environment and secure benefit. The values so expressed are economic values in the sense we have defined. Slightly more complex are values expressed through *options* to use the environment, that is, the value of the environment as a potential benefit as opposed to actual present use value. Economists refer to this as *option value*. It is essentially an expression of preference, a willingness to pay, for the preservation of an environment against some probability that the individual will make use of it at a later date. Provided the uncertainty concerning future use is an uncertainty relating to the availability, or 'supply', of the environment, the theory tells us that this option value is *likely* to be positive (see below). In this way we obtain the first part of an overall equation for total economic value. This equation says:

Total user value = Actual use value + Option value

Intrinsic values present more problems. They suggest values which are in the real nature of the thing and unassociated with actual use, or even the option to use the thing. Chapter 1 drew attention to one meaning of 'intrinsic' value, namely a value that resides 'in' something *and that is unrelated to human beings altogether*. Put another way, if there were no humans, some people would argue that animals, habitats, etc. would still have 'intrinsic' value. We drew attention in Section 1.10 to a separate, but not wholly independent concept of intrinsic value, namely value that resides 'in' something but which is captured by people through their preferences in the form of non-use value. For the rest of this chapter it is this second definition of intrinsic value that we use. That is, values are taken to be entities that reflect people's preferences, but those values *include* concern for, sympathy with, respect for the rights or welfare of non-human beings and the values of which are unrelated to human use. The briefest introspection will confirm that there are such values. A great many people value the remaining stocks of blue, humpback and fin whales. Very few of those people value them in order to maintain the option of seeing them for themselves. What they value is the *existence* of the whales, a value unrelated to use although, to be sure, the vehicle by which they secure the knowledge for that value to exist may well be film or photograph or the recounted story. The example of the whales can be repeated many thousands of times for other species, threatened or otherwise, and for whole ecosystems such as rainforests, wetlands, lakes, rivers, mountains, and so on.

These *existence values* are certainly fuzzy values. It is not very clear how they are best defined. They are not related to vicarious benefit, i.e. securing pleasure because others derive a use value. Vicarious benefit belongs in the class of option values, in this case a willingness to pay to preserve the environment for the benefit of others. Nor are existence values what the literature calls *bequest values*, a willingness to pay to preserve the environment for the benefit of our children and grandchildren. That motive also belongs with option value. Note that if the bequest is for our immediate descendants we shall be fairly confident at guessing the nature of their preferences. If we extend the bequest motive to future generations in general, as many environmentalists would urge us to, we face the difficulty of not knowing their preferences. This kind of uncertainty is different to the uncertainty about availability of the environment in the future which made option value positive. Assuming it is legitimate to include the preferences of as yet unborn individuals, uncertainty about future preferences could make option value negative. Provisionally we state that:

Intrinsic value = Existence value

where, for now, existence values relate to values expressed by individuals such that those values are unrelated to use of the environment, or future use by the valuer or the valuer on behalf of some future person.

In this way we can write our formula for total economic value as:

Total economic value = Actual use value + Option value + Existence value

Within this equation we might also state that:

Option value = Value in use (by the individual) + Value in use by future individuals (decendant and future generations) + value in use by others (vicarious value to the individual)

The context in which we tend to look for total economic values should also not be forgotten. In many of those contexts three important features are present. The first is *irreversibility*. If the asset in question is not preserved it is likely to be eliminated with little or no chance of regeneration. The second is *uncertainty*: the future is not known, and hence there are potential costs if the asset is eliminated and a future choice is foregone. A dominant form of such uncertainty is our ignorance about how ecosystems work: in sacrificing one asset we do not know what else we are likely to lose. The third feature is *uniqueness*. Some empirical attempts to measure existence values tend to relate to endangered species and unique scenic views. Economic theory tells us that this combination of attributes will dictate preferences which err on the cautious side of exploitation. That is, preservation will be relatively more favoured in comparison to development.

There is no particular agreement on the nature of the equation for total economic value. Some writers regard intrinsic value as part of existence value rather than as its equivalent. Others regard intrinsic value as being inclusive of option value. To a considerable extent the variations in definition appear to relate to what is meant by 'use'. Thus if it means actual current use by the individual expressing the preference, bequest values are not use values. The view taken here, however, is that the issue of when use occurs and by whom cannot be regarded as differentiating characteristics: all uses, whenever they occur and whoever they are by, give rise to use values. Equally, all use values are conceptually distinct from the intrinsic value of the environment which we currently equate with existence value. It is clear that the concepts of option and existence value need further investigation.

9.5 OPTION VALUE

The willingness to pay for an environmental good, e.g. wildlife preservation, a national park, improved water or air quality, is related to the consumer surplus that the individual expects to receive from that good. We saw that gross WTP was made up of the intended expenditure on the good plus the consumer surplus (CS). The benefit to the individual will therefore be the excess WTP over what is actually paid out, since the latter is the cost to the individual. This excess is CS. Since decisions are made on the basis of what is *expected*, we can say that the relevant CS is *expected* CS, which we write as E(CS).

If we are sure of our capability of buying the good, and of our future preferences, and of the availability of the good when we want it, E(CS) is a proper measure of the benefit of the good. It is this that we would wish to put into our cost–benefit assessment. If it costs an amount C to preserve a wildlife habitat, for example, we can say that it is worth preserving it if $C <$ E(CS). However, the idea that we are certain of both the factors influencing our *demand* for the wildlife habitat, and the factors influencing its *supply* is not realistic. On the demand side we might be unsure of our income and unsure of our preferences in the future. On the supply side, we may be unsure that the habitat will be there for us to enjoy. It is this presence of *uncertainty* that requires us to modify the use of E(CS) as our measure of benefit.

We can illustrate the required modification by considering supply uncertainty. This is very relevant in the real world because natural environments are everywhere being reduced in size and number. We cannot be sure that a given environment will be available to us in the future. The basic idea is that, given this supply uncertainty, and given the fact that most people do not like risk and uncertainty (they are said to be *risk averse*), an individual will be willing to pay *more* than the expected CS in order to ensure that he or she can make use of the environment later on. The total WTP is called *option price* (OP) and it comprises the expected consumer surplus *plus* 'option value' where option value (OV) is the extra payment to ensure future availability of the wildlife habitat; that is:

Option price = Expected consumer surplus + Option value

or

OP = E(CS) + OV

On this basis, simply estimating future use of the wildlife habitat will give us only E(CS) and will ignore OV. We will have underestimated the true value of the habitat.

Once different attitudes to risk are introduced and the uncertainty is extended to the 'demand side', we cannot be sure OV is positive. Indeed, even with supply side uncertainty, there is ambiguity over the sign of OV. The analytical basis for these judgements is complex (see the notes on further reading for this chapter) but the general outcome is as shown in Table 9.4, although the reader is warned that the signs shown for supply uncertainty require certain technical assumptions to be fulfilled.

A further source of value is *quasi option value*. Imagine a development that threatens to destroy the wildlife habitat we have

Table 9.4

	Sign of OV		
	Risk loving	Risk neutral	Risk averse
Demand uncertainty			
Income	+ve	0	–ve
Preferences	?	?	?
Supply uncertainty	–ve	0	+ve

been hypothesising. The development has a certain value in terms of people's willingness to pay for its outcome. An illustration might be a tropical forest which contains a rich range of diverse species which may have future value for scientific and commercial purposes. Many experts argue this, for example, with respect to plant species for pharmaceuticals and for crop breeding. There are uncertain benefits from the preservation of the habitat, but these benefits could become more certain through time as information grows about the uses to which the forest species can be put. But if the development takes place, this source of genetic information is lost for ever. Quasi option value (QOV) is the value of preserving options for future use given some expectation of the growth of knowledge. If QOV is positive it would tend to support the view that the development should be postponed in order to make a better decision later.

The literature suggests that if the expected growth of information is independent of the developments, i.e. we do not need the development to generate the information, then QOV will always be *positive*. If, on the other hand, the information depends on the development, QOV could be positive or negative: positive when the uncertainty is about the benefits of preservation, and negative when the uncertainty is about the benefits of the development. It seems fair to say that the types of information growth in question in the real world are *not* related to development. Hence the presumption must be that QOV is always positive.

9.6 EXISTENCE VALUE

Existence value is a value placed on an environmental good and which is *unrelated to any actual or potential use of the good.* At first sight this may seem an odd category of economic value for, surely, value derives from use? To see how existence values can be positive consider the many environmental funds and organisations in existence to protect endangered species. The subject of these campaigns could be a readily identifiable and used habitat near to the person supporting the campaign. It is very often a remote environment, however, so much so that it is not realistic to expect the campaigner to use it now, or even in the future. Nonetheless, many people support campaigns to protect tropical forests, to ban the hunting of whales, to protect giant pandas, rhinoceros, and so on. All are consumable vicariously through film and television, but vicarious demand cannot explain the substantial support for such campaigns and activities. This type of value, unrelated to use, is existence value.

Existence value provides one of the building bridges between economists and environmentalists, for it is not readily explained by the conventional motives. Economists have suggested a number of motives, all of which reduce to some form of *altruism* – caring for other people or other beings:

1. *Bequest* motives relate to the idea of willing a supply of natural environments to one's heirs or to future generations in general. It is no different to passing on accumulated personal assets. As noted above, however, we prefer to see bequest motives as part of a *use* value, the user being the heir or future generation. It is possible, of course, to think of a bequest as relating to the satisfaction that we believe will be given to future generations from the mere existence of the asset, but the very notion of bequest tends to imply that the inheritor makes some use of the asset.
2. *Gift* motives are very similar but the object of the gift tends to be a current person – a friend, say, or a relative. Once again, gift motives are more likely to be for use by the recipient. We do not therefore count the gift motive as explaining existence value – it is one more use value based on altruism.
3. *Sympathy* for people or animals. This motive is more relevant to existence value. Sympathy for animals tends to vary by culture and nation, but in a great many nations it is the norm, not the exception. It is consistent with this motive that we are willing to pay to preserve habitats out of sympathy for the sentient beings, including humans, that occupy them.

Much of the literature on existence value stops here. The reason for this is that altruistic motives are familiar to economists. They make economic analysis more complex but, by and large, altruism can be conveniently subsumed in the traditional model of rational economic behaviour. In terms of the idea that individuals maximise utility, or welfare, what we can say is that altruism gives utility to the giver, and the giver's utility depends on the utility of other people, or other beings. This interpretation fits neatly into the rational economic man concept, and avoids facing up to still other motives that may be relevant to explaining existence value.

What might these other motives be? One suggestion is that non-human beings have rights, and that when people express an existence value unrelated to their own or anyone else's use of the environment, they are, as it were, voicing those rights because the beings in question cannot do so. But if this is a motive for existence value, then it appears to cause problems for the model of rational economic man, or so some economists fear. It means, for example, that actions may be motivated by factors other than maximising utility. In turn, this means that we will not be able to explain the world (completely anyway) in terms of utility maximisation. Nor, if the rights of others have superior moral standing over utility maximisation, can we prescribe policy on the basis of maximising utility (or benefits). Given the powerful superstructure that economists have built up on the basis of utility maximisation, it is very understandable that they should be unwilling to sacrifice its generality. But the idea that behaviour often *is* motivated by the respect for the rights of others is hardly surprising. It is a fact of life. Why, then, is it any more odd to think of valuations reflecting the rights of other beings? We are, after all, used to the idea that we can pursue our own pleasure only within limits set by society, limits that attempt to embody rights.

The issue here may then be one of deciding when it is, and is not, proper to take account of existence values. If the aim of society is to allocate resources so as to maximise, as far as possible, the utility of individuals in society, then it will be correct to take account of existence value if it is altruistically based. If, on the other hand, existence value relates to a rights motive, and we do not wish such motives to be relevant to the design of policy, it will be improper to take account of it. The reader must decide for himself or herself. For the record, we see no inconsistency in taking account of existence value – whatever its basis – because the values in question are of people and because social policy typically does reflect both wants and rights.

A second motive for existence value unrelated to altruism is *stewardship*. We might also refer to this motive as *Gaian*, after the Greek goddess of the earth, Gaia. A Gaian motive might be based on the idea that the Earth is something far greater and more important than the multitudes of people it supports, and that its population has a responsibility to see that it survives. The implication, of course, is that individual wants may have to be sacrificed to some greater good but, again, we should not be surprised at this idea. Families engage in such activities frequently. There exists also a modern Gaian movement based on a scientific hypothesis that the Earth is a living organism which adjusts in a self-regulating manner to external shocks. An ironic twist to the Gaian motive for existence value, however, is that, in this view, humans are rather unimportant in the self-regulation.

9.7 EMPIRICAL MEASURES OF OPTION AND EXISTENCE VALUE

It is possible to secure empirical estimates of option and existence value by the use of procedures which adopt a questionnaire approach to the WTP for benefits. This approach, the *contingent valuation* approach, is described in Chapter 10. In this section we report several studies which have attempted to obtain actual measures.

David Brookshire, Larry Eubanks and Alan Randall (1983) measured the *option price* (option value plus expected consumer surplus) and *existence value* of grizzly bears and bighorn sheep in Wyoming, both species being subject to threats to their existence. By asking hunters for their WTP in a context where the probability of there being adequate supplies of these species was variable, the authors were able to uncover different types of economic value. A hunter who was certain of his own intentions nonetheless faced uncertain supply. The pattern of bids is shown in Figure 9.3. The U refers to respondents who were uncertain if they would hunt, the C to respondents who were certain they would. This captures an element of demand uncertainty. The subscripts 5 and 15 refer to the number of years before a programme of protection would permit the hunting to take place, the programme being hypothetically paid for by the licences for which the respondents were bidding.

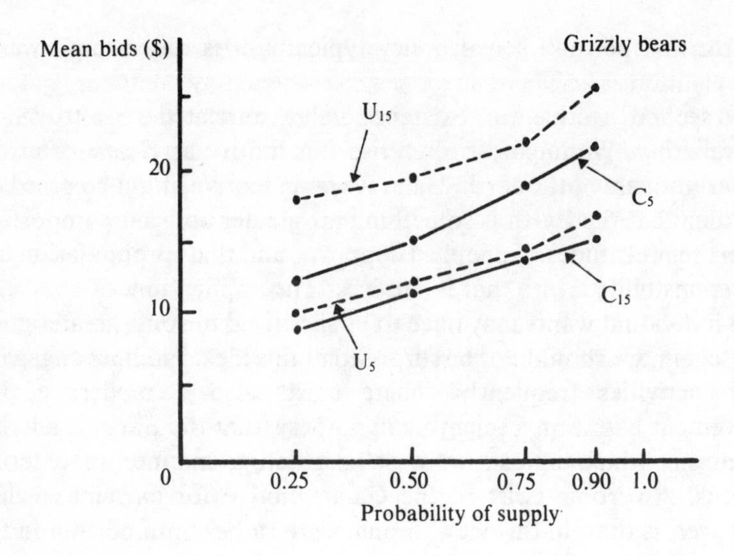

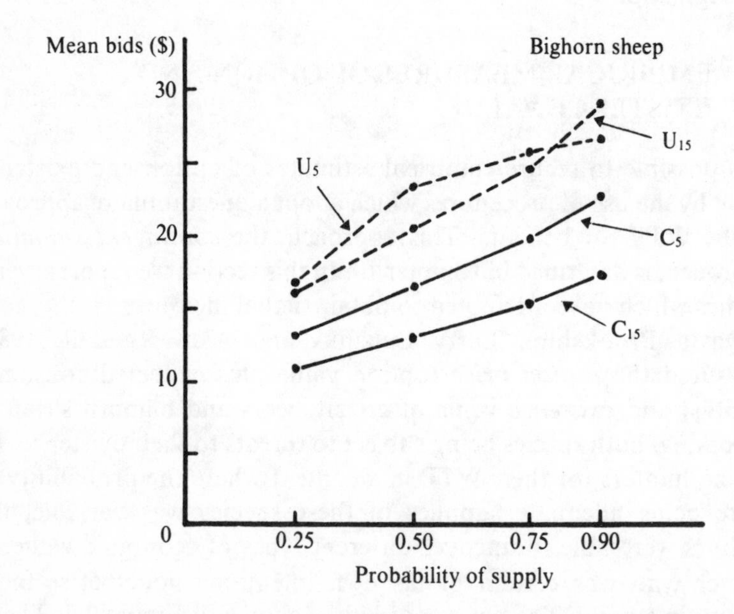

Figure 9.3 Mean grizzly and bighorn bids for certain (C) and uncertain (U) hunting demands over alternative time horizons (5 and 15 years). (*Source*: D. Brookshire, L. Eubanks and A. Randall, 'Estimating option prices and existence values for wildlife resources', *Land Economics*, **59**(1), February 1983.)

The overall option price should increase as the probability of supply increases. This was the result predicted by the theory and it is seen to hold in this case. One might also expect the bids based on certain demand to exceed those based on uncertain demand, but the diagrams show that there is no systematic relationship. Respondents who indicated they would never hunt the bears or sheep were asked what they would nonetheless pay to preserve the species. They were further divided into *observers* (a form of use value) and *non-observers* ('pure' existence value). The results provided estimates of 'observer option price', i.e. the option price associated with keeping the species for recreational observation, and existence value. The results are shown in Table 9.5. Clearly, these are significant sums. To see this compare them to the average option prices for hunting under, say, 90 per cent probability of future supply. For grizzly bears and the five-year time horizon the sequence would be $21.50 option hunting price compared to $21.80 option observer price and $24.00 for existence value. Average existence value is on a par with the bids to maintain the population for hunting and observation.

In a later paper, Brookshire *et al.* (1985) detail findings relating to the Grand Canyon. By looking at the bids made by respondents to experience improved visibility (regardless of whether visits take place or not), the authors find that the total 'preservation bid' for the Grand Canyon's visibility was $4.43 per month, compared to a 'user bid' of $0.07 per month. Interpreting existence value as the difference between total preservation value and use value, the finding is thus that existence value dominates preservation in this case. Existence value stands in the ratio of 66:1 to user value (note that what is being preserved is visibility, not the site itself). The explanation for such a

Table 9.5 Existence values and option prices for grizzly bears and bighorn sheep.

	Bears		Sheep	
	5 years	15 years	5 years	15 years
Average observer option price ($)	21.8	21.0	23.0	18.0
Average existence value	24.0	15.2	7.4	6.9

Source: Brookshire *et al.* (1983).

large ratio is that the resource in question is unique – it has no substitutes. Where substitutes exist one would expect existence values to be lower, and this tends to be the picture in other studies on existence value.

Jon Strand (1981) reports a CVM-type study of acid rain for Norway. After indicating the nature of the environmental problem – damage to freshwater fish from acid rain – respondents were given a starting point figure for the global cost of stopping acid pollution which was translated into a special income tax. They were then asked if they were willing to pay this sum. The approach was thus of the 'take-it-or-leave-it' kind rather than one involving iterative bids in which respondents could vary their bid according to different levels of clean-up. But the hypothetical tax rates were varied across the four samples of respondents interviewed, i.e. the tax rate was the same for each sample but varied between samples. The 'yes' responses were found for the lower taxes. Strand then estimates 'bid curves' using this information in a conditional probability framework, i.e. estimating the probability that a respondent would pay a particular tax given a certain income. Strand estimates that the average bid was 800 Norwegian krone per capita. Given a population of 3.1 million, this translates to a 'national' annual benefit of 2.5 billion krone. Earlier work by Strand suggests that user values are about 1 billion krone, so that subtracting this from the implied total preservation value of 2.5 billion krone gives an existence value of 1.5 billion krone. In 1982 terms this translates to some $270 million or about 1 per cent of the Norwegian GNP. Note that, by asking for WTP, the Strand study probably underestimates the true value of benefits of reduced aquatic acidification. The reason for this is that a good deal of the acidity arises from 'imported' pollution and respondents will generally have been aware of this. Accordingly, they may well have had the attitude that others besides themselves should pay for the clean-up.

10 · MEASURING ENVIRONMENTAL DAMAGE II: VALUATION METHODOLOGIES

10.1 TOTAL ECONOMIC VALUE AND DECISION-MAKING

Chapter 9 showed that the relevant concept when measuring the benefit of an environmental improvement is total economic value (TEV). In the same way, if we wished to measure the damage done to the environment, say by a development project, we would want to calculate the TEV that is lost by the development. Damage and benefit are obverse sides of the same concept.

The relevant comparison when looking at a decision on a development project is between the cost of the project, the benefit of the project, and the TEV that is lost by the development. More formally, we can write the basic rules as:

(i) proceed with the development if

$$(B_D - C_D - B_P) > 0$$

and

(ii) do not develop if

$$(B_D - C_D - B_P) < 0$$

where B_D refers to the benefits of development, C_D refers to the costs of the development and B_P refers to the benefits of preserving the environment by not developing the area.

TEV is in fact a measure of B_P, the total value of the asset left as a natural environment. The benefits and costs of the development will be relatively simple to measure, primarily because they are likely to be in the form of marketed inputs and outputs which have observable prices. This is clearly not going to be the case with TEV, so we need now to investigate ways in which we can measure the component parts of TEV.

10.2 DIRECT AND INDIRECT VALUATION

The approaches to the economic measurement of environmental benefits have been broadly classified as *direct* and *indirect* techniques. The former considers environmental gains – an improved scenic view, better levels of air quality or water quality, etc. – and seeks directly to measure the money value of those gains. This may be done by looking for a *surrogate market* or by *experimental* techniques. The surrogate market approach looks for a market in which goods or factors of production (especially labour services) are bought and sold, and observes that environmental benefits or costs are frequently attributes of those goods or factors. Thus, a fine view or the level of the air quality is an attribute or feature of a house, risky environments may be features of certain jobs, and so on. The experimental approach simulates a market by placing respondents in a position in which they can express their hypothetical valuations of real improvements in specific environments. In this second case, the aim is to make the hypothetical valuation as real as possible.

Indirect procedures for benefit estimation do not seek to measure direct revealed preferences for the environmental good in question. Instead, they calculate a 'dose-response' relationship between pollution and some effect, and only then is some measure of preference for that effect applied. Examples of dose-response relationships include the effect of pollution on health, the effect of pollution on the physical depreciation of material assets such as metals and buildings, the effect of pollution on aquatic ecosystems and the effect of pollution on vegetation.

However, indirect procedures do not constitute a method of finding the willingness to pay, WTP, for the environmental benefit (or the willingness to accept, WTA, compensation for environmental damage suffered). What they do is to estimate the relationship between the 'dose' (pollution) and the non-monetary effect (health impairment, for example). Only then do they apply WTP measures taken from direct valuation approaches. Accordingly, we do not discuss indirect procedures further in this chapter.

10.3 THE HEDONIC PRICE APPROACH

The value of a piece of land is related to the stream of benefits to be derived from the land. Agricultural output and shelter are the most obvious of such benefits, but access to the workplace, to commercial amenities and to environmental facilities such as parks, and the environmental quality of the neighbourhood in which the land is located are also important benefits which accrue to the person who has the right to use a particular piece of land. The property value approach to the measurement of benefit estimation is based on this simple underlying assumption. Given that different locations have varied environmental attributes, such variations will result in differences in property values. With the use of appropriate statistical techniques the hedonic approach attempts to (a) identify how much of a property differential is due to a particular environmental difference between properties and (b) infer how much people are willing to pay for an improvement in the environmental quality that they face and what the social value of improvement is. Both the identification and the inference activities involve a number of issues which are discussed in some detail below.

The identification of a property price effect due to a difference in pollution levels is usually done by means of a *multiple regression* technique in which data are taken either on a small number of similar residential properties over a period of years (time series), or on a large number of diverse properties at a point in time (cross section), or on both (pooled data). In practice, almost all property value studies have been cross-section data, as controlling for other influences over time is much more difficult.

It is well known of course that differences in residential property values can arise from any sources, such as the amount and quality of accommodation available, the accessibility of the central business district, the level and quality of local public facilities, the level of taxes that have to be paid on the property, and the environmental characteristics of the neighbourhood, as measured by the levels of air pollution, traffic and aircraft noise, and access to parks and water facilities. In order to pick up the effects of any of these variables on the value of a property, they *all* have to be included in the analysis. Hence such studies usually involve a number of *property* variables, a number of *neighbourhood* variables, a number of *accessibility* variables and finally the *environmental* variables of interest. If any variable that is relevant is excluded from the analysis then the estimated effects on property value of the included variables could be biased. Whether the bias is upward or downward will depend on how the included and excluded variables relate to each other and to the value of the property.

On the other hand if a variable that is irrelevant is included in the analysis then no such systematic bias results, although the estimates of the effects of the included variables are rendered somewhat less reliable. This would suggest then that we include as many variables as possible. However, doing so creates another difficulty. Typically many of the variables of interest are themselves very closely correlated. So, for example, accessibility to the town centre is often closely related to some measures of air pollution, and one measure of air pollution, such as total suspended particulate matter, is very closely correlated to other measures such as sulphur dioxide. To overcome this, many studies use only one 'representative' measure of pollution.

The first stage in the hedonic price approach, then, is to estimate an equation of the form:

property price = f(property variables, neighbourhood variables, accessibility variables, environmental variables)

or, symbolically,

$$PP = f(PROP, NHOOD, ACCESS, ENV)$$

where f(.....) simply means 'is a function of' (depends upon). The actual specification of this equation is a matter of professional choice. A familiar one is:

$$\ln PP = a \ln PROP + b \ln NHOOD + c \ln ACCESS + d \ln ENV$$

where 'ln' simply refers to logarithm. By feeding in the observed values for property prices, the property variables, the neighbourhood, accessibility and environmental variables, a simple computer program will generate the values of *a, b, c* and *d.* In this case, the value of *d* will tell us by how much property prices vary if we alter the value of the environmental variable. Provided we can relate the property price to the willingness to pay, we have nearly solved the problem of valuing environmental damage (or improvement).

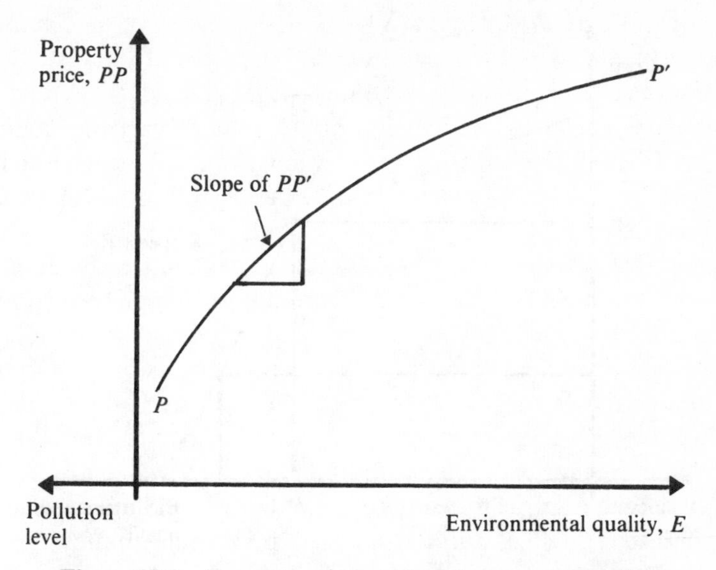

Figure 10.1 Property prices and environmental quality.

Figure 10.1 shows a typical relationship between pollution and property values that might be uncovered by the hedonic price techniques. It shows that as the pollution level decreases, so property values rise, but at a declining rate. Figure 10.2 plots the *slope* of the relationship in Figure 10.1 against the level of pollution. This is shown as AB. Hence it gives, for each level of pollution, the amount by which property values would fall if pollution levels were to be increased by a small amount.

If we are to obtain an estimate of the demand for environmental quality we would like to know how much households are willing to pay for given levels of environmental quality. In Figure 10.2 consider an individual or household who is living in an environment with an ambient pollution level P^0. It is assumed in the hedonic methodology that this choice has been arrived at in a rational manner. That is to say, the household concerned has weighed the benefits of living in alternative locations against the costs and on balance has chosen location P^0. To arrive at this decision it must have concluded that the extra payment required in higher property prices for an improvement in the environment from a pollution level slightly higher than P^0 to P^0 is just equal to the benefits of that improvement. Hence we can define the amount W^0 as that household's willingness to pay for the last unit of environmental quality. But such a willingness to pay is a

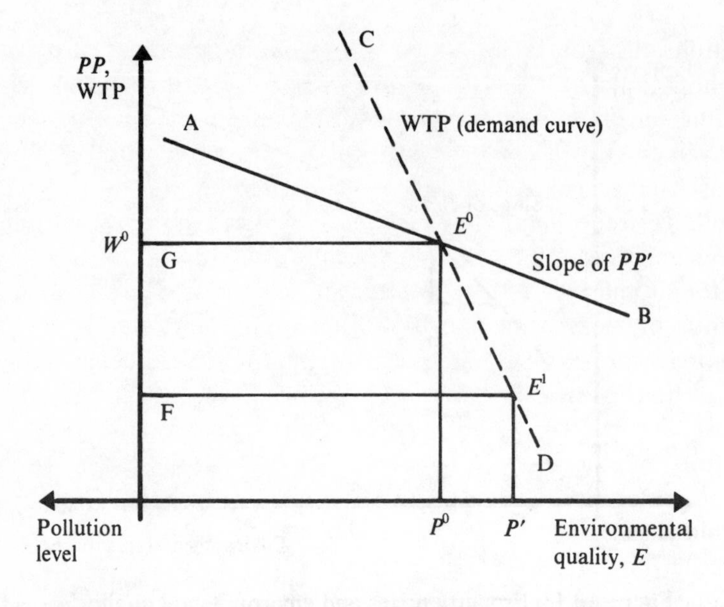

Figure 10.2 Willingness to pay and hedonic property prices.

point on the household's demand curve, and other such points are indicated by the broken line CD through E^0.

What this shows us is that the estimated hedonic price relationship can be used to obtain a point on each household's demand curve, and that the slope of the estimated relationship is a locus of points on the demand curves of many different households. If all such households were identical in every respect then the derived curve AB in Figure 10.2 would also be the demand curve for environmental quality. Each household's willingness to pay for a small improvement at every level of pollution P must also be every other household's willingness to pay if they are all identical, and the locus of willingness-to-pay points defines the demand curve. In general, however, households will differ in income and preference for environmental quality. When that is the case the hedonic approach as outlined so far only gives us partial information on the demand structure. What is now required is to see how this marginal willingness to pay varies with household income and other household characteristics. This involves a further statistical exercise which would then estimate the demand function for environmental quality.

In order to value any environmental improvement we would now use the estimated inverse demand functions CD. Suppose that pollution falls from P^0 to P'. Then the gain in consumer surplus to *each* household at P^0 is the area E^0E^1FG. By adding up all such consumer surpluses we obtain the overall value of the environmental improvement. In fact, most empirical studies work with schedules such as AB, i.e. this second stage is not carried out.

Table 10.1 reports the results of hedonic price air pollution studies where significant effects of air pollution on property values have been found and where these effects can be expressed irrespective of the units of measurement of pollution or property values (i.e. in percentage terms). As stated earlier, many such studies find it difficult to distinguish between different forms of air pollution because of their strong inter-correlation. In these cases the one pollution measure included inevitably picks up the effects of all forms of air pollution with which it is strongly correlated. The results in Table 10.1 suggest that a 1 per cent increase in sulphation levels will result in falls in property values between 0.06 and 0.12 per cent. A similar increase in particulates lowers property values by between 0.05 and 0.14 per cent. Where the pollution variable is picking up more that one measure of air pollution, property value falls of

Table 10.1 Impact of air pollution on property values.

City	Year of: (a) property data (b) pollution measure	Pollution	Percentage fall in property value for a percentage increase in pollution
St Louis	(a) 1960	Sulphation	0.06–0.10
	(b) 1963	Particulates	0.12–0.14
Chicago	(a) 1964–67	Particulates	
	(b) 1964–67	and	0.20–0.50
		sulphation	
Washington	(a) 1970	Particulates	0.05–0.12
	(b) 1967–68	Oxidants	0.01–0.02
Toronto-Hamilton	(a) 1961	Sulphation	0.06–0.12
	(b) 1961–67		
Philadelphia	(a) 1960	Sulphation	0.10
	(b) 1969	Particulates	0.12
Pittsburg	(a) 1970	Dustfall and	0.09–0.15
	(b) 1969	sulphation	
Los Angeles	(a) 1977–78	Particulates	
	(b) 1977–78	and	0.22
		oxidants	

Source: D. W. Pearce and A. Markandya, *The Benefits of Environmental Policy*, OECD, Paris, 1989.

between 0.09 and 0.5 per cent are recorded. Again we should note that the fall in property values per unit increase in pollution could vary with the level of pollution.

10.4 CONTINGENT VALUATION

The contingent valuation method (CVM) uses a direct approach – it basically asks people what they are willing to pay for a benefit, and/or what they are willing to receive by way of compensation to tolerate a cost. This process of 'asking' may be either through a direct questionnaire/survey, or by experimental techniques in which subjects respond to various stimuli in 'laboratory' conditions. What is sought are the personal valuations of the respondent for increases or decreases in the quantity of some good, contingent upon a hypothetical market. Respondents say that they would be willing to pay or willing to accept if a market existed for the good in question. A contingent market is taken to include not just the good itself (an improved view, better water quality, etc.), but also the institutional context in which it would be provided, and the way in which it would be financed.

One major attraction of CVM is that it should, technically, be applicable to all circumstances and thus has two important features:

- it will frequently be the *only* technique of benefit estimation
- it should be applicable to most contexts of environmental policy.

The aim of the CVM is to elicit valuations – or 'bids' – which are close to those that would be revealed if an actual market existed. The hypothetical market – the questioner, questionnaire and respondent – must therefore be as close as possible to a real market. The respondent must, for example, be familiar with the good in question. If the good is improved scenic visibility, this might be achieved by showing the respondent photographs of the view with and without particular levels of pollution. The respondent must also be familiar with the hypothetical means of payment – say, a local tax or direct entry charge – known as the payment *vehicle*.

The questioner suggests the first bid – the 'starting point bid (price)' – and the respondent agrees or denies that he/she would be willing to pay it. An iterative procedure follows: the starting point price is increased to see if the respondent would still be willing to pay it, and so on until the respondent declares he/she is not willing to pay the extra increment in the bid. The last accepted bid, then, is the maximum willingness to pay (MWTP). The process works in reverse if the aim is to elicit *willingness to accept* (WTA): bids are systematically lowered until the respondent's minimum WTA is reached.

A very large part of the literature on CVM is taken up with discussion about the 'accuracy' of CVM. Accuracy is not easy to define. But since the basic aim of CVM is to elicit 'real' values, a bid will be accurate if it coincides (within reason) with one that would result if an actual market existed. But since actual markets do not exist *ex hypothesi* (otherwise there would be no reason to use the technique), accuracy must be tested by seeing that:

- the resulting bid is similar to that achieved by other techniques based on surrogate markets (house price approach, wage studies, etc.)
- the resulting bid is similar to one achieved by introducing the kinds of incentives that exist in real markets to reveal preference.

There are various ways of classifying the nature of the biases that may be present in the CVM. A classification is shown in Table 10.2.

Table 10.2 Sources of bias in CVM.

Strategic	Incentive to 'free ride'?
Design	(a) starting point bias (b) vehicle bias (c) informational bias
Hypothetical	Are bids in hypothetical markets different to actual market bids? Why should they be?
Operational	How are hypothetical markets consistent with markets in which actual choices are made?

The concern with *strategic bias* is long-standing in economics and emanates from the supposed problem of getting individuals to reveal their true preferences in contexts where, by not telling the truth, they will still secure a benefit in excess of the costs they have to pay. This is the *free rider problem*. For example, if individuals are told that a service will be provided if (a) the total aggregated sum they are willing to pay exceeds the cost of provision, and (b) that each will be charged a price according to their maximum WTP, then the presumption is that each individual will understate his or her true demand. The context is one in which the good in question is a 'public good', or has features of a public good. Such goods are difficult to provide in a way that excludes anyone from enjoying them, and the consumption of the good by each individual tends not to be at the cost of consumption to other individuals. Environmental quality has these features. Hence the relevance of the 'free rider' problem. Typically, however, CVM studies have not found strategic bias to be significant.

The potential for *design bias* arises from various sources. The first of these is *starting point bias*. It will be recalled that the interviewer suggests the first bid, the starting point. It is possible that this will influence the respondent in some way, perhaps by suggesting the range over which the 'bidding game' would be played by the interviewer, perhaps by causing the respondent to agree too readily with bids in the vicinity of the initial bid in order to keep the game as short as possible.

CVM studies have attempted to test for this source of bias, usually by offering different starting bids, and sometimes by letting the respondent make the first bid. Statistically, then, it is possible to see if the mean (average) bid is affected by the choice of starting bid. The results are not conclusive, some studies finding no correlation between starting bids and mean bids, others finding that mean bids were very much affected by starting bids.

Vehicle bias

This arises from the choice of the 'vehicle', or instrument of payment, used in the approach. Such vehicles include changes in local taxes, entrance fees, surcharges on bills (e.g. electricity bills), higher prices for goods, and so on. Respondents may be 'sensitive' to the vehicle, perhaps regarding $1 paid through taxes as being more costly to him than $1 paid through an entrance fee.

The tests for vehicle bias are conceptually very simple. The average bid should not differ significantly between type of vehicle, e.g. the value of an improvement to the environment should be roughly the same whether the hypothetical payment is a tax increase or an entrance fee to the area, etc. If mean bids do vary by type of vehicle, vehicle bias may be said to exist. There are exceptions to this basic rule, but tests of the rule – by seeing how mean bids do vary with choice of instrument – seem to suggest some source of bias. The research issue that arises is then how to choose a 'neutral' vehicle.

Information bias
This may arise from various aspects of the CVM. Starting point bias, for example, could be regarded as a form of information bias since it is the interviewer who 'informs' the respondent of the first bid. The sequence in which information is supplied may also influence respondents, e.g. indicating the 'importance' of a feature before explaining the nature of the choice. The general amount and quality of information is also of significance, particularly if the total cost of the environmental improvement is included in the information. The tests for such bias are difficult and usually involve either withholding information from one group and supplying it to another, or measuring the degree of information thought to be held by respondents. Various studies suggest no effect, while others derive measured differences in WTP according to information differences.

Hypothetical bias
The basic idea of CVM is to elicit hypothetical bids that conform to actual bids if only actual markets exist. The basic difference between actual and hypothetical markets is that in actual markets purchasers will suffer a cost if they get it wrong – regret at having paid too much, for example. One obvious test is to carry out the CVM using hypothetical *and* actual payments. What work there has been suggests that hypothetical bias is still a problem in the CVM approach. Both information and hypothetical bias problems seem to produce random variation in study results and are therefore more properly to be regarded as reliability rather than bias problems.

Operational bias
This may be described in terms of the extent to which the actual 'operation conditions' in the CVM approximate actual market conditions. This had led researchers to suggest various 'reference operating conditions' (ROCs) which should be met. The lists vary but all would include the requirement that respondents be familiar with the good they are being asked to value, and that they have either prior experience of varying the quantities of the good, or can 'learn' how to do this through repeated bids. One might add to the list the requirement for the general absence of uncertainty, but it is worth

Table 10.3 Comparisons of CVM with other techniques.

	CVM results		Indirect market study	
Study	Commodity	Value[a]	Method	Value[a]
Knetsch and Davis (1966)	Recreation days	$1.71 per household/day	TCM	$1.66 per household/day
Bishop and Heberlein (1979)	Hunting permits	$21 per permit	TCM value of time = 0 value of time = ¼ median inc. value of time = ½ median inc.	 $11.00 $28.00 $45.00
Desvousges *et al.* (1983)	Water quality improvements:	User values:[b] average (across question format)	TCM	User values
	(a) loss of use	$21.41		$82.65
	(b) boatable to fishable	$12.26		$ 7.01
	(c) boatable to swimmable	$29.64		$14.71
Seller *et al.* (1984)	Boat permit to:	Close-ended consumer surplus:	TCM	Consumer surplus
	Lake Conroe	$39.38		$32.06
	Lake Livingston	$35.21		$102.09
	Lake Houston	$13.01		$13.81
Thayer (1981)	Recreation site	Population value per household per day: $2.54	Site substitution	Population value per household per day: $2.04
Brookshire *et al.* (1982)	Air-quality improvements:	Monthly value[c]	HPM (property values)	Monthly value:
	(a) poor to fair	$14.54		$45.92
	(b) fair to good	$20.31		$59.09
Cummings *et al.* (1983)	Municipal infrastructure in:	Elasticity of substitution of wages for infrastructure	HPM (wages)	Elasticity of substitution of wages for infrastructure; 29 municipalities:
	(a) Grants, NM	–0.037		
	(b) Farmington, NM	–0.040		–0.035
	(c) Sheridan, WY	–0.042		
Brookshire *et al.* (1984)	Natural hazards (earthquakes) information	$47 per month	HPM (property values)	$37 per month

Source: Cummings, Brookshire and Schulze (1986), p. 125.

[a] Mean values amongst respondents.

[b] Values apply to post-iteration bids for users of the recreation sites.

[c] Value for sample population.

noting that this automatically raises problems for the use of CVM in eliciting option values which arise precisely because of uncertainty.' It was noted above that the concept of 'accuracy' is a little elusive when considering benefit measurement techniques. Validity is a multidimensional concept and no one test will prove definitive. But some reassurance is likely to derive from any discovery that differing techniques secure similar valuations. Table 10.3 summarises several studies that have attempted comparisons of CVM and other valuation approaches. The studies compared CVM with one or other of the travel cost methods (TCM) (see below), hedonic property price approach (HPM = house price method), and site substitution approach (not discussed here). The *ranges* of values all overlap if accuracy is expressed as ±60 per cent of the estimates shown, and overlap in thirteen of the fifteen comparisons if the range is ±50 per cent. These are familiar ranges of error in estimates of demand functions in economics. This does not mean that the CVM is 'correct' since, as noted above, we have in turn to make some judgement as to how correct the comparator techniques are. But it does tend to be reassuring.

One significant feature of the CVM literature has been its use to elicit the different kinds of valuation that people place on environmental goods. In particular, CVM has suggested that existence values may be very important. Schulze *et al.* (1983), for example, have suggested that the benefits of preserving visibility in the Grand Canyon are of the order of $3.5 billion per year, and some $6.2 billion per year if the visibility is extended to the southwestern parklands of the USA. Making allowance for future population trends, annualised benefits rise to $7.4 billion. These compare with the control costs of some $3 billion per year. Existence value estimates are as yet few in number and should be treated with caution. CVM results are best interpreted as indicating respondent total economic valuation i.e. a global assessment encompassing use, option and existence values.

10.5 TRAVEL COST APPROACHES

Travel cost models are based on an extension of the theory of consumer demand in which special attention is paid to the value of time. That time is valuable is self-evident. What precisely its value is, remains a question on which there is some disagreement, as will become clear later. However, as a starting point let us imagine a household consisting of a single person who works as a driver. He can work

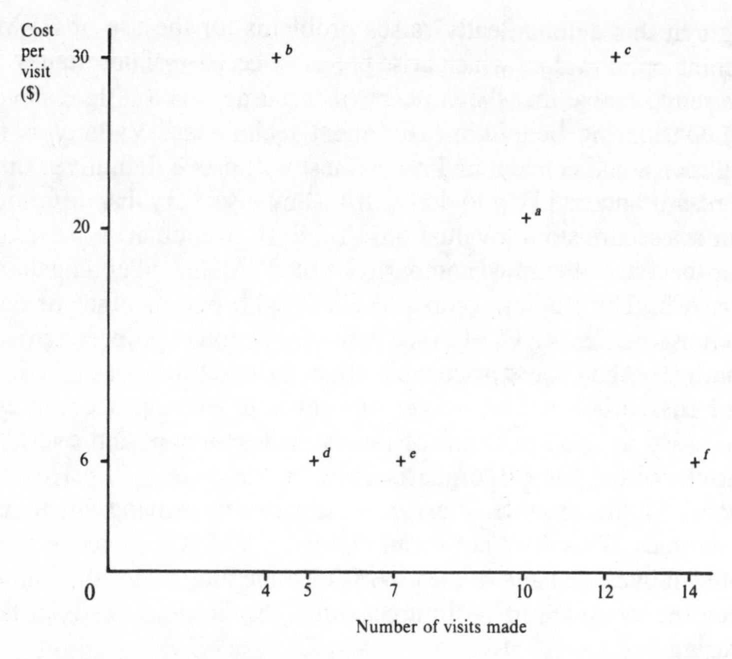

Figure 10.3 Observations of recreational visits and visiting costs.

as many or as few hours as he wishes and he earns $5 an hour. He is fortunate enough not to pay taxes, and enjoys (or dislikes) driving for work or for recreation equally much. On a particular day he can either drive to a park that takes an hour to get to, and spend some time there, or he can go to work. In these circumstances he is faced with possibly two decisions. The first is whether to go to the park or to go to work. The second is, if he goes to the park, how much time to spend there. Suppose that the cost of the journey in terms of petrol and wear and tear is $3 and there is an entry fee of $1. If he goes to the park and spends a couple of hours there, then it will have cost him $4 in cash *plus* the loss of income of $20. The true cost of the visit consists of the entry fee, plus the monetary costs of getting there, plus the foregone earnings. If we had information on all these variables and we could obtain it for a large number of individuals, along with the information on the number of visits that each had made (and would make) during the season, then we could attempt to estimate the household's willingness to pay for a given number of visits. However, at first glance the data would not look very orderly. Figure 10.3 shows the kind of data that we might find.

Our single earner household, for example, could be represented by the point *a*: he makes ten visits at a cost per visit of $20. Points *b* and *c* represent two households, each of whom face very high costs ($30). Of these *b* makes very few visits because it is a poor household living far from the recreational site, and *c* is a high earning household located near the park that makes a lot of short visits (being a high earner it has a high foregone-earnings component to its costs). Points *d*, *e* and *f* also represent households with the same costs per visit. Whereas both *d* and *e* make few visits, *d* does so because it has no attraction to the facilities offered, but *e* does so because it has access to another park close to its residential location. Household *f*, on the other hand, makes a lot of visits. Although it is identical to *e* in every other respect, it is not located close to another recreational area.

It is clear from the above that if we are to trace out how a particular household, such as *a*, would react to changes in the cost per visit, then we need to group together households that are similar to *a*. The locus of points linking such households would then constitute their demand curve for the recreational facilities that that site has to offer. Similarity here means grouping our observations according to income, preference for recreation and access to other recreational facilities. Given the demand curves we can calculate the benefits of the site by taking the area under these curves to obtain the consumer surplus as indicated in Chapter 2. Adding up the consumer surpluses for different categories or households gives us the overall benefit of the site.

If the model developed here is to be used to evaluate the benefits of environmental improvements, then further work has to be done. It is no longer enough to separate out the groups according to what other recreational facilities they may have access to. We now need to know how much of the willingness to pay of a category of households will increase if the facility at a particular site is improved to allow, for example, the possibility of fishing in a lake where none was possible before. This in turn requires knowledge of how much of the willingness to pay for each site is due to each of its specific facilities. Then by looking across sites we will be able to trace out changes in this wilingness to pay as facilities change. The data required for such an exercise would include the facilities of each site and the location of each household relative to all the sites. This is clearly a very large amount of information and so some simplifying assumptions will be necessary in many cases. What these are and

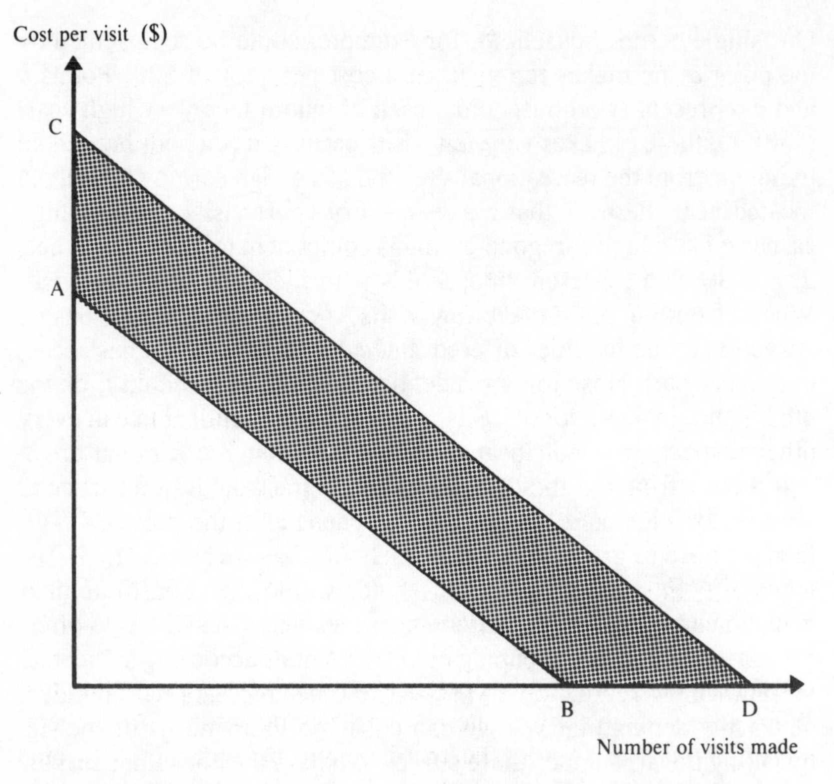

Figure 10.4 Benefits of improving a recreational site.

how they affect the results is discussed later in this section.

If we can derive the demand curve for recreation for a particular category of households defined by household characteristics such as income, education and the liking for recreational facilities, and we can show how this demand curve would shift if facilities improved, then the benefit of the improvement can be derived as shown in Figure 10.4. AB is the curve prior to the change and CD is the curve after the change. The benefits of this group of consumers are given by the area ABCD. Adding across groups gives the total benefit.

10.6 WILLINGNESS TO PAY VERSUS WILLINGNESS TO ACCEPT

The CVM has been particularly instrumental in a debate over the relationship between WTP and WTA measures of environmental change. It will be recalled that WTP is generally elicited when

Table 10.4 Disparities between WTP and WTA (in year-of-study dollars).

Study	WTP	WTA
Hammack and Brown (1974)	$247.00	£1,044.00
Banford *et al.* (1977)	43.00	120.00
	22.00	93.00
Sinclair (1976)	35.00	100.00
Bishop and Heberlein (1979)	21.00	101.00
Brookshire *et al.* (1980)	43.64	68.52
	54.07	142.60
	32.00	207.07
Rowe *et al.* (1980)	4.75	24.47
	6.54	71.44
	3.53	46.63
	6.85	113.68
Hovis *et al.* (1983)	2.50	9.50
	2.75	4.50
Knetsch and Sinden (1983)	1.28	5.18

Source: Cummings *et al.* (1984).

considering the valuation of a potential environmental *benefit*, whereas WTA seems more appropriate if we are asking someone to 'accept' a *cost*. If we take a given state of the environment, then, we could ask for the WTP to improve the environment still further and the WTA to reduce environmental quality from the initial position. Economic theory tells us that these two values should not differ significantly. Yet the CVM studies tend to suggest quite major disparities. Table 10.4 shows some examples of the kinds of differences that have been found. How are the differences to be explained? There are various options. The main ones are as follows:

1. Economic theory is wrong and people value gains and losses 'asymmetrically', attaching a lot more weight to a loss compared to the existing position than to a gain.
2. The relevant CVM studies are flawed and no reliance can be placed in the disparate estimates.
3. CVM studies tend to deal with large, discrete changes and 'instant' valuations. These cannot be compared to the context in which economic theory concludes that WTA and WTP must be very similar.

The literature is divided on which explanation is correct, proposition

1, in particular, exciting considerable controversy. Psychologists suggest that *prospect theory* explains a good deal of the difference. In prospect theory individuals' values relate to gains and losses in comparison to some 'reference point'. This contrasts with the economic assumption that individuals maximise 'utility'. What matters is the point from which the gains and losses are measured. This may, for example, be the status quo. Second, prospect theory suggests that values for negative deviations from the reference point will be greater than values placed on positive deviations. Gains will be valued less than losses, just as the CVM studies suggest. Third, the manner in which the gains and losses are to be secured matters a great deal. An 'imposed' loss, for example, will tend to attract a much higher value than a voluntarily secured gain of equal quantity. Thus, it is suggested that a loss of something that is already owned is regarded as more important than the gain of something not yet possessed; the regret attached to going without something one never had is less than the cost of losing what one already has.

Other writers suggest that the disparity between WTP and WTA disappears when proper incentives are established for people to tell the truth in response to questions about their valuation of environmental quality. These 'proper' responses, it is argued, are present in the market-place. Moreover, markets contain in-built mechanisms whereby buyers and sellers learn about the commodity they are trading, whereas hypothetical market situations do not. Brookshire and Coursey 61987), for example, report experiments in valuation of tree-growing in a city area and conclude that

> when the market-like elicitation process is repeated even a small number of times, values for the public good are more consistent with the traditional economic notions of diminishing marginal utility. Although individuals may initially exaggerate their preferences for the public good, they modify their stated values as functions of the incentives, feedback, interactions and other experiences associated with the repetitive auction environment (Brookshire and Coursey, 1987, p. 565).

A great deal more research is needed to investigate these issues. The issue is important, for if WTA and WTP really do differ by multiples of three or more (as some of the empirical literature suggests) then the kind of values placed on the environment are similarly affected.

11 · POLLUTION CONTROL POLICY IN MIXED ECONOMIES

11.1 POLLUTION CONTROL: THEORY VERSUS PRACTICE

In Chapters 4 – 8 we analysed the economic approach to pollution and pollution control policy. We noted that, from the point of view of economic theory, market-incentive policy instruments (e.g. effluent charges and rights) can be shown to be the least-cost solution to the problem of attaining ambient environmental standards. In the context of pollution abatement it appears that economic-incentive instruments possess the inherent advantages of cost-effectiveness and also provide inducements to technological innovation. Economists therefore have often been critical of the direct regulation and standards approach to pollution control, despite the fact that this 'command and control' strategy has been universally favoured by governments and their regulatory agencies. The use of charges and rights (permits) is not entirely absent in operational control strategies, but their adoption and implementation has been quite restricted and they have been limited to a supplementary role.

A number of practical or political reasons for the lack of widespread acceptance of market-incentive instruments in pollution control were briefly examined in Chapter 6. It may also be the case that some of the advantages claimed for effluent charges and some rights schemes exist only under unrealistic or restrictive assumptions. We noted in Chapter 8 that a number of real pollution situations involve a range of pollutants and a complex 'receiving' environmental system or set of systems. Quite complicated charging schemes may be required in such circumstances with significant informational requirements. The bargaining (over charge levels, timing of

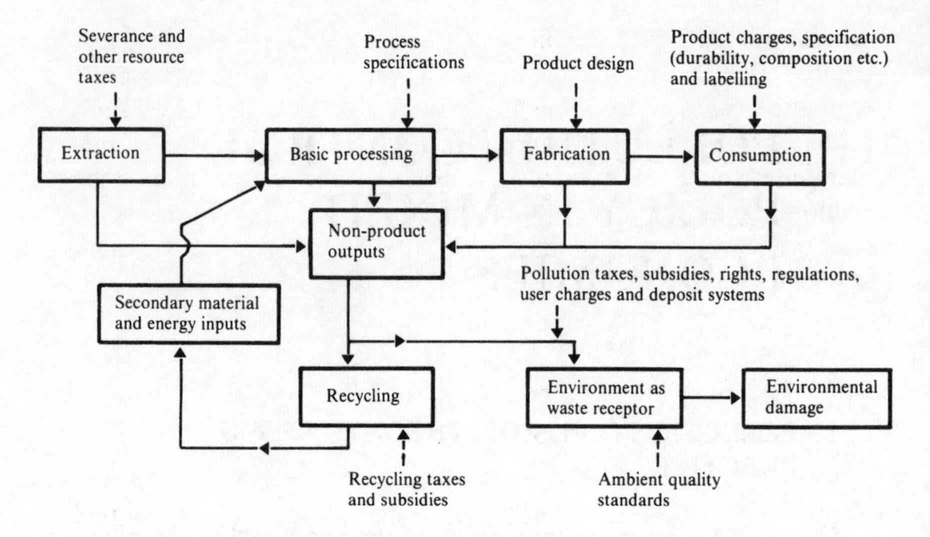

Figure 11.1 Pollution control policy instruments.

application and effluent measurement) that has to take place between the control agency and the polluter may not therefore be significantly reduced by replacing direct regulation with an effluent charge.

More empirical data and individual case-study analysis is required in order to clarify a number of the 'direct regulation versus incentives' arguments. A limited number of UK studies relating to a narrow water pollution context (i.e. trade-effluent discharging and public treatment) indicated that up to 30 per cent of dischargers do not understand simple pricing systems related to changing effluent strength/volume compositions. These same dischargers also lacked any clear idea of alternative treatment methods and costs, recycling opportunities or other waste-reduction options. In these circumstances the regulatory agency's 'field' role in discretionary standard-setting (i.e. providing information, persuading and advising the polluter) is still required under an economic-incentive system.

Given the inherent uncertainties involved, pollution control in practice is best viewed as an iterative search process. Policy-makers are simply trying to discover and arrive at a set of pollution control arrangements which make enough people better off, so that those circumstances are preferable to current circumstances. Ideally the appropriate set of policy instruments required to achieve the 'socially acceptable' environmental quality goals should be based on the criteria of political feasibility, cost-effectiveness, flexibility and equity. The impact of a particular set of instruments (direct regulations and/or economic incentives) will depend on its location in the economic system. Figure 11.1 illustrates the processes of raw material extraction, processing, manufacture and consumption in the economy, and the points where various policy instruments to discourage waste and encourage waste reduction, recycling and more stringent residuals management could be introduced.

11.2 DIRECT REGULATIONS: A 'COMMAND AND CONTROL' PHILOSOPHY

Traditionally, in all the mixed industrialised economies regulatory instruments have formed the basic foundations of environmental protection policy. Environmental legislation has been implemented via regulatory instruments coupled with systems of monitoring and sanctioning of non-compliance. This 'regulative tradition' has involved the setting of ambient air and water quality standards or objectives/targets, and the imposition of emission/discharge limits and/or product or process standards, through a system of licensing and monitoring. Polluters' compliance is mandatory and sanctions often exist for non-compliance.

Pollution control usually involves a process of bargaining and negotiation before the exact content of the regulatory package is determined. This negotiatory aspect improves the likelihood of polluter compliance, offers a degree of flexibility and can potentially reduce the perceived uncertainties of both regulator and polluter. In Europe and especially within the European Community (EC) there has been a general trend in pollution control away from a case-by-case facilitative process to a much more interventionist activity involving broad considerations of quality levels and emission/discharge standards on a regional scale. Distinctions are also made between existing sources of pollution and potential new sources.

Different countries have adapted different approaches to the problem of relating emission/discharge standards to the ambient environment goals. The following specifications cover most of the approaches that have been taken:

- limits in terms of maximum rate of discharge from a pollution source
- specification of a given degree of pollution control, e.g. percentage removal of all particulates from the emission
- requirement to implement some variant of the 'best practicable means', or 'best available control technology' for pollution reduction
- pollution density limits related to discharges/emissions
- discharge bans related to pollution concentration measures or damage costs
- discharge limits set by reference to the use of specified inputs to or outputs from the production process.

11.3 TECHNOLOGY-BASED POLLUTION CONTROL STRATEGIES: THE USA AND THE UK

In the UK and the USA effluent standards have been based on the principles of Best Practicable Means (BPM) and Best Available Control Technology (BACT) respectively. Conceptually, both approaches involve the balancing of costs and benefits. This involves assessments of industrial plant financial viability as well as environmental assimilative capacities and pollution abatement technological options. The practice of pollution control in the UK and the USA, however, has been markedly different.

The USA

In the USA, air pollution control policy, for example, is based on achieving ambient air quality standards through strict formulation of technology standards (technology-forcing) applied by the private discharger/emitter. Under the Clean Air Act (1969) the USEPA set National Ambient Air Quality Standards (NAAQSs) which were to be implemented by states under State Implementation Plans (SIPs). Plans were developed for 247 air quality control areas, which determined emission levels for existing pollution sources. More stringent limits were set for new and modified sources on an industry-by-industry basis.

Nationwide ambient concentration standards for ozone (hydrocarbons emissions) and other 'criteria' pollutants were set, together with New Source Performance Emission Standards (NSPSs). Still more stringent case-by-case emission limits for processes subject to preconstruction new source review (NSR) were also formulated. So state control of existing sources (under SIPs) is the primary means for meeting ambient air quality objectives; NSPSs and NSRs are then meant to ensure that attainment will not be undermined as new plants add emissions. The system should also offer safeguards against new plants being located in relatively lax 'pollution haven' areas.

Over the period 1970–76 SIPs fell behind schedule as polluters and regulators fought a series of lawsuits against a general background of uncertainty over pollution damage impacts and abatement options. When it was clear many states could not meet SIP deadlines, the EPA started to formulate its offset policy (see Chapter 8). New and modified sources were allowed in 'non-attainment areas' when Lowest Achievable Emission Rate Technologies (LAERTs) were applied and when any additional emissions were offset.

In 1977 the Clean Air Act was amended to allow extension of deadlines for achieving NAAQSs and the formulation of new technology standards. In non-attainment areas, existing sources were allowed to apply Reasonably Available Control Technology (RACT – in principle less stringent than BACT), taking into account technological and economic feasibility. SIP deadlines continued to be problematic for non-attainment areas and the EPA was obliged to issue 'conditional approvals', based on states' forecasts of when NAAQSs would be met.

In 1978 attainment area programmes were buttressed by Prevention of Significant Deterioration (PSD) regulations. In PSD areas, modified existing sources had to satisfy BACT, although netting was allowed. Modified sources must meet BACT, unless net emissions are equal to the level before modification, other sources having abated their emissions.

By 1981 the Clean Air Act regulations had achieved a 58 per cent reduction in total annual particulates emissions and a 25 per cent reduction in sulphur oxides emissions. These large reductions were achieved despite a growth in coal-fired electricity generation and in general industrial activity.

Nevertheless, because of the 'slippage' in the regulatory programme and the revision of SIPs, the EPA began to search for ways and means of promoting more cost-effective pollution control.

Concern had been growing about the efficiency of a control approach which relied on standardised, end-of-pipe control technology to facilitate rule-making and enforcement. Uniform state or nation-wide emissions limits for discrete industrial processes seemed to be becoming increasingly costly as emission levels were tightened (as predicted by economic theory – see Chapters 4 – 6). The bubble policy (see Chapter 8) was formulated over the period 1977–1986 with the basic objective of increased economic efficiency in mind. During the 1980s many trading policies for different types of sources were initiated, resulting in a final Emissions Trading Policy Statement which was published in the Federal Register of December 1986.

The control of ground-level ozone in the USA has developed into quite a controversial issue which strikes at the heart of the regulatory approach to pollution control. Some 68 out of a total of 216 mostly urban areas have missed the December 1987 deadline for attaining the NAAQS for ozone. This standard requires that the peak daily 1-hour ozone concentration in an area does not exceed 0.12 parts per million (ppm) more than three times in three years. According to the EPA, during 1987 the average number of days that an area exceeded the standard increased over that of the previous three years.

Despite widespread non-attainment, the EPA has indicated that, after a review of toxicological, clinical and epidemiological evidence about the possible effects of ozone on human health, the current standard is not stringent enough. As we have already noted, ambient quality standards are, in principle, set according to a primary criterion – the protection of public health with a margin of safety. The legislative history of the US Clean Air Act and subsequent court decisions indicated that issues of technical feasibility and costs are not to be considered in setting the standard. Widespread non-compliance is not permitted to influence the choice of the standard. This approach may become less and less tenable (in cost terms) as more sophisticated experiments show very small effects, but effects nevertheless, at lower and lower levels of ozone exposure.

The ozone standard debate highlights a fundamental social question: *should human health be protected at all costs*? It seems clear that short-term ozone exposure does cause short-term, reversible reductions in lung function and increases in the frequency of respiratory symptoms. Long-term exposure may be linked to premature ageing of the lung and an increased risk of developing chronic respiratory disease. Ozone can also damage plants, organic materials and reduce visibility. Economists would seek to introduce the additional criterion of efficiency into the debate. They would say that the benefits of further protection against ozone should be balanced against the costs of obtaining it.

Such a balancing procedure does have a place in the policy process, because US Executive Order 12291 mandates that cost–benefit analysis be applied to all major regulations. The US Congress has implicitly recognised the high cost of attaining the ozone standard by allowing deadline extensions for compliance. EPA has commissioned researchers to investigate both the costs and benefits of ozone legislation. A county-level model to estimate the health benefits of a variety of national ozone control policies has been developed. Health benefits of air pollution reduction are measured as individual willingness to pay (see Chapter 9) to move from their current health state to a better one. The model uses baseline ozone concentrations for each day of the year, 1984 population data, the results of epidemiological and clinical (controlled laboratory) studies of the effect of ozone on health, and data on willingness to pay for reduced respiratory distress (see Chapter 10).

The aggregate dollar benefits using the epidemiological studies ranged from $153 million to $1,785 million annually. Clinical studies data resulted in benefits of $51 million to $4.7 billion annually. There was also a skewed distribution of the health benefits, with Southern Californian benefits exceeding half the total national benefits. People in this area only comprise some 11 per cent of the total population living in non-attainment areas. The benefits estimates are subject to some important caveats – they exclude benefits from reducing the rate of ageing of the lung; willingness-to-pay values exhibited a very broad range, with significant numbers of outliers; and non-health impacts of ozone were also ignored.

Given that different areas of the USA will experience very different levels of health benefits from ozone standard attainment, it is likely that benefit–cost ratios will also differ greatly across areas. Deadlines for compliance could therefore be linked to the size of the benefit–cost ratios, encouraging efficiency while still permitting discretion on grounds of equity or safety margins.

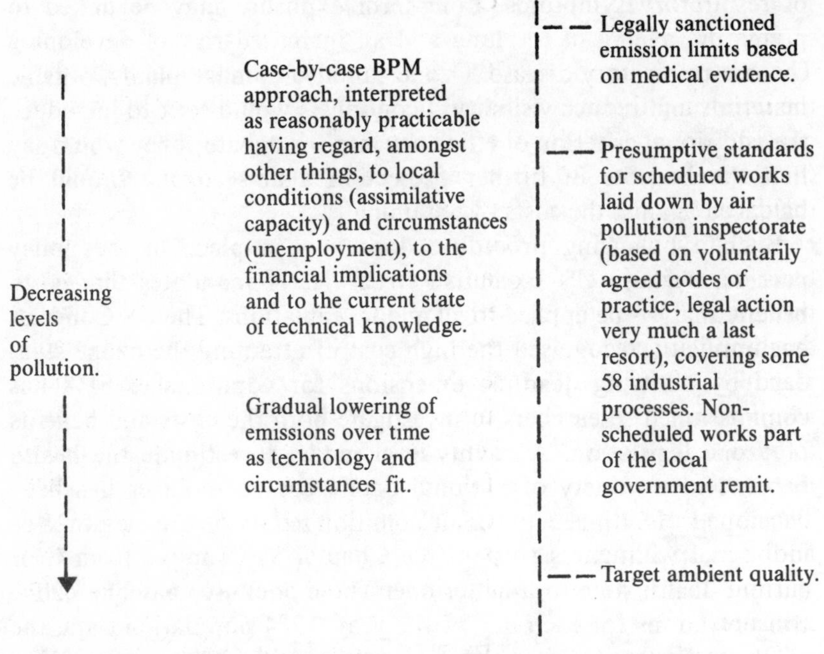

Figure 11.2 UK approach to air pollution control, pre-1987

The UK

In the UK pollution control policy has been implemented to a large degree on the basis of, and through the process of, negotiations and bargaining with, individual polluters. Proponents of this BPM approach assert that its advantages lie in its flexibility, its adaptability to particular circumstances and its gradualness (see Figure 11.2). Critics have argued that BPM is too flexible, that there are insufficient guidelines to help regulatory officials to judge the merits of a particular case and that all the discussions are too confidential.

Significant on-going changes to the UK's system of pollution control began to be introduced in 1987. A shift in the basic philosophy of control is under way because of pressure for change mandated by policy strategies and directives (carrying the force of law) favoured at the EC level.

11.4 EC POLLUTION CONTROL POLICY

The central component of the EC's pollution abatement strategy is the *uniform* effluent emission limit based on BACT. The EC has also consistently pressed for the inclusion of a *precautionary approach* (i.e. an anticipatory and preventative approach) in any controls over the discharge of potentially hazardous substances. This approach is consistent with the use of uniform emission limits or even discharge bans depending on how dangerous the discharged substances are thought to be (often whether or not full scientific evidence is available) rather than the traditional flexible, ambient quality standards of the UK. Polluters would also be required to undertake a *residual duty* under EC pollution legislation. Dischargers would have to operate their process and carry out other functions not specified in the consent to discharge effluent, in accordance with best practice. The aim of this requirement is, at least in part, to try to stimulate the adoption of cleaner processes and operational practices, thereby shifting the focus on pollution control away from end-of-pipe measures.

The passing of the 1976 EEC Directive on dangerous substances marked the beginning of a decade of dispute within the EC about water pollution control policy. On the one side, the UK has consistently favoured a policy based on a case-by-case approach with effluent limits set according to the receiving capacity of each water body and related environmental quality objectives (EQOs). But the rest of the Member Countries and the Commission have increasingly favoured discharge limits in the form of uniform end-of-pipe standards based on BACT.

The initial 'framework' Directive provides for both approaches, but each time succeeding Directives relating to specific target substances have been put forward, the controversy over control principles (discharge limit values versus EQOs) has re-emerged. In 1982 the Commission published a list of 129 priority 'black list' substances (requiring special attention because of their toxicity, persistence and capacity for bioaccumulation) but over the period 1976–1986 only four Directives dealing with mercury, cadmium and lindane had been adopted. By 1988 some twelve blacklist substances were subject to control via discharge limits or EQOs. The EC also intends to identify 'grey list' substances (e.g. chromium, copper, nickel, lead and zinc) and control their discharge. Both the selection procedure and the discharge limits or EQOs for such substances are proving extremely controversial.

During 1987–1988 it became clear that the UK was in the process of accepting a new approach, more consistent with EC thinking, to controlling discharges to water of dangerous substances taken from its own 'red list'. The new system is to be developed in parallel with a new integrated pollution control policy utilising technology-based controls (i.e. limit value approach) on solid, liquid and aqueous residues from about 3,000 industrial processes. We examine the integrated pollution control approach below.

The red list approach was a major break with tradition – it was an implicit acceptance that the policy of allowing discharges of all substances up to the point where ostensibly they are safely assimilated by the receiving waters is inappropriate for the most hazardous substances. The red list system identifies dangerous substances in terms of the following criteria: toxicity, persistence, bioaccumulation, physico-chemical properties and production levels. A carcinogenicity criterion may also be included. Only some forty-nine substances on the EC black list seem to be candidates for the UK red list. Once on the list 'strict' EQOs will be set for all such substances. Industrial processes discharging 'significant' quantities of these materials either directly to rivers or into sewers are likely to be 'scheduled' by the new integrated pollution inspectorate (formed in 1987), HMIP. Discharge limits will be set in terms of Best Available Technology Not Entailing Excessive Cost (BATNEEC). This new approach applies initially only to new or significantly modified plant, but will be extended gradually to existing installations.

In 1984 the EC passed a 'framework' Directive on air pollution from industrial plants, which required 'authorisation' of emissions from nineteen types of works. This legislation has forced the UK to put forward new air pollution control plans. The main feature of the new proposals (first issued in late 1986) would be to extend to all processes under local authority control the prior approval powers which, with minor exceptions, had been available only to the Industrial Air Pollution Inspectorate (and after 1987 to HMIP). A new two-part schedule will be required listing the processes under local and central control, together with emission consents for each plant. The consents would be linked to national BPM guidelines for each process. These changes will require new legislation which is expected to be introduced in 1989/90.

The UK government has also agreed, as part of the 1987 International Conference on the Protection of the North Sea, to apply BATNEEC principles to a wider range of dangerous substances entering rivers and estuaries bordering the North Sea.

11.5 INTEGRATED POLLUTION CONTROL: THE UK BPEO CONCEPT

Integrated Pollution Control (IPC) was first recommended in a UK Royal Commission on Environmental Pollution (RCEP) report published in 1976. The Commission had initially been investigating air pollution and the concept of BPM, but ended up recommending a new unified inspectorate to control pollution from complex industrial processes, regardless of the receiving media. The unified inspectorate was the Commission's partial remedy for what it perceived as ineffective pollution regulation delivered by the cumbersome and uncoordinated pollution control system in England and Wales. Clearly the inspectorate is only the first step on the way towards a more coherent and comprehensive regulatory system.

The integrated inspectorate (HMIP) was eventually established in 1987 and one of its recommended basic guiding principles, Best Practicable Environmental Option (BPEO), was not defined until the publication of the twelfth report of the Royal Commission in 1988. According to the RCEP, a BPEO is:

> the outcome of a systematic consultative and decision-making procedure which emphasises the protection and conservation of the environment across land, air and water. The BPEO procedure establishes, for a given set of objectives, the option that provides the most benefit or least damage to the environment as a whole, at acceptable cost, in the long term as well as in the short term.

IPC will therefore involve HMIP in considering the aqueous, solid and gaseous residues from industrial processes in the context of their potential aggregate environmental impact, thereby breaking down the barriers imposed by having separate enforcement bodies each looking after a single environmental medium (see Figure 11.3).

The BPEO framework will have several key components. First, processes to be subject to the new approach will be scheduled in regulations. Some 2,500 – 3,000 processes will be covered, including

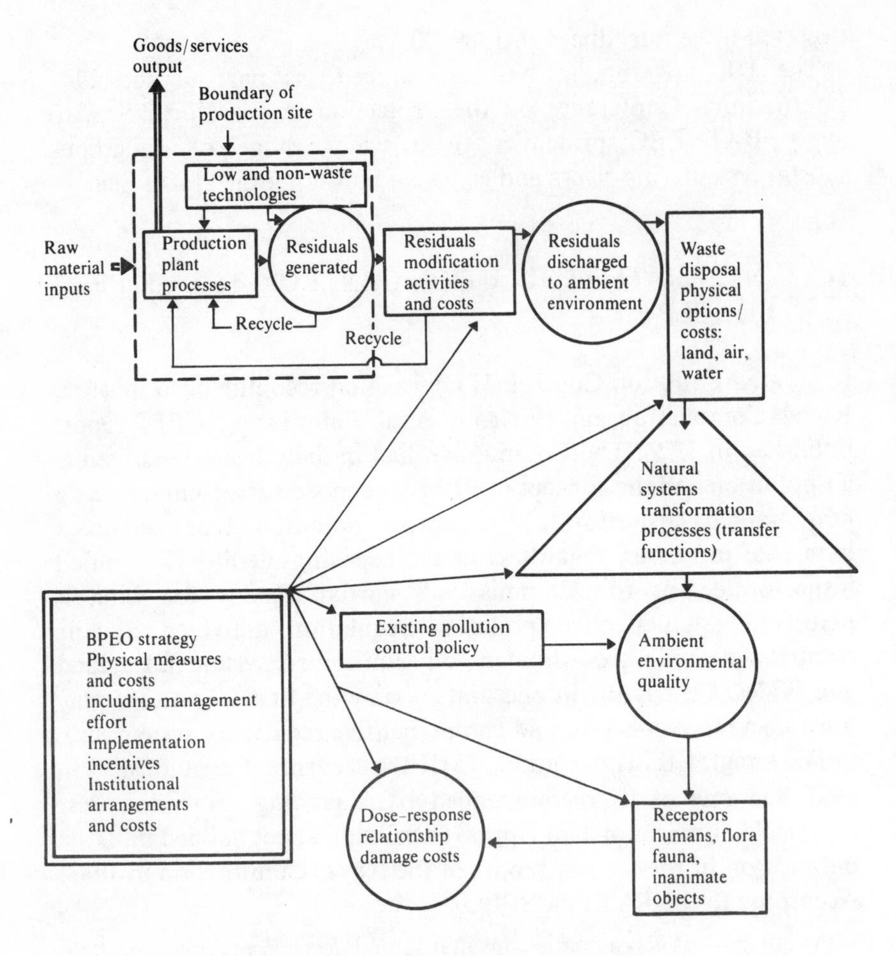

Figure 11.3 Basic BPEO model schema.

the 2,000 works already scheduled under air pollution legislation plus other processes, some of which will be discharging red list substances to water. Second, BATNEEC-type limits will be applied to the solid, liquid and gaseous residues from these processes. Third, controls applicable to scheduled processes will be set out in a series of IPC notes itemising preferred technologies and discharge limits. Fourth, operators will be obliged to seek prior approval from HMIP before operating a new scheduled process or substantially modifying an existing one. HMIP authorisations will be subject to periodic review and will be linked to a 'residual duty' provision.

As it is currently envisaged the BPEO approach has a number of shortcomings. In particular it is not clear how IPC will be applied to sites at which processes and products are frequently changed (e.g. fine chemicals). Only 'overall site criteria' may be practicable and acceptable in these cases; there is an obvious link to the US bubble concept here. The exclusion of most solid waste disposal facilities from the IPC regime is another problematic feature. Toxic waste disposal sites, for example, are not to be scheduled. At a more fundamental level there is no methodology laid down or even guidance on how risks in different environmental media might be compared during a BPEO study.

11.6 ECONOMIC-INCENTIVE POLLUTION CONTROL INSTRUMENTS

Some of the basic principles and recent changes in the direct regulatory approach to pollution control, both in Europe and the USA, have been outlined above (p. 167). The shifts in regulatory practice are the result of changing economic and policy contexts. During the 1980s European and North American countries have become increasingly affected by one or more of the following tendencies:

1. A move towards reduced direct government intervention in society, both in a financial (privatisation and the enterprise culture) and regulatory sense ('deregulation').
2. A move towards policy integration and particularly integrated pollution control, combined with a growing recognition of the need for increased cost-effectiveness of control.
3. A gradual transition away from end-of-pipe pollution abatement to preventative measures such as process control and a greater emphasis on the 'precautionary' (anticipatory) approach.

The combined effect of all these tendencies had led some countries to reassess existing economic instruments (e.g. revenue-raising charges) that have so far played a relatively minor role in pollution control, and to at least consider the virtues of new types of economic instruments.

The distinctions between economic and regulatory instruments are not particularly sharp in reality, as combinations of instruments are

Table 11.1 Typology of economic-incentive pollution control instruments.

Type of instrument	General description
Charges	
Effluent charges	Paid on discharges into the environment and are based on the quantity and/or quality of the effluent
Incentive effluent charges	Revenue collected via the charge is not returned to the polluter
Distributive effluent charges	Revenue collected via the charge is returned to the polluter, in the form of subsidies for new pollution control equipment
User charges	Payments for the cost of collective or public treatment of effluents
Product charges/tax differentiation	Additions to the price of products, which are polluting or are difficult to dispose of, the former have a revenue-raising feature
Administrative charges	Control and authorisation fees
Subsidies	
Grants	Non-repayable forms of financial assistance, contingent on the adoption of pollution abatement measures
Soft loans	Loans linked to abatement measures and carrying below-market rates of interest
Tax allowances	Allows accelerated depreciation, tax or charge exemptions or rebates if certain pollution abatement measures are adopted
Deposit–refund system	Systems in which surcharges are laid on the price of potentially polluting, a refund of the surcharge is given on the return of the product or its residuals
Market creation	Artificial markets in which actors can buy and sell 'rights' for actual or potential pollution
Emissions trading (bubbles, offsets, netting and banking)	Within a plant, within a firm or among different firms
Market intervention	Price intervention to stabilise markets, typically secondary materials (recycled) markets
Liability insurance	Polluter liability leading to insurance market

frequent. A number of possible classifications of economic instruments therefore exists. Table 11.1 presents a general typology.

The actual mix of pollution control instruments used varies substantially from country to country. Any given country's choice of policy instrument may be governed by a variety of considerations, including some assessment of how well the instruments meet both 'conformity criteria' and 'optimality criteria'. The former include consistency of aims and means, legal acceptability of means, complementary nature of the means *vis-à-vis* the general system of environmental policy and administrative culture. The latter include effectiveness, efficiency, distributional implications and political acceptability. The choice of instruments will therefore reflect inter-country differences in environmental policy principles, differences in relative power between various interest groups and differences in political structures (e.g. centralised versus federal systems).

In order to get a rudimentary assessment of the 'success' of operational economic-incentive instruments and forecast the future prospects for a more widespread adoption of such instruments, a selection of instruments will be evaluated in terms of the following criteria:

- Environmental effectiveness (i.e. the impact on environmental quality, either through pollution abatement by target groups, or collective/public actions)
- Economic efficiency (more strictly, cost effectiveness, i.e. the achievement of policy objectives at minimum cost)
- Administrative efficiency and practicability (information requirements, management costs and propensity to induce opposition), compatibility with institutional framework (in particular with the 'Polluter-Pays Principle').

The pollution control strategies that have evolved and continue to be adapted reflect choices not between regulations and one or more economic-incentive instruments, but between various combinations of regulation and incentives. The rationale for such combined packages encompasses the need for revenue to finance environmental protection, the incentive provided by economic instruments to better implement associated regulations and the potential stimulatory impact on technical change.

In Chapters 4–8 we examined the properties of economic instruments (assuming that they all attempted to internalise a more appropriate environmental behaviour by means of financial incentives) and showed that such instruments would be expected to provide effective and efficient control over pollution. Table 11.2 surveys the actual use of such instruments in mixed economies. In general, it seems to be the case that only a limited number of these instruments (e.g. emissions trading) are explicitly designed to achieve an economically efficient solution. The financial revenue–generating

Table 11.2 Types of economic instrument: past or current usage by country.

Country	Effluent charge				User charge	Product charge	Administrative charge (licensing and control)	Tax differen-tiation	Subsidies (including grants, soft loans and tax allowances)	Deposit–refund	Market creation	
	Air	Water	Waste	Noise							Emission trading	Market intervention
Australia		×	×		×		×					
Belgium			×		×		×					
Canada					×				×	×		
Denmark					×		×	×	×	×		
Finland					×	×	×		×	×		×
France	×	×		×	×	×	×		×			
Fed. Rep. of Germany		×		×	×	×	×	×	×			
Italy		×			×	×	×					
Japan	×			×	×							
Netherlands			×	×	×		×	×	×	×		×
Norway					×	×	×		×	×		
Sweden					×	×	×	×	×	×		
Switzerland				×	×			×				
UK				×	×		×	×				
USA			×	×	×	×			×		×	

Adapted from Opschoor and Vos (1989).

capacity of charge is important on a number of actual charging systems especially in the light of general public expenditure constraint policies. Whenever economic instruments require complex modes of operation they meet with resistance, both from polluters and regulatory agencies. Most economic instruments operate as adjuncts to the primary direct regulations and this is unlikely to change in the foreseeable future.

Given the limited actual use of economic incentives in pollution control policies in all mixed economies, it is interesting to note that in 1985 the Organisation for Economic Cooperation and Development (OECD) member countries all adopted an environmental declaration which included a re-affirmation of the *Polluter-Pays Principle* (PPP). Countries agreed to seek to introduce more flexible, efficient and cost-effective pollution control measures through a consistent application of the PPP and economic instruments.

PPP dates back to OECD recommendations in 1972 and 1974 and although the principle is well established, it is far from unambiguous. The 1972 Standard PPP interpretation has it that PPP means that the polluter should bear the expenses of carrying out the pollution prevention and control measures decided by public authorities to ensure that the environment is in an acceptable state. The 1974 Extended PPP interpretation says that, if a country decides that above and beyond the costs of controlling pollution the polluters should compensate the polluted for the damage which would result from residual pollution, this measure is not contrary to the PPP, but the PPP does not make this additional measure obligatory.

Other ambiguities relate to the use of tax advantages and subsidies to assist polluters to 'clean up' over a transitional period of time. There is no agreement over the length of the transitional period or over whether subsidies are, or are not, consistent with PPP. Both in France and the Federal Republic of Germany, for example, subsidies are considered compatible to PPP as long as, in principle, the polluter remains fully responsible for his pollution and if subsidies would aid the implementation of PPP or enable stricter environmental controls.

Overall, the practice of pollution control suggests that efficiency and equity properties of control instruments are less important than environmental effectiveness, political/bureaucratic compatibility and revenue-raising properties. PPP in this context provides a moral and economic rationale to enable governments to facilitate the finance-finding function that some instruments contain.

As far as the individual instruments themselves are concerned actual *effluent charge rates* are set at too low a level to provide an incentive role. They do, however, fulfil successfully a revenue-raising function. Higher charge rates are likely to prove acceptable only if the charge base is detailed and therefore complex. Administrative efficiency is then likely to conflict with effectiveness and efficiency. *Product charges* have generally lacked an incentive impact because they have been set at too low a rate. Most effective to date have been a Finnish charge on beverage containers and a Norwegian charge on food containers. Administrative efficiency is improved if charges are linked to existing tax or excise systems. Thus in the USA, a 'feedstock' charge has been introduced in order to create funds (Superfund) to help clean-up or removal of inactive or abandoned hazardous waste sites. *Subsidies* are not very compatible with the PPP, but are nevertheless widely applied. They have helped in old plant renewal, in the introduction of clean technologies and generally fulfill a facilitating role during transition periods. PPP has therefore been reinterpreted along national lines and seems not to be a crucial factor in decision-making on pollution control instruments.

Emissions trading, as applied in the US air pollution context, does seem to have worked in terms of stimulating economic efficiency, although its environmental effectiveness is less apparent. Substantial cost savings by polluters have been reported and emissions trading has facilitated continuous economic growth in non-attainment (poor air quality) areas. Administrative costs have also been high.

We conclude by noting that while the 'deregulation' trend in mixed economies could lead to a stronger role for economic-incentive instruments this is most likely to be because of their revenue-raising properties rather than their economic efficiency properties *per se*. The shift in pollution control philosophy away from remedial end-of-pipe measures towards a more precautionary/anticipatory approach may result in more applications of product charges, subsidies aimed at process charges and deposit–refund systems. Even so these types of instrument will still be only supplementary tools in the pollution-abatement policy kit.

12 · POLLUTION CONTROL POLICY IN CENTRALLY-PLANNED ECONOMIES

12.1 SOVIET IDEOLOGY AND ENVIRONMENTAL IMAGE

Although socialist central plan economies differ in terms of the degree of concentration and centralisation that their systems operate under, we will concentrate on just one, the Soviet Union, as our example. Such economies are distinguished from market-based systems in that the central economic questions of what to produce, how to produce it and to whom it should be distributed are settled, in principle, via a central hierarchical planning process. The conventional official Soviet viewpoint has always been that central plan socialism is capable, without significant restructuring, of adequately coping with environmental degradation. It is argued that the Soviet system, built on 'scientific' Marxian principles, is subject to environmental problems qualitatively different from those that afflict capitalist market economies. This position is founded on a distinct Marxian perspective on the natural environment, which we outlined in Chapter 1. In reality, the Soviet Union and all other socialist systems are, like their capitalist counterparts, far from immune from environmental despoliation.

During the late nineteenth century, in the decades before the Bolshevik revolution which established the Soviet state, two broad intellectual currents emerged – Slavophile rurality (i.e. small-scale rural communities and their associated lifestyles offer the best prospect for civilised living) and Western scientific materialism. Lenin, the founding father of the Soviet Union, rejected the ecocentric Slavophile view in favour of the technocentric 'scientific socialist' position, intellectually dependent on the writings of Marx and Engels. There are two basic strands in the Marx/Engels attitude to nature. The first emphasises the need to humanise nature via science so that inherent value can be turned into use value. The second warned against the predatory exploitation of nature said to be typical of capitalist systems. Engels in particular advocated recycling waste and hinted at ecological limits and pollution-free industrial production (there is a link here to the modern Soviet interest in so-called 'Low and Non-Waste Technological Processes').

Overall, however, the basic ideological belief in inevitable progress outweighs the Engels caveats and results in the modern Soviet rejection of zero growth or steady-state economies. Instead, Soviet people should develop progressively and take on the task of 'optimising the biosphere' aided by a belief in the inevitability of material and technological progress.

Marxist ideology encourages the view that the natural resource base would be sufficient to support any population level, given advanced forms of production, provided that society is organised on socialist principles. In particular, goods should be 'rationally valued' (in terms of labour power) so that use value and exchange value are equivalent and alienation and exploitation absent. In practice of course the Soviet Union's economy is no less reliant on ecological foundations than any other economy and in some respects faces unique environmental difficulties, as well as having vast potential untapped resources.

The Soviet Union encompasses some 8.6 million square miles of territory (a sixth of the planet's land surface), around a third of which is forested and nearly half of which lies in the permafrost zone. Thus although many areas are richly endowed with natural resources, the exploitation of such assets will be costly in both financial and ecological terms. Much of Soviet territory is ecologically fragile and, although rich in energy and other mineral resources, is also remote from the established centres of population and industry in European Russia. Siberia (60 per cent of the Soviet land mass) exemplifies both the future resource exploitation opportunities and inherent costs facing succeeding generations. The massive new oil and gas fields of Western Siberia will be costly to work because of remoteness, the hostile climate and difficult terrain which has hindered the development of an adequate infrastructure.

The Soviets face a serious problem in the general lack of correspondence between the distribution of water resources and new mineral and energy resource deposits on the one hand, and population, economic activity and agricultural potential on the other: 84 per cent of the country's river flow enters the Arctic and Pacific oceans; only 16 per cent crosses the southern and western sections, much of which has an arid or semi-arid climate. But it is in these latter areas where one finds 75 per cent of the population, 80 per cent of industrial activity and over 80 per cent of Soviet crop land. Food and crop production has grown increasingly dependent on irrigation as a means of increasing and stabilising harvests, which have been a persistent problem since the inception of the Soviet state. Only around 10 per cent of the Soviet land area is suitable for agriculture and only 1 per cent of this 10 per cent gets an annual rainfall of 70 cm or more, which is required for stable harvests (the equivalent US figure is 50 per cent).

Scientific socialism has also not been able to bypass the laws of thermodynamics. Water pollution is severe in the industrialised areas of European Russia and the Ukraine, with the Volga and Don rivers as well as the Black, Caspian, Aral, Azov and Baltic Seas and Lake Baikal suffering heavy effluent loading. Air pollution is a more regionalised problem and urban areas have ambient quality levels no better than Western cities, despite the less widespread private automobile ownership pattern; this latter feature is offset by the Soviet planners' traditional strategy of locating heavy industry and worker housing and amenities in close proximity to each other. History is important in this context, as the Bolshevik party initially drew its support from the minority urban population (which the Party needed in order to expand) living in an overwhelmingly rural peasant-based country. A growing number of Soviet writers have acknowledged that the process of heavy industrialisation and technocentrist materialism have generated ecological disruption under socialism as well as capitalism. But pollution and related problems have occurred, it is argued, only because socialism in the Soviet Union has not yet achieved its full potential. Pollution then is only a temporary anomaly.

Nevertheless, the Soviet Union shares common borders with fourteen other nations. Environmental problems do not respect these borders, as was dramatically emphasised by the 1986 Chernobyl nuclear power station accident. Thus, the Soviet environmental management policies, or lack of them, are of continuing worldwide importance.

12.2 ECONOMIC SELF-SUFFICIENCY AND THE TRANSFORMATION OF NATURE

A core factor in the official Soviet environmental world view is the sheer size of the country, which has spawned a somewhat complacent attitude toward resource depletion and waste-assimilation capacities. Political necessity has forced successive leaders into industrial and agricultural expansion programmes based on long-term self-

Table 12.1 Soviet net import dependence of minerals as a percentage of consumption (1983).

Metals	Main suppliers	Percentage
Aluminium		44
Antimony	Yugoslavia	6
Bauxite and alumina	Greece, Guinea, Yugoslavia, Hungary, India, Jamaica	48
Chromium		17
Cobalt	Cuba	47
Copper		0
Diamond, gem		280
Gold		29
Iron ore		21
Lead		0
Manganese		12
Mercury		0
Molybdenum	Mongolia	15
Nickel		23
PGM		64
Silver	Switzerland, UK	10
Tin	Malaysia, Singapore, UK	33
Titanium		10
Tungsten	People's Republic of China, Mongolia	43
Zinc	Australia, Finland, Peru, Poland	2
Non-metals		
Asbestos		43
Fluorspar	Mongolia, People's Republic of China	53
Phosphate		16
Potash		26
Fuels		
Crude oil		23
Natural gas		13

Sources: (1) US Department of the Interior, Bureau of Mines, 1983, Minerals Yearbook. (2) P.R. Bellinger, 'Probability of continued Soviet mineral self-sufficiency', *Resources Policy*, 1985, **11**, 3.

sufficiency goals. The growth ethic was reinforced in the leadership's mind by a 'catching up with the West' mentality which led to heavy investment in basic industries and defence. Thus, for example, the USSR is now the world's leading producer of oil, natural gas, coal, asbestos, iron ore, manganese, mercury, nickel, platinum group metals, potash, titanium and vanadium. It ranks second in aluminium, chromite, gold, lead, phosphate, tin, tungsten and zinc. Overall, it has a lack of net import dependence in minerals (see Table 12.1).

The leadership has consistently emphasised a need to transform or modify the environment in order to augment nature and improve human living conditions. Both the collectivisation drive in the agricultural sector and the heavy industrialisation process during the Stalinist period (1928–1953) spawned a propensity for massive schemes. The culmination of this trend was Stalin's 'Plan for the Transformation of Nature' initiated in 1948. The plan had a dual form: to transfer part of the northerly rivers stream flow southward to the more arid regions of Central Asia and Southern Russia, and to plant massive windbreaks and shelter belts of trees in the Steppe regions of European Russia. The windbreak scheme has at least been partially successful. The river-diversion plan has, on the other hand, continued to be a source of controversy and debate up to the present. We look in more detail at this plan in a later section (p. 184).

Other examples of gigantic projects are the White Sea-Baltic Sea Canal and Khrushchev's Virgin Lands Campaign to boost agricultural output. In more recent decades, Stalin's pathological determination to achieve 'final' victories over nature, regardless of cost, has been replaced by a somewhat cautious approach based on a cost-effectiveness criterion and incorporating environmental impact information. Within certain limits, experts are now allowed to voice concerns and advocate a diversity of approaches toward environmental protection.

One long-standing environmental conservation (if not preservation) measure in the USSR has been the establishment of an extensive network of nature preserves, *zapovedniki*. The first was approved in 1919 and by 1986 there were some 130 in existence (see Table 12.2). Within the preserve boundaries, plants and animals and habitats are fully protected. Many threatened species have also been successfuly propagated in them. *Zapovedniki* have been further supplemented by national parks, biosphere reserves and hunting

Table 12.2 Preservation/conservation areas in the USSR.

Type of area	1950	1960	1970	1980	1986
Zapovedniki	128	90	100	120	130
Biosphere reserves	0	0	0	7	17
Hunting preserves	0	0	6	7	8
National parks	0	0	0	7	10

Source: Adapted from P.R. Pryde, 'Soviet Management of Biosphere Resources', Paper IIUG pre 87–4, International Institute for Environment and Society, 1987, Berlin.

preserves. The primary purpose of these restricted-access areas is not tourism but rather environmental monitoring and research. Most have been created in areas of typical or significant landscapes and ecosystems, or in areas of particular biotic importance. In recent years, references to genetic preservation have begun to appear much more frequently in specialist Soviet journals. Endangered species have been recorded in volumes called *Red Books* since 1978. A 1985 edition lists some 135 endangered plant species and seventy species of endangered wildlife. We examine the economics of species extinction in Chapter 17.

If all types of partially and fully protected areas are counted, only some 8 per cent of the total area of the country is covered. The system needs to be expanded if it is to play a significant role in genetic preservation, and a unified management agency for all the *zapovedniki* is also required.

12.3 RIVER-DIVERSION SCHEMES AND THE LAKE BAIKAL CONTROVERSY

Some Western analysts think it fruitful to conceive of the Soviet policy process in environmental protection in terms of a 'state corporatist' model, which has evolved since Stalin's time. Corporatism, like pluralist models representative of market democracies, takes into account the role of groups in the policy process. But state corporatism contains a much greater role for the state (planner, not individual preferences) in the recognition of problems, placing issues on the public agenda and adjusting policies to suit changing circumstances. Responsibility for the environment, control of information and the generation of new ideas must flow through the Party-State structure.

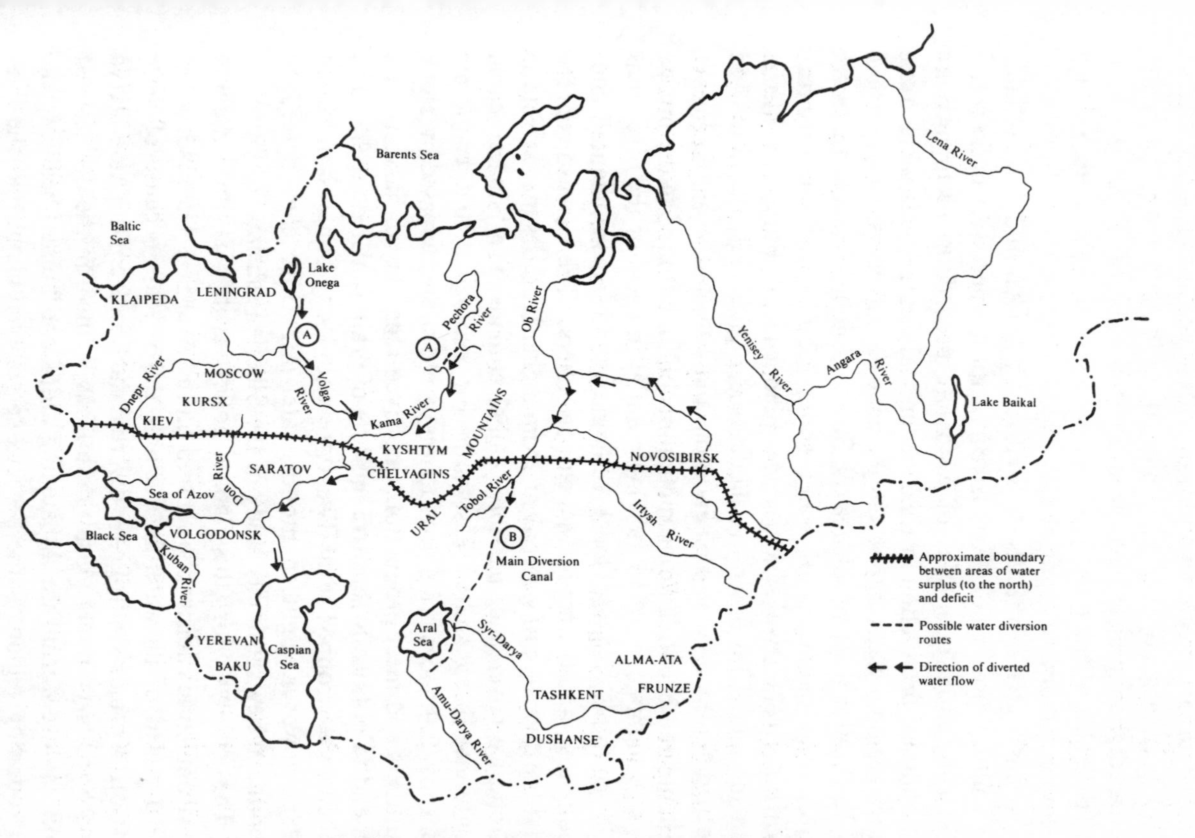

Figure 12.1 Soviet river-diversion schemes.

The river-diversion schemes, for example, have involved over 120 institutions in environmental impact assessments. In the course of the debate local, regional and transnational ecological consequences have been argued over and internal nationality issues have also surfaced. Economists have been especially critical about the cost-effectiveness of the plan. The main issues in the debate are outlined below in relation to Figure 12.1.

A major goal of the USSR is currently greater and more reliable agricultural production. Extensive irrigation is necessary in the south and substantial quantities of water are needed. Plans to increase agricultural production in Kazakhstan and Central Asia will require extra-regional water imports in the next 10 – 20 years.

To counter these problems, the Soviet Union proposes to transfer substantial quantities of water south from its northern Arctic-flowing rivers (two schemes are planned, (*a*) European Russia scheme and (*b*) a Siberian scheme, see Figure 12.1; the latter had been deferred until implementation and assessment of the former scheme had occurred (construction by 1990), but in 1984 the project was brought forward to 1987). Potential impacts of the diversions will certainly include local ecological and environmental consequences, and the net impact will be adverse. There is also concern that the diversion of water away from higher latitudes may affect the Arctic ice cover and, possibly, global climate. The 'Sibaral' canal route has been shifted eastwards for ecological and economic reasons. Enhanced grain production would allow increased livestock numbers and meat production, improved reliability of harvests, and an end to politically sensitive imports of grain is also desirable. The scheme was dropped at the Twenty-seventh Party Congress in March 1986; however, eighteen months later Mr Gorbachev agreed to conduct a new feasibility study of the Siberian scheme.

The catalyst for this latest policy switch is an on-going environmental catastrophe centred on the Aral Sea. The sea has lost 60 per cent of its water in the past thirty years because of large irrigation projects which have diverted the flow from the Amu Darya and Syr Darya rivers. A salt desert has formed on the dry sea bed and its dust storms are threatening crops and human health in the surrounding region. Some Soviet experts even claim that the Aral Sea is a decisive factor in the climate of the whole of Asia. There is a potential threat to rainfall and food supplies in the Indian sub-continent.

Environmental protection policy came of age with the widely publicised controversy over pollution in Lake Baikal. The lake is 700km long and is both the deepest and the most voluminous freshwater lake in the world. It supports over 1,700 species of plants and animals. Most significantly, some 75 per cent of these are endemic, and make Lake Baikal a biospherically significant asset. The main threat to Lake Baikal came from two wood-processing plants which emitted effluent either directly into the lake or into a feeder river. An additional threat was posed by logging activities on the steep hillsides surrounding the lake, which carried with them a soil-erosion potential.

Defenders of Baikal legitimised and strengthened their environmental case by identifying their 'interest group' goal with a goal of part of the Soviet leadership circle, namely surmounting the waste and inefficiency of departmentalism. Individual Soviet ministries had a tendency to pursue narrow 'departmental' interests at the protected zone and in 1971 increased effluent clean-up was ordered along with improved logging management. It remains to be seen whether the deterioration in water quality that has occurred is significant and irreversible.

By the 1970s, environmental protection was an issue which decision-makers had to take seriously. However, one should not over-emphasise this rise of environmentalism in the USSR. Official macroeconomic concerns of growth and efficiency legitimised discussion, by experts, of wider environmental issues, but no great surge of public participation. The Lake Baikal debate stimulated the passing of several high-level environmental protection resolutions, in addition to expenditure allocations for pollution control in the state plan and the upgrading of an environmental monitoring committee (*Gidromet*). Between 1981 and 1986 over 53 billion roubles were allocated to environmental protection.

12.4 COMPARATIVE ENVIRONMENTAL EFFECTIVENESS

Materials-balance models demonstrate that in any industrialised, growth-orientated economy, natural resources and ecological foundations of the system would be heavily stressed. Environmental convergence theory maintains that environmental disruption will tend to be uniform between countries at similar levels of economic development, regardless of their political/institutional form. Environmental effectiveness does not therefore vary between such economies. It has been pointed out, however, that convergence theory is methodologically incorrect in that it fails to distinguish between the actual amount of environmental disruptions observed in an economy and the conceptually distinct environmental efficiency of an economy. The latter point can be defined, in principle, as the point at which marginal pollution damage costs are just equal to the costs of preventing that additional unit of pollution (abatement costs) (see Chapter 4).

Comparisons can be made of the appropriateness of the institutional forms and incentive structures which determine environmental performance. Such *a priori* evaluations are, however, frequently inconclusive. The reason for this is that assessments of environmental effectiveness are complicated by problems of data compatibility (the lack of comparable national pollution indices) and the influence of so-called non-systemic factors. Comparing, for the sake of argument, the USSR and the USA, one set of non-systemic influences – the more frequent inversions, colder climate, factory location pattern and sectoral composition of GNP – require that the USSR allocates more resources to air pollution control than the USA would to achieve the same ambient quality. On the other hand, another set of non-systemic factors – the lower overall level of GNP, relative paucity of automobiles and large land mass – reverses the logic of the argument. If the latter set of factors dominates and if the two countries had similar attitudes to the environment, one would expect the USSR to devote fewer resources to environmental protection. If this is what is observed (empirically this seems likely), it does not necessarily mean that the Soviet system is less effective environmentally.

Concentrating just on the institutional forms employed in different countries to decide upon and implement pollution control strategies, it would seem that superficially the socialist system has some advantages. It fosters a collectivist approach in which limitations are placed on individual choice for the sake of the wider society. According to a number of the minority economic doctrines (see Chapter 1) just such an emphasis on group identity, social responsibility and public forms of ownership in the economy is required if a sustainable system is to evolve. Further, some writers have argued that Soviet-style central planning actually confers an informational advantage which could be utilised to achieve more efficient environmental management. What is being claimed is that highly centralised decision-making processes in the existing central planning structures are capable, in principle, of generating much of the information required to achieve acceptable ambient quality levels at the least possible abatement costs.

The structural advantage argument in favour of socialist systems has been disputed in principle and on grounds of operational reality. It has been argued that an economy such as that in the USSR is not innately superior to a decentralised market economy in terms of the maximisation of the flow of information, simply because a planned economy on the scale required functions more as an administrative than an economic organisation. It must also be the case that the planning system in question must have as its primary organisational principle the maximisation of environmental protection.

In practice, the USSR has failed to realise fully any formal structural advantage (if in fact it exists) because of a range of dysfunctional forces. Both the formal constitution and the legislation relating to the environment that has been passed appear impressive. Article 18 of the New Constitution (1977) reads as follows:

> In the interests of the present and future generations, the necessary steps are taken in the USSR to protect and make scientific, rational use of the land and its mineral and water resources, and the plant and animal kingdoms, to preserve the purity of air and water, ensure reproduction of natural wealth, and improve the human environment.

Article 67 addresses the individual directly: it reads:

> Citizens of the USSR are obliged to protect nature and conserve its riches.

So the Constitution places an emphasis on obligations to both past and future generations, particularly the immediately succeeding generations. In the past this has been partly a device to justify a policy of surplus re-investment in capital projects rather than diverting resources to current consumption. But it is also encapsulated in Marxist ideology that portrays any period or generation as a mere point in time along a political–economic continuum.

No less than six major environmental laws have been passed in recent years in the USSR – Principles of Water Legislation (1970 and revised in 1979); Principles of Land Use (1968); Principles of Mineral

Legislation (1975), Principles of Forestry Legislation (1977); and laws on air quality (1981) and the animal world (1980).

The Constitution and all this formal legislation is still to a significant extent offset by in-built economic, political and technical forces in the Soviet system:

1. The planning system still gives top priority to meeting quantitative norms (production quotas, etc.), and the 'taut planning' system implicitly downgrades the rank–order status of environmental protection activities at lower levels in the planning hierarchy; laws, national or republican, will conveniently be ignored when 'economic' interests are at stake.

2. The incentive bonus scheme imparts a perverse incentive structure which results in resource conservation not being practicable at plant level. Managers tend to adopt a risk-averse strategy and concentrate on meeting output targets: fines for pollution are low and enforcement not uniformly strict.

3. Departmentalism (ministerial empire building) means a lack of coordination in the planning systems. There are few mechanisms and no agency to deal with cross-media (and therefore cross-ministry) environmental problems. For example, serious administrative problems exist concerning the protection of endangered plants. There is no formal law on botanical resources and a large number of agencies are responsible for the protection of plants – the Ministries of Agriculture and Health, USSR Academy of Sciences, All Union Society for the Protection of Nature and Local Soviets, or administrative organs. As a result inter-agency communication and cooperation is often very cumbersome or non-existent.

4. The lack of an economic resource pricing structure meant that, in the past, resources were often treated as virtually free goods. Resource users were not resource owners thus often precluding incentives for long-term sustainable management practices. In the past it was also the case that resource exploiters could not independently develop uses and customers for the by-products of their operations. Only the forest legislation and the revised law on water usage embody charges for the use of natural resources.

5. Official environmental agencies have a lower status than industry interests. Embryonic Soviet 'pressure groups' do not operate

like their Western counterparts. Groups of scientists/academics have successfully appealed to the top echelons of the Soviet system on environmental conservation issues, such as the protection of Lake Baikal. But such groups are closed out of high-level deliberations and diffusion of information is slow and cumbersome compared to pluralist societies.

6. Soviet laws which embody the strictest environmental standards and emission norms are exceedingly difficult (costly) to meet and are therefore often ignored in practice. Maximum permissible norms for waste discharges/emissions are most often based on concentrations of pollutants in the ambient environment, rather than on amounts of pollutants discharged/emitted at source. This makes the establishment of legal liability for pollution violations very difficult. For water pollution, the focus on concentrations is open to abuse if plant directors dilute waste-water effluent with clean water, thereby formally complying with pollution norms. The air quality legislation introduced in 1981 has introduced a system of maximum allowable emission standards.

In the USSR institutional and/or attitudinal changes tend to take place only slowly; nevertheless, progress in terms of increased environmental protection measures has been made, especially in the last ten years or so. Problem areas requiring much more immediate official attention and/or new management strategies include the commercial nuclear power programme (especially the role of graphite reactors), soil conservation, reclamation of contaminated land, the need for an expanded network of biotic reserves and the lack of a single national agency to manage protected areas.

Glasnost and environmental protection

Under *glasnost*, Soviet scientists and academics may be able to voice their concerns increasingly about the environment. Many regions and towns now have nature protection societies, and small active environmental groups are beginning to appear in Soviet society. Public bodies such as local Soviets and official inspectorates are finding that under *glasnost*, and with media support, they can increase the stringency of environmental protection policy. In the summer of 1988, for example, the Soviet of Archangel region forced the closure of part of the Voloshka cellulose plant because of excessive pollution. The Akhangaran cement works in Uzbekistan has been shut down by the atmospheric air inspectorate, and the water conservation inspectorate has closed four sections of the Estonian phosphite works in Maardu.

The democratisation of Eastern Europe which began in 1989 has spread rapidly. It has served to bring to light the massive extent of environmental degradation in these formerly centrally planned economies. The market economies will be required to channel multilateral aid into Eastern Europe in order to create a more modern industrial base. Over the long term, this should lead to a higher level of ambient environmental quality, as old inefficient, highly polluting plants, are replaced. However, there is a requirement for this aid to be explicitly 'environmentally sensitive'. This is a lesson that aid agencies are only just beginning to learn in the context of developing countries (see chapter 22).

13 · GLOBAL POLLUTION POLICY

13.1 INTRODUCTION

In this chapter we look at selected issues of pollution in a context wider than the confines of national boundaries. Two contexts are under consideration.

First, pollution arising from one country can cause damage in another country. This *transboundary pollution* takes on the features of an *externality* between the 'emitter' and the 'recipient'. Such externalities include pollution of the river Rhine in Europe which serves several different countries in terms of supply water, acting as a receiving medium for wastes, and as a recreational and commercial watercourse. But perhaps the most celebrated transboundary pollutant is the so-called 'acid rain' – air pollutants that travel across boundaries and cause damage in other countries. The USA is an emitter in this respect and Canada is the recipient. In Europe, there are several major emitters and generally the northern countries, especially Scandinavia, are the recipients. As we shall see, the nature of the externality is complex, for not only do the emitters cause damage to the recipients in other countries, but may simultaneously be recipients both of others' pollution and their own. In this respect, transboundary pollution has some features in common with our other context.

The second pollution context is also one of externality but it has the features of *mutual externality* in that polluters damage themselves and others because what they pollute is the *global commons*. The global commons refer to resources that are shared as common property by all nations – the atmosphere, the stratosphere and the world's oceans. These resources are treated as open access resources but they are common property resources as well because the group of individuals using them comprises everyone. The main examples of global commons pollution are carbon dioxide (CO_2), which gives rise to a global warming of the earth (the 'greenhouse effect'), and chlorofluorocarbons (CFCs) which are associated with damage to the ozone layer round the earth.

13.2 ACID RAIN

Acid rain gets it name from the solution of sulphur oxides (SO_2 and SO_3) and nitrogen dioxide (NO_2) in cloud and rain droplets, causing the resulting sulphuric and nitric acids to fall as an acid 'rain'. Henceforth, for convenience, we shall refer to the pollutants as 'SOX' and 'NOX' – sulphur oxides and nitrogen oxides. Acid rain is something of a misnomer because the damage done by this form of *wet deposition* should not be singled out from the particles of NOX and SOX that fall as *dry deposition:* both cause damage. The pollutant *ozone* is also implicated in the acid rain story. Ozone is formed by photochemical reaction in the atmosphere, and also exists in 'natural background' levels.

Table 13.1 Sources of SOX and NOX emission (percentage).

SOX/NOX	USA	Canada	Germany	UK
SOX emissions				
Power stations	66	16	56	66
Industries	22	32	28	24
Smelters	6	45	–	–
Households and commercial	3	4	13	9
Transport	3	3	3	1
Total	100	100	100	100
NOX emissions				
Transport	44	61	45	32
Power stations	29	13	31	46
Industry	22	20	19	19
Households and commercial	4	5	5	3
Other	1	1	–	–
Total	100	100	100	100

Sources: For USA, West Germany and Canada: Sandra Postel, *Air Pollution, Acid Rain, and the Future of Forests*, World Watch Institute, Washington DC, 1984. For UK: House of Commons Environment Committee, *First Report – Air Pollution* Vol. 1, HMSO, London, 1988.

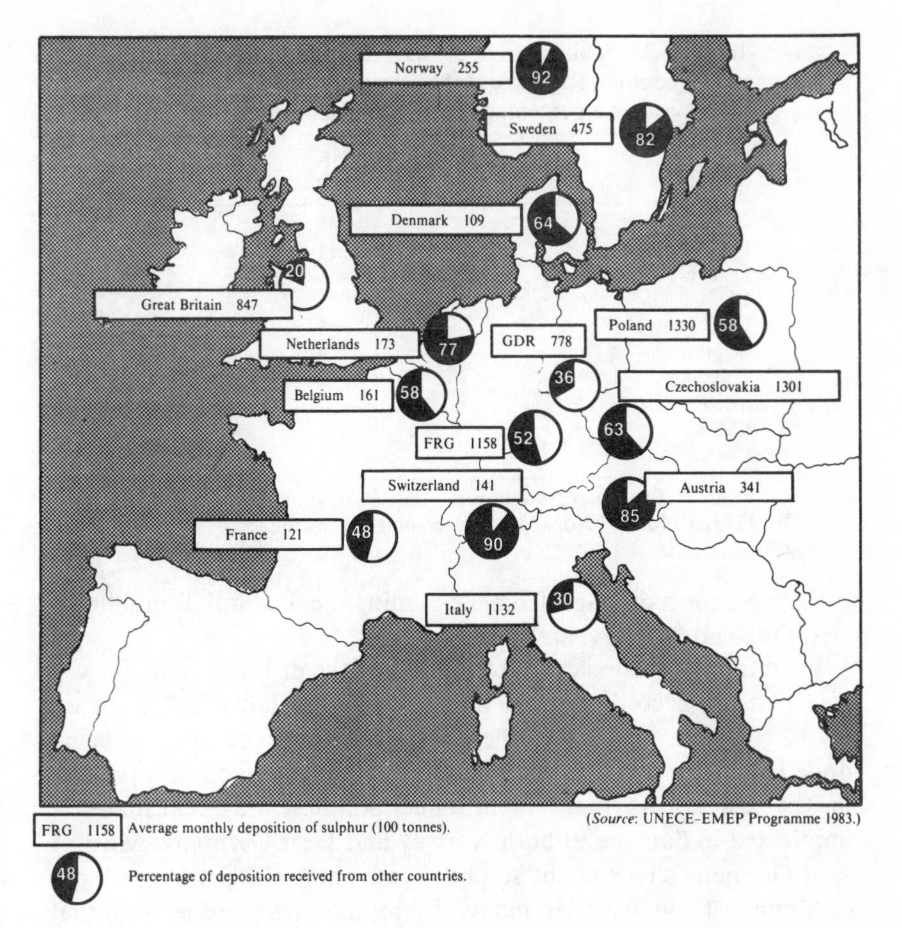

(*Source*: UNECE–EMEP Programme 1983.)

Figure 13.1 Sulphur deposition in Europe 1980.

The sources of the SOX and NOX emissions are varied. In the USA and Europe the main source of emissions is power stations that burn coal and oil with high sulphur content. In Canada the main source is the smelting industry. Table 13.1 shows the relevant proportions for SOX and NOX.

Figure 13.1 shows the general emitter–recipient picture for Europe. Thus the UK received some 84,700 tonnes of sulphur per month in 1980 of which 20 per cent came from other countries and 80 per cent was 'self-inflicted'. In Norway, however, 255 tonnes per month were received and virtually all of it (92 per cent) was 'imported'. A more detailed analysis shows who pollutes whom. The approximate proportions for the early 1980s for two of the countries

Table 13.2 Main sources of SOX deposition received in the Federal Republic of Germany and Norway (expressed as a percentage of all SOX deposition).

Sources	Federal Republic of Germany	Norway
Germany	48	4
East Germany	11	6
France	8	2
UK	5	8
Czechoslovakia	5	3
Belgium	4	1
Poland	3	4
Sweden	–	3
USSR	–	5
Unknown	8	51

Source: European Monitoring and Evaluation Programme (EMEP) Model, Norwegian Metereological Institute, Oslo.

suffering heavy damage from acid rain, the Federal Republic of Germany and Norway, are shown in Table 13.2.

Even though the exact source of half of the deposition in Norway is not known according to the model source for Table 13.2, it is easy to see that Norway holds others responsible for the damage being done by acid rain and that it puts great emphasis on control policies in the UK as its main identifiable polluter. East Germany is implicated in damage to both Norway and West Germany. Most of East Germany's receipts of sulphur are from itself, but about 9 per cent comes from West Germany. There may therefore be a mutual benefit in bargaining over reductions at source, whereas Norway, not being a significant source of emissions to the sources of its own pollution, does not have this option.

Acid rain is implicated in several types of damage. It is known that pine and spruce forests in the USA, Canada and some parts of Europe have been damaged. West Germany's *Waldsterben* ('forest death') is widely thought to be related to acid rain. Longer-term growth of forests may also be at risk. Damage to crops is less certain but work in the USA suggests that a number of tobacco, wheat and soya crops may be damaged by sulphur depositions. Other land-based flora and fauna have also been affected, notably near to urban centres, reducing local biodiversity. Buildings suffer corrosion from acidic deposits, causing extra maintenance to be carried out. SOX and NOX are also implicated in the acceleration of morbidity, especially respiratory diseases. A major cause for concern is the acidification of lakes and the effects on fish populations. In southern Norway, salmon populations have declined rapidly. Local soils have not neutralised the acids, and soil run-off has acidified watercourses, also releasing aluminium from the soil which causes fish death. The exact mechanisms at work are complex, and not all scientists would agree that acid rain is even the main cause of the acidification.

Controlling acid rain is essentially an issue of international bargaining and cooperation because, as noted above, emitters and recipients differ. To some extent, self-interest may assist in the solution of the problem. For example, if West Germany and the UK are mainly 'self-polluters', they have an incentive to reduce their own pollution provided they are satisfied that the pollution is causing damage. A comparison of the costs of control and the benefits that would ensue, would be sufficient to determine whether a clean-up is worthwhile. But, suppose control costs outweigh benefits? Should the individual nation not then implement a control strategy? If there are benefits to *other* countries of an individual country undertaking control measures, then these should be included in the cost–benefit comparison. But this requires that the controlling nation takes an international view of the benefits and not a purely nationalistic one. Moreover, recipients have very limited scope for action in terms of ameliorating damage – lakes can be limed, for example, in an attempt to neutralise the acids but, clearly, control at the source of the pollution is to be preferred. Hence the need for international action.

The international community has responded to the problem. Throughout the 1970s various attempts were made to obtain an international convention to control emissions, although they were mainly symbolic and lacked a clear control strategy. The main development in Europe was the Convention on Long Range Transboundary Air Pollution, drawn up by the Economic Commission for Europe (a United Nations organisation, not to be confused with the Commission of the European Communities – the 'Common Market'), and this was eventually ratified in 1983. In 1982 a conference on the acidification of the environment in Stockholm called on European nations to agree to an action programme to reduce SOX emissions first, with NOX emissions being on a longer-term agenda. Particular criticism was levelled at the 'tall chimneys' policy of a number of countries, for example the UK. Under this policy sulphur depositions were reduced *within* the country of origin by building tall chimneys which carried the emissions away. But, as we saw at the beginning of this book (p. 36), it is not possible to destroy matter. The SOX must therefore have been transported elsewhere. Essentially, tall chimneys converted a domestic pollution problem into an international one.

In 1983 Norway proposed a '30 per cent Club', the idea being that all parties should agree to a 30 per cent reduction in SO_2 by 1993 compared to the 1980 level. The problems of securing international cooperation in pollution control were quickly apparent. Several countries, including France, the UK, the USA and the Eastern Bloc, rejected the proposal. But other countries did agree and in 1984 the 30 per cent Club was formed. NOX was also on the agenda for future consideration. Meanwhile the European Commission (the 'Common Market') was also urging member states to agree to significant SOX and NOX reductions. The UK's repeated opposition to all the measures proposed for significant reductions earned it the name of the 'dirty old man of Europe'. In 1986, under pressure, the UK agreed to 'retrofit' three major coal-fired power stations with flue gas desulphurisation equipment. In 1988 the UK agreed to a European Community 'Directive' (which has the force of law) on emissions from large combustion plants requiring a 20 per cent reduction by 1993 on the 1980 emissions level, a 40 per cent reduction by 1998 and a 60 per cent reduction by 2003. These targets relate to the UK, but similar targets exist for other member countries. Similarly, NOX reductions have been brought into the agreement. For the UK these are 15 per cent by 1993 and 30 per cent by 1998, and this will involve fitting low NOX burners to power stations. Various 'let-out' clauses exist in the agreement, relating, for example, to unexpectedly high costs of compliance and technical difficulties.

The European negotiations illustrate the role which international pressure, bargaining and negotiation must play in securing changes in transboundary pollutants. Even then, it has taken some fifteen years to secure those agreements which do exist.

13.3 CHLOROFLUOROCARBONS

The ozone layer is concentrated some 6–30 miles above the earth in the stratosphere. It absorbs much of the ultraviolet radiation from the sun which would otherwise be harmful to mankind and other living species. Certain trace gases can however break down the ozone layer. One of these, chlorofluorocarbons (CFCs), has been demonstrated to have caused significant damage to the ozone layer. CFCs are chemically very stable until they reach the stratosphere where the ultraviolet radiation causes the chemical to break up and release chlorine which then impacts on the ozone layer. Two types of impact seem to occur. One is the general reduction of the layer and the other is a punching of 'holes' in the layer. In both cases the ultraviolet radiation from the sun is increased in terms of its receipt on earth. The evidence for the 'holes' is persuasive. How far there is a general depletion of the ozone layer is less certain.

CFCs have widespread uses. A major use is in aerosol cans which are familiar to everyone in the form of hairsprays, deodorants, insect sprays and so on. CFCs provide the propellant in these sprays, but can be easily substituted by other gases. CFCs are also used in refrigeration equipment, as a foam blowing agent in the manufacture of foam fillings, and as a solvent. Substitution for these uses is also possible but is less easy. Table 13.3 shows the structure of sales in the European Community in 1985.

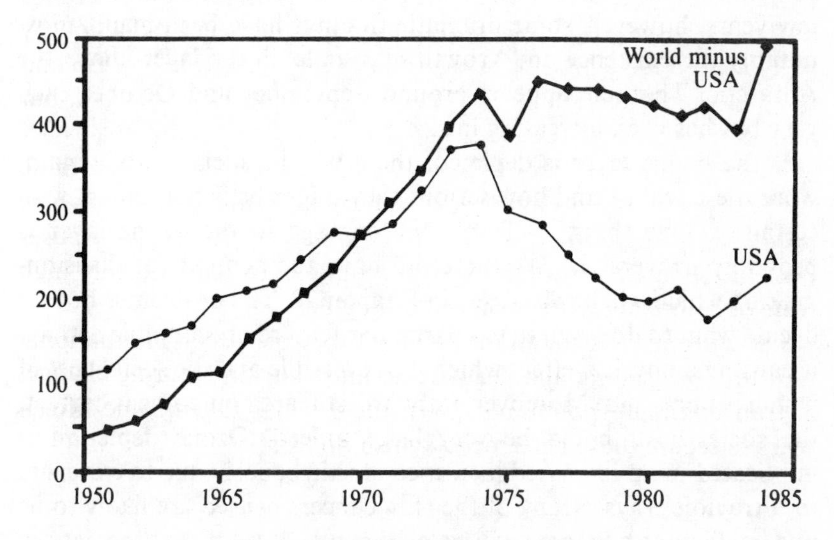

Figure 13.2 World CFC production.

Source: CMA, 'Production, Sales, and Calculated Release of CFC-11 and CFC-12 Through 1984', October 1985 and US International Trade Commission, *Synthetic Organic Chemicals*, Annual Series.)

Table 13.3 European sales of CFCs in 1985

Total sales within the EEC	228,000	tonnes
Of which (per cent):		
Aerosols	51	
Refrigeration	11	
Foam plastics	33	
Other	5	
Exports from EEC	107,000	tonnes
Total sales (domestic plus exports)	335,000	tonnes

Figure 13.2 shows world production of CFCs. Particularly notable is the growth in production outside the USA compared to the dramatic reduction in USA production in the 1970s. The change in USA production is due to a ban on the use of CFCs in aerosol production, a measure taken because of concern over the ozone layer effects.

Quite how the ozone layer is changing because of these trace gases is not known with certainty because of the complex nature of the chemical interactions that take place within the stratosphere and the fact that research into these interactions is fairly recent. In the past few years, however, some dramatic findings have been made, most notably the existence and growth of a 'hole' in the layer above the Antarctic. The hole appears around September and October each year but has been increasing in size.

If the ozone layer is depleted, there will be social costs. Again, what these will be and how serious they might be is not known with certainty. One thing is clear – the damage to the ozone layer is probably irreversible. We therefore have the context for decision-making which will be discussed in Chapter 20. The problem is how to decide what to do given that a particular feature of economic activity is causing a physical effect which is irreversible and the social cost of which is uncertain. Moreover, if the worst feared outcome is correct, the social costs could be very large indeed. Ozone depletion is implicated in an increased incidence of skin cancers due to exposure to ultraviolet rays. Many of the skin cancers caused are likely to be non-malignant but some will be malignant. A 1 per cent increase in ultraviolet radiation is likely to be associated with a 5 per cent increase in non-malignant cancers, and perhaps a 1 per cent increase in malignant cancers, in both cases with an associated death rate.

One estimate suggests that if no controls were put on CFCs, there might be 3 million deaths from non-malignant skin cancers and 200,000 from malignant skin cancers in the USA population born before the year 2075.

CFCs are more generally implicated in the global warming phenomenon caused by carbon dioxide emissions (see below). Ozone depletion may also result in the suppression of immune systems, leading to an increase in parasitic and viral infections. There may also be effects on plant yields, and on aquatic systems.

International action has been taken on CFCs, although its adequacy is debated. In 1978 the USA banned the use of CFCs in aerosols. In 1985 a Convention for the Protection of the Ozone Layer was drawn up in Vienna. In 1987 the Montreal Protocol on Substances that Deplete the Ozone Layer was signed by all major CFC-producing countries. This Protocol took effect at the beginning of 1989 and by mid-1990 consumption of CFCs should be 'frozen'. By mid-1994 consumption should be reduced by 20 per cent, and a further 30 per cent reduction should take place by mid-1999, i.e. an overall 50 per cent reduction in consumption is to be achieved by the turn of the century. Consumption differs from production by the amount of net imports or exports that a country may undertake. *Production* will be allowed to increase by 10 per cent (on the 1986 level) to 1990, but must thereafter fall to 90 per cent of the 1986 level by 1994 and to 65 per cent by 1999. The reason for distinguishing between production and consumption limits is to allow the developing countries flexibility in expanding their consumption, but within limits. Otherwise the controls would discriminate against the developing nations by imposing possible larger costs of substitute technologies.

How far this Protocol is adequate is debated. Some authorities argue that substantially larger cuts in emissions need to be secured in order just to stabilise the ozone depletion situation. But it is significant that the Montreal Protocol stands out as an international agreement about a pollutant that has yet to impose its damage. In other words, it illustrates *anticipatory* policy rather than 'reactive' policy which is based on waiting to see what damage has actually been done. The irreversibility of the damage is the main justification for this policy stance.

During 1988 the UK government announced that after consideration of its Department of the Environment's Stratospheric

Ozone Review Group (SORG) report, emissions of CFCs must be reduced by at least 85% as soon as possible. It has called for the review of the Montreal Protocol to be brought forward to 1989. The SORG report indicates that the ozone hole was more severe, and lasted longer in 1987 than in any previous year. The stratospheric ozone loss reached 95 per cent at times, while ozone in the full depth of the atmosphere was reduced by 60 per cent. CFCs are almost certainly to blame. Rising concentrations of the so-called greenhouse gases, such as carbon dioxide, will also tend to increase the area of the ozone hole. In contrast to their warming effect on the troposphere (see below), these gases lead to a cooling at higher altitudes.

SORG's calculations suggest that the concentrations of CFC-11 and CFC-12, the main ozone-depleting CFCs currently in industrial use, will almost double by 2050 even if the Protocol is observed fully. Merely to stabilise atmospheric concentrations at their present levels would require emissions of CFC-11 and CFC-12 to be cut by 77% and 85% respectively.

The crucial question now is how fast the reduction should be pursued, and whether it should encompass chlorine-containing compounds which do not at present fall within the scope of the Protocol. A recent USEPA report contains a projection that chlorine levels will treble by 2075 if emissions of methyl chloroform (a solvent widely used for degreasing in the metal and electronics industries as well as in dry cleaning) grow annually by 3.5–5 per cent. According to EPA, chlorine-containing chemicals not covered by the Protocol will account for 40 per cent of the projected growth in stratospheric chlorine by 2075. Methyl chloroform will contribute 80 per cent of this figure if its usage grows as projected. The report's conclusion is that an immediate total phase-out of the substances already covered by the Protocol, together with a freeze on methyl chloroform, are needed to hold chlorine levels in the atmosphere at present levels.

Thus it would seem that the pace of anticipatory ozone depletion policy change is going to quicken appreciably over the next few years. The complexities and interrelatedness of global pollution problems are well illustrated by an issue raised in the SORG report – that while the substitutes for CFCs which are now being developed will have a lower or zero depletion potential they may still have an undesirably large greenhouse effect. The exact magnitude of the effect is not, however, known at present. We now turn to an examination of the greenhouse effect.

13.4 CARBON DIOXIDE AND THE GREENHOUSE EFFECT

Human economic activity is causing the release of certain pollutants (atmospheric trace gases – mainly carbon dioxide, methane, nitrous oxide and CFCs) which tend to block the emission of heat from the earth's surface. These gases are transparent to short-wave radiation from the sun, but absorb long-wave radiation from the Earth, thus trapping heat. In theory, increasing their concentration in the atmosphere causes the earth's surface and lower atmosphere to warm – like a greenhouse. Unlike the ozone effect (in the upper atmosphere) the greenhouse effect operates mostly in the lower atmosphere. The progressive and even accelerating accumulation of these gases, if allowed to continue unchecked is, many scientists believe, likely sooner or later to induce a 'significant' warming trend and climate modification. Great uncertainty, however, surrounds the magnitude of the warming effect, the rate of warming and its geographical and seasonal pattern. These uncertainties in turn make predictions of the likely impacts of climate modification on human systems and ecosystems subject to wide error ranges.

As yet, the evidence linking temperature trends and the enhanced greenhouse effect is ambiguous. Observed warming evidence is not inconsistent in direction and magnitude with greenhouse expectations and analysis of the palaeoclimatic records is also encouraging but not conclusive. The underlying real long-term temperature trend is obscured by the inherent spatial and temporal variability of the global weather. In the absence of conclusive proof, scientists have been employing theoretical models – general circulation models (GCMs) – to assess the processes known to occur and their possible interactions. Projections are then made into the future so as to forecast potential effects over time.

Until recently, carbon dioxide and methane emissions were largely responsible for potential changes in climate via the greenhouse effect. Concentrations of carbon dioxide in the atmosphere have risen by 25 per cent over the last two hundred years and the trend is continuing upwards. Currently, fossil fuel burning is accounting for about 80 per cent of annual carbon dioxide emissions, with deforestation (particularly tropical forest loss) also playing an important part. Future atmospheric carbon dioxide concentrations will depend primarily on fuel mix policy and the aggregate energy demand for fossil fuels. It is likely that even with policy switching to alternative fuels, increased energy use efficiency and extensive energy conservation measures, carbon dioxide concentrations will continue to increase over the next fifty years or so. Developing country economies will also significantly affect future carbon dioxide emissions as they attempt to foster increased rates of economic growth, thereby increasing their demand for energy.

The other trace gases (methane, nitrous oxide and CFCs) are present in the atmosphere (due to fossil fuel burning, forest burning, agricultural land conversion, fertiliser use and manufacturing) in much smaller concentrations than carbon dioxide, but their concentrations are increasing at a relatively fast rate. Their combined influence may become quantitatively comparable to that of carbon dioxide in the future. It is important to stress, however, that all projections of the future emissions of greenhouse gases and the rate at which they accumulate in the atmosphere are highly uncertain. Prior knowledge of trends in, for example, energy demand and the extent and type of expansion in industry and agriculture are all required for accurate predictions.

Several teams of scientists have been active in modelling global climate, using GCMs, in order to understand climate dynamics and to quantify the consequences of the enhanced greenhouse effect. How warm the world is going to get and at what rate, will depend on the sensitivity of climate to given changes in gas concentration and on the time required for the climate system to approach equilibrium. GCM experiments have indicated that, overall, the warming effect is enhanced by climate system feedbacks. The current state-of-the-art estimate is that a doubling of pre-industrial carbon dioxide concentration (here taken to represent the combined effective concentrations of all the greenhouse gases) can cause a global mean warming (i.e. global mean equilibrium surface air temperature) of 1.5–4.5°C. To put these numbers in perspective, one need only note that the warmest period in the past 100,000 years was only 1°C, on average, warmer than today!

An important distinction exists between the *equilibrium climate response* (i.e. the steady-state change in climate that would eventually occur for a given concentration of greenhouse gases and allowing for a lag caused by the thermal inertia of the oceans) and the *transient response* (i.e. actual expected mean temperature change, usually some 40–80 per cent of the projected equilibrium warming). Both climate responses have important implications for policy strategies designed to account for the social and environmental impacts of climate change.

The equilibrium warming concept implies that there is considerable unrealised greenhouse warming 'in the pipeline' but not yet evident. Thus, even if increases in trace gas emissions were to stop immediately, considerable warming should still occur. This amount of climate change has been irreversibly committed to future generations by the current generation. So anticipatory policy decisions on trace gas emissions and on impact adaptation measures will incur costs to the present generation and a stream of benefits stretching out into the future over several generations. It is also likely that the longer the delay in decision-making, the greater the costs, if climate change proves to be significant. Project or policy appraisal over long periods of time involves the issue of discounting whereby impacts in the far future are discounted more heavily than those in the near future. It also involves potential ethical concerns over fairness to future generations (intergenerational equity). Chapters 14 and 15 analyse both these issues in detail.

The policy implications of transient warming are determined by the magnitude of the impacts of climate change and how quickly they occur. So if global average temperatures rise fairly slowly then it is plausible to think that via 'natural adaptations' and/or some 'forced adaptations', socio-economic systems and their supporting eco-systems will have time to adjust. If the climatic warming turns out to be on the high side of the estimate range then anticipatory policy measures will be required (i.e. emission reductions or bans, fuel switching, etc.), in order to slow the rate of change and to reduce societal and environmental vulnerability.

GCMs, despite shortcomings such as poor modelling of ocean circulation processes and hydrological processes as well as lack of knowledge on cloud formation and feedbacks and poor regional climate forecasts, indicate that warming will be greater in high latitudes than in low latitudes; warming will be greater in winter than in summer; and global precipitation, on average, will increase (particularly in the mid-to-high latitudes). The models also show considerable seasonal and spatial variations. They are unable to provide accurate details of regional climate when simulating present-day conditions. This raises serious questions about their ability to forecast the regional details of any future change in climate and therefore provide guidance on regional impacts and possible adaptive strategies. It is also likely that one of the first tangible signs that a substantial climatic change is underway will be an apparent increase in the frequency of certain types of extreme events.

We highlighted earlier the fact that any assessment of the possible impacts of the greenhouse effect is hindered by a lack of reliable projections of climatic change on the regional level. Potential direct impacts of the greenhouse problem are, however, many and varied. The direct impacts will occur as a result of three factors – climatic change, sea-level rise and higher levels of atmospheric carbon dioxide. The greatest effects are likely to result from climatic change, although in coastal areas rising sea level may prove to be a significant hazard. Higher atmospheric levels of carbon dioxide could affect the productivity of ecosystems.

It is probably more correct to think of a series of impact interactions being stimulated by the initial first-order biophysical impacts. Secondary impacts will be felt at the local enterprise level (farm, industrial firms, etc.) and lead on to tertiary impacts at the regional and national level. Additional complexity is introduced if interactions of the same order occur (intra-sectoral, e.g. between different farming systems; and inter-sectoral, e.g. effects of climate change on agriculture, forestry, soils and water resources). The most widely discussed impacts are:

- agricultural change, in terms of changes in productivity from existing land use, changes in location, crop mix, agricultural technology and management practice, plant disease, etc.
- coastal resource change: loss of wetlands, salinisation of land, coastal erosion and land loss due to sea-level rise
- forestry impacts such as increased die-back
- other land-use changes due to migration of population
- water resources impacts: river flow change, ground water impacts
- effects on natural ecosystems and biological diversity, land and water-based, including fisheries, endangered species
- effects on navigation, e.g. due to ice-pack change, coastal erosion, etc.
- infrastructure damage: property losses, especially in coastal towns and cities, damage to ports and harbours, airports, etc.
- effects on land-based transport systems
- effects on human health
- impacts on the energy sector.

The net impact will vary from region to region depending on:

- the rate of change in the environment and its nature
- the sensitivity of the ecological and social systems that are affected
- the degree to which these systems can naturally adjust or adapt (without deliberate human intervention)
- the human response.

A number of policy responses are possible and each policy option will have associated with it a probable stream of costs and benefits. Socio-economic impacts will thus be policy-dependent not only in the sense of varying with the level of climate change, but also in that the differing costs of control and adaptation will themselves have economic impacts, e.g. on the balance between investment and consumption, on fiscal revenues and expenditure, international trade, etc.

A convenient if somewhat simplified classification of policy scenarios might be:

1. *Do nothing.* In this option no anticipatory policy is undertaken, and the resulting trace gas concentrations are therefore dependent upon some forecast of how the world and regional economies will behave, together with some assumption about 'natural' adaptation by humans as global warming increases. The benefits of this policy are the avoided control costs. The costs of the policy occur in the form of the damage done, plus any costs involved in 'natural adaptation'. The 'do nothing' scenario can be treated as the base case. Other scenario impacts are then measured relative to this one.

2. *Anticipatory prevention.* In this option, various measures are undertaken to prevent the global warming from being as great as it otherwise would be. This involves reducing the projections of trace gas emissions in the 'do nothing' scenario. The costs of this policy consist of the investments that would have to take place (e.g. in energy efficiency measures, scrubbing of CO_2, switches in CFC technology, etc.), together with any losses due to changes in consumption behaviour, relocation of economic and social activity, etc. The benefits consist of the avoided damage compared to the 'do nothing' scenario.

3. *Adaptation.* In this option the emissions of greenhouse gases are not reduced compared to the 'do nothing' scenario, but the *consequences* are mitigated by investments and lifestyle changes –

e.g. switches in crops, consumption changes, etc. The costs of adaptation can then be compared to the benefits which will consist of the damage avoided, using the 'do nothing' scenario as a base case.

The distinction between these scenarios is somewhat artificial in that 'natural' adjustments will occur in all three, and anticipatory action may well include adaptive expenditures, as well as perhaps reducing the level of such expenditures.

One impact of the greenhouse effect which would be felt globally and does represent a reasonably tractable research problem is the effect on *global-mean sea level.* There are four possible reasons why global warming may cause sea level to rise: thermal expansion of the oceans; melting of mid- and high latitude alpine glaciers; melting of the Greenland and Antarctic ice sheets; and the possible disintegration of the West Antarctic ice sheet. All projections of changing global sea level indicate that the net effect will be positive and the rate of change significantly faster than that which occurred during the past one hundred years. Estimates range from 20 cm to 1.5m by 2030 to 2050. If global warming stopped abruptly in 2030, global sea level would continue to rise for another thousand years (again a major intergenerational impact).

Some 3 billion people live in coastal regions, and their cities, agricultural land, water resources, beaches and coastal wetlands may be at risk from inundation or storm-induced flooding (the latter representing the changing risks of extreme events). The projected rise in sea level may exceed the capacity of some in-place coastal defences; for example, a 1m rise in sea level would substantially reduce the effectiveness of the Netherlands flood defences. The level of protection provided by the system could fall from a 1 in 10,000 year standard to a 1 in 100 year standard.

Taking the problem of sea level rise in the context of those countries that partly rely on engineering solutions to mitigate saline inundation hazard (e.g. the UK, Netherlands and USA), what policy responses are possible?

Depending on the type of coastal environment, the following options are possibilities:

1. *Do nothing*. This could literally be just that, i.e. abandon defences and let nature take its course.
2. *Do nothing extra.* Because of the uncertainties, a 'wait-and-see' strategy might seem politically attractive. Coastal and sea

defences would be kept at their current levels and maintained/re-placed as required, in the expectation that storm frequencies will not change significantly. Effectively, the degree of 'acceptable risk' may need some adjustment.

3. *Adaptation.* This could involve: improving/reinforcing existing 'engineered' protective structures; enhancing 'natural' protection (e.g. improving beach levels through nourishment, with the aim of providing a wide beach in front of a wide belt of dunes); compensating property owners in the hazard zone to avoid the need for future protection; flood-proofing individual properties, or changing their use.
4. *Adaptation/prevention.* Anticipatory responses which could involve strategic retreat – moving existing defences landward to allow roll-back of the active coastal zone (e.g. returning agricultural land to salt-marsh which then acts as a stabilising factor, keeping pace with sea level rise as long as mud supplies are adequate); coastal hazard zone designation and planning controls; de-intensify land uses.

In summary, a key feature of the climatic change debate is pervasive *uncertainty*. There is uncertainty about trace gas projections, compounded by the fact that the state of climate modelling is not such that the response of climate to the gases can be known with certainty. The regional variation in climate responses is known with even less certainty, and yet detailed regional impacts are necessary if there is to be any broadly accurate assessment of socio-economic impacts. The response of people to the change in climate is also uncertain. Possible responses include doing nothing, anticipatory behaviour, on-going adaptive behaviour and crises reaction. Uncertainty about when an impact will occur is potentially important if the methodologies for calculating costs and benefits adopt a discounting approach.

In conclusion, it can be argued that whether society responds to the greenhouse problem in a largely anticipatory or adaptive way will determine the scale of the required international cooperation. If the greenhouse effect is perceived to be significant then anticipatory measures (i.e. emission reductions) make sense and will require international cooperation. It is arguable that at a certain point it will be more efficient, aside from questions of equity, for Western industrialised economies to abate the greenhouse effect through providing Third World economies with the best available control technologies and capital resources. Policy actions in developing countries could be facilitated via a package of measures linked to bilateral aid flows and to multilateral development bank loan programmes.

Since a certain amount of climatic warming will occur regardless of actions taken by governments now, there is a strong case for compensation of those who will lose. Both Third World people alive now and global future generations seem to be deserving candidates. In the case of the greenhouse effect and future generations the issue appears to be that we cannot avoid harming future generations, if only by imposing risk. So the choice really is between the greenhouse effect with compensation and the greenhouse effect without compensation. Furthermore, if compensating investment is not made in the present, compensation is not an option in the future.

The discounting approach to future costs and benefits, the notions of intergenerational equity and irreversibility are all taken further in Chapters 14, 15 and 20.

PART III

ETHICS AND FUTURE GENERATIONS

14 · DISCOUNTING THE FUTURE

14.1 THE PROBLEM

Economic analysis tends to assume that a given unit of benefit or cost matters more if it is experienced now than if it occurs in the future. This lowering of the importance that is attached to gains and losses in the future is known as *discounting*. The rationale for discounting is discussed below, but it will be immediately apparent that, if discounting has a logic of its own, it will still create problems when applied to environmental issues. To see this, consider a development that yields immediate and near-term benefits but which has fairly catastrophic environmental consequences for future generations. Examples might be the storage of highly radioactive waste, the emission of chlorofluorocarbons (CFCs) that damage the ozone layer, acid rain, carbon dioxide and the greenhouse effect, and so on. So long as the weight we attach to the future gets less and less the further into the future we go which, as we shall see, is what discounting does, then the less important such catastrophic losses will be. In other words, discounting contains an in-built bias against future generations.

Later chapters will also show that discounting affects the rate at which we use up natural resources. The higher the discount rate – the rate at which the future is discounted – the faster the resources are likely to be depleted. Once again, the effect appears to be one of discriminating against future generations.

Another way of looking at this 'discrimination against the future' is to say that discounting appears to be inconsistent with the philosophy of 'sustainability' discussed in Chapter 3. This is because the higher the discount rate the lower the importance attached to the future, and hence the less likely we are to honour the idea of conserving the natural capital stock. If high discount rates lead to a depletion of the capital stock, then sustainability is jeopardised.

For all these reasons, then, discounting is important. Many environmentalists, and some economists, even regard it as immoral simply because it does appear to be inconsistent with the ideas of conservation and sustainability. Yet discounting turns out to be an everyday occurrence. We therefore need to understand its basis.

14.2 THE RATIONALE FOR DISCOUNTING

The existence of interest rates explains discounting. One pound in year 1 would accumulate to £$(1 + r)$ in year 2 if the interest rate is r per cent (r is typically expressed as the corresponding decimal, e.g. 5 per cent would be 0.05, 12 per cent would be 0.12, and so on). *Looked at from the standpoint of year 1*, we can ask the question: 'How much is £1 in year 2 worth to us in year 1?' The answer will be that it is worth £1/$(1 + r)$, for the simple reason that if we had this sum in year 1 we could invest it at r per cent and obtain in year 2

$$\frac{\pounds 1}{(1+r)} \times (1+r) = \pounds 1$$

In the same way, we see that £1 in year 3 can be expressed as a value to us in year 1 as follows:

$$\pounds 1/(1+r)^2$$

since in year 3

$$\frac{\pounds 1}{(1+r)^2} \times (1+r) \times (1+r) = \pounds 1$$

We now have the general formula for discounting. A benefit B in any year t can be written as B_t, and from the above procedure we know that this benefit will have a value to us in year 1 of

$$\frac{B_t}{(1+r)^t}$$

Notice the procedure whereby we look at future benefits (and costs; the procedure is the same) from the standpoint of the present. This is why expressions such as the one above are called *present values*. The procedure for finding a present value is known as *discounting* and the rate at which the benefits or costs are discounted is known as the *discount rate*.

How then do positive interest rates arise? There are two underlying reasons for positive discount rates. First, people discount the future because they simply prefer their benefits now to later. We say they have *time preference*: they are impatient. The underlying value judgement in welfare economics is that people's preferences matter – they should count in whatever social decision-rule we devise. The alternative to letting preferences count is to override them, to say that we know best what is good for other people. This overriding of preferences is something that societies commonly do, but it is clearly only to be done when there are very good reasons for it. If we accept that preferences matter, we are logically obliged to accept that people's preferences for the present over the future must be allowed to count as well.

The second source of interest rates is the *productivity of capital*. The basic observation about capital is that if we divert some resources for investment (capital formation) rather than consumption, those resources will be able to yield a higher level of consumption in a later period than if we consumed them now. Clearly, it is worth waiting for these extra future benefits provided the cost in terms of impatience (the time preference cost) is less than the future benefits. We therefore see that there is a strong link between capital productivity and time preference: we are not likely to undertake investment unless the future benefits outweigh the time preference interest rate.

The STPR

If we now translate these two ideas to the level of *society* we can argue that positive interest rates have two sources. The first is *society's* time preference rate. This we call the *social time preference rate* (STPR). The STPR will reflect more than the underlying level of impatience which is known as *pure time preference*. It is likely also to reflect a social judgement to the effect that since future societies are likely to be richer than current ones, an extra \$1 or £1 of benefit to them is worth less, has less utility, than \$1 or £1 to the current society. This can be formalised by saying that we should discount the future because of *diminishing marginal utility of consumption.* In fact we can present a formula for the STPR as follows:

$$s = \text{STPR} = c\,e + p$$

where c = the rate of growth of real consumption per capita, e = the elasticity of the marginal utility of consumption function (see below) and p = the pure time preference rate of interest.

We do not derive the formula here (see Further Reading to this chapter for sources). The measure e shows the relationship between the utility that we think is derived from extra units of consumption, and, for analytical convenience, the relationship is expressed as an elasticity, i.e. the percentage change in utility that would arise from a percentage change in consumption. The component ce in the formula thus accounts for the idea that as future societies are likely to be richer we should attach less weight to their gains, i.e. we should discount those future gains. The component p reflects impatience. In each case we would get a percentage, so that adding the two together will give us s in a percentage form.

SOC

The second source is the productivity of capital. This is easier to understand because all we need to do is to express the future flow of benefits obtainable from investment as a flow of consumption. If we invest 100 units and get back 110 in consumption in the future, the net productivity of capital is 10 units; 10/100 = 10 per cent. Suppose that this 10 per cent covers all the benefits so that we can ignore any external effects, then 10 per cent is a social rate of return to capital investment. If we are considering a use of resources that yields us 8 per cent while we know that by investing elsewhere we could get 10 per cent then the proper decision is not to invest in the 8 per cent project – we could do better elsewhere. The 10 per cent is thus the social opportunity cost; or its SOC, of the project under consideration.

We can show the underlying analytics with the aid of Figure 14.1. This shows consumption in two years, t and $t + 1$. The function TT' is a *transformation function*, or *production possibility curve*. It shows the possible configurations of production between two *years*. It says, for example, that if resources are wholly devoted to year t, output will be OT. If they are wholly devoted to year $t + 1$ output would be OT'. Notice that OT' is greater than OT. This is intuitively

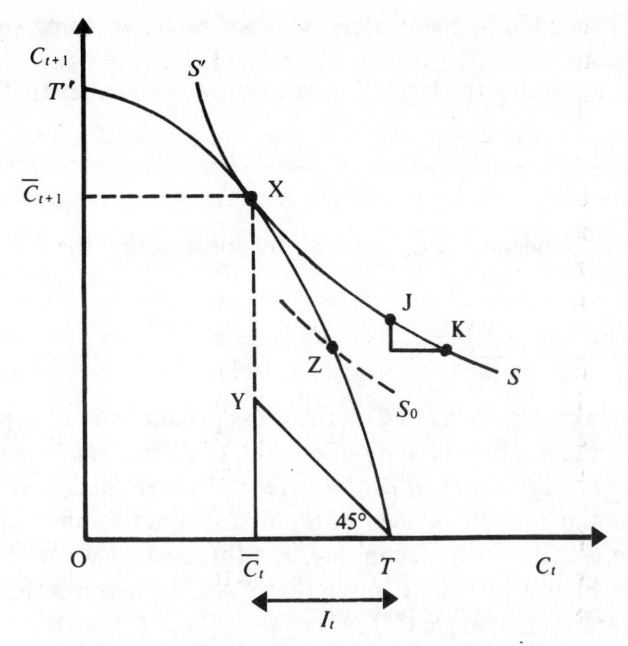

Figure 14.1 Deriving the social rate of discount

acceptable because devoting resources to production in year $t + 1$ means, given that we have measured the transformation in terms of units of consumption, that OT' can only come about by investing all the resources that would have been consumed in year t. That is, if the economy is at position T' it means that OT was invested in year t and the consumption goods resulting are OT' in year $t + 1$.

Also shown in Figure 14.1 is an indifference curve. This is a *social* indifference curve indicating the combinations of consumption in period t (C_t) and consumption in period $t + 1$ (C_{t+1}) between which society is indifferent. This is denoted by SS'.

Clearly, society will be in an optimal position if it located at point X, for then society is able at this point to climb on to the highest possible social indifference curve given the constraints set down by the function TT'. In fact we can find out just how much investment and consumption occur if the economy settles at X. We can read off C_t and C_{t+1} immediately and we see that these are given by $\overline{C}_t$ and $\overline{C}_{t+1}$. But TT' embodies the investment that also takes place, i.e. the difference between $\overline{C}_t$ and OT must be the level of real investment in year t, I_t. In turn we see that it is I_t that generates the consumption level C_{t+1}. Using these ideas we can establish some important equations.

First, observing the level of investment I_t, we can see that

$$\frac{\bar{C}_{t+1}}{I_t} = \frac{Y\bar{C}_t + XY}{I_t} = \frac{Y\bar{C}_t}{I_t} + \frac{XY}{I_t} \quad (14.1)$$

But $Y\bar{C}_t = I_t$ because $Y\bar{C}_t$ is drawn by constructing the 45° line YT. Hence

$$\frac{\bar{C}_{t+1}}{I_t} = 1 + \frac{XY}{I_t} \quad (14.2)$$

The first expression in (14.2) is the gross productivity of capital, and XY/I_t is the *net* productivity of capital, or its *internal rate of return* (or *marginal efficiency of capital*). Yet this latter concept is precisely the discount rate introduced as the first of the two alternatives, i.e. the interest rate in the economy, r. Moreover, if we make I_t very small in Figure 14.1, we can see that $\bar{C}_{t+1}/I_t$ measures the *slope of* TT′ (the tangent of angle $\bar{C}_t TX$ is $\bar{C}_t X/\bar{C}_t T$ and $\bar{C}_t X = O\bar{C}_{t+1}$). Hence we can rewrite (14.2) as:

$$\text{Slope of } TT' = 1 + r \quad (14.3)$$

where r is now the marginal rate of return on capital.

Turning our attention to SS' we can proceed in a similar way. Consider points J and K in Figure 14.1. These are points on the same social indifference curve, so that the utility *lost* by moving from K to J would be $\Delta C_t MU_t$, i.e. the change (Δ) in C_t multiplied by the marginal utility associated with C_t. The utility *gained* would be $\Delta C_{t+1} MU_{t+1}$. Since J and K are on the same indifference curve, we can write

$$-\Delta C_t MU_t = \Delta C_{t+1} MU_{t+1} \quad (14.4)$$

and hence

$$\frac{-\Delta C_{t+1}}{\Delta C_t} = \text{Slope of } SS' = \frac{MU_t}{MU_{t+1}} \quad (14.5)$$

The slope of SS' is simply the ratio of the two marginal utilities of consumption (which is what we would expect from our knowledge of indifference curves in general).

Now, as we move along SS' in the direction K to J, society will tend to require more and more C_{t+1} to compensate for a unit loss of C_t. Very simply, $\Delta C_{t+1}/\Delta C_t > 1$. Hence we now have, from (14.5),

$$\frac{MU_t}{MU_{t+1}} > 1 \qquad (14.6)$$

Writing the excess of this ratio over unity as s we have

$$\frac{MU_t}{MU_{t+1}} = 1 + s = \text{Slope of } SS' \qquad (14.7)$$

We now define s as the *social rate of time preference.*

The discussion so far suggests that there are two possible sources for the discount rate, r. It could be SOC or STPR. In a world in which there are no taxes and where there are perfectly functioning capital markets, it turns out that SOC = STPR: we would not have to worry about which approach we took. In a less than perfect market situation, however, a choice has to be made. A voluminous literature exists on this issue, but it does not concern us here. Suffice it to say that some people argue for SOC, some for STPR and some for a sort of average of the two.

Whatever, the school of thought in question, however, the discount rates that emerge are all positive. That is, the literature has a consensus that the fact that people prefer the present to the future and the fact of capital productivity imply positive discount rates. We need to see whether this conclusion is likely to be modified by criticisms advanced by environmentalists.

14.3 A CRITIQUE OF DISCOUNTING

Pure time preference

The objections to allowing the component p (reflecting impatience) to influence the discount rate are several. First, economists would point out that individual impatience is not necessarily consistent with maximising an individual's lifetime welfare maximisation. The proof is complex but is a variant of the idea expressed by a number of economists that 'impatience discounting' is fundamentally irrational. It can lead to decisions which are incompatible with long-run welfare. Second, what individuals want carries no necessary implications for public policy. We overrule individual preferences in many contexts. Why not overrule their impatience? Third, even if we elevate want-satisfaction to moral status (which is what the welfare economics approach does), it is satisfaction of wants as they arise that matters. But this means that tomorrow's satisfaction matters, not today's assessment of tomorrow's satisfaction.

How compelling these arguments are against allowing pure time preference to influence STPR is clearly going to depend a great deal on one's view of the underlying value judgement about letting preferences count. It becomes an issue of letting preferences count however they are formed, or modifying the rule to allow only some preferences to count. Either route is morally hazardous. Others will find the conceptual point about the logic of want–satisfaction implying no discounting to be worth more consideration.

Risk and uncertainty

It has been argued that a benefit or cost is valued less the more uncertain is its occurrence. Since uncertainty is usually expected to increase with time from the present, this declining value becomes a function of time and hence is formally expressible in the form of a discount rate for risk and certainty.

The types of uncertainty that are generally regarded as being of relevance are as follows:

1. Uncertainty about the presence of the individual at some future date (the 'risk of death' argument).
2. Uncertainty about the preferences of the individual even when his existence can be regarded as certain.
3. Uncertainty about the availability of the benefit or the existence of the cost.

The objections to using uncertainty to justify positive discount rates are several. First, uncertainty arising from not being sure that the individual will be present to receive a distant benefit – the 'risk of death' argument – ignores the view that society is 'immortal' in contrast to the individual's mortality. Second, uncertainty about preferences is clearly relevant if we are talking about certain goods and perhaps even aspects of environmental conservation, but is handled by the idea of option price as the sum of expected consumer surplus and option value (Chapter 9). Third, uncertainty about the presence or scale of benefits and costs may be unrelated to time, and certainly appears unlikely to be related in such a way that the scale of risk obeys an exponential function as is implied in the use of a single rate in the discount factor:

$$e^{-rt} \text{ or } 1/(1+r)^t$$

What is being argued here is not that uncertainty and risk are irrelevant to the decision-guiding rule, but that their presence should not be handled by adjustments to the discount rate, for such adjustments imply a particular behaviour for the risk premium which is hard to justify.

Diminishing marginal utility of consumption

It will be recalled that in addition to impatience, discounting was justified by the assumption of diminishing marginal utility of consumption. The objections to this are twofold. First, economists dispute whether there is any meaningful way to measure the value of e, the elasticity of the marginal utility of consumption function. The debate is too complex to be reviewed here but revolves round the measurability of utility, the nature of comparisons of utility of different people at different times (a problem of the 'interpersonal comparison of utilities'), and even the nature of utility comparisons of the same person over time. Second, the argument assumes that real consumption (c in the equation) will grow over time. This may seem reasonable given the past experience of the wealthier countries of the world. It is not reasonable for many of the poorest, nor can we assume that the value of c is independent of the level of environmental degradation. That is, high discount rates, based in part on the utility of consumption argument, can lead to environmental degradation (because of the discrimination against sustainability practices) which in turn brings about a fall in real consumption per capita. Put another way, the discount rate is not independent of environmental quality.

Opportunity costs

The environmental literature has made some limited attempts to discredit discounting due to opportunity cost arguments. This literature is, however, confusing since most of the objections arise because the implication of opportunity cost discounting is that some rate greater than zero emerges and this is then held to be inconsistent with a concept of intergenerational justice. There do, however, appear to be two criticisms which are generally, but not wholly, independent of this wider concern.

The first arises because the discount *factor* arising from a constant discount *rate* takes on a specific exponential form. This is because discounting is simply the reciprocal of compound interest. In turn, compound interest implies that if we invest £100 today it will compound forward at a particular rate, provided we keep not just the original £100 invested but also re-invest the profits. Now suppose the profits are consumed rather than re-invested. The critics suggest that this means that those consumption flows have no opportunity cost. What, they say, is the relevance of a discount rate based on assumed re-invested profits if, in fact, the profits are not re-invested but consumed? If the argument is correct it provides a reason for not using a particular rate – the opportunity cost rate – for discounting streams of consumption flows as opposed to streams of profits which are always re-invested. But, in that context, it would not provide a reason for rejecting discounting altogether, since consumption flows should be discounted at a social time preference rate. That is, the critics have not seen that a future benefit *is* worth less than a current benefit if we admit any of the arguments for a social time preference rate. As it happens, the particular critics in question would not admit to believing the arguments for time preference rates either, so their position would be consistent.

The second argument relates to intertemporal compensation. Consider an investment which has an expected environmental damage of £X in some future time period T. Should this £X be discounted to a present value? The argument for doing so on opportunity cost grounds is presumably something like the following: if we debit the investment with a social cost now of £$X/(1 + r)$, then that sum can be invested at r per cent now and it will grow to be £X in year T and can then be used to compensate the future sufferers of the environmental damage. The critics argue that this has confused two issues. The first is whether future damage matters less than current damage of a similar scale. The second is whether we can devise schemes to compensate for future damage. The answer to the first question they suggest, is that it does not matter less than current damage, or if it does, it matters less only because we are able to compensate the future as shown. If we are not able to make the compensation, the argument for being less concerned, and hence the argument for discounting, become irrelevant.

Part of the problem here is that actual and 'potential' compensation are being confused. As typically interpreted economic analysis requires only that we could, hypothetically, compensate losers, not that we actually do. In this case, the resource cost to the current generation of hypothetically compensating a future generation is, quite correctly, the discounted value of the compensation. What the critics are objecting to, we suggest, is the absence of built-in actual compensation mechanisms in the analysis. We have considerable sympathy with that view, but it is not relevant to the issue of how to choose a discount rate.

These particular arguments against opportunity cost related discounting are not persuasive. It seems fair to say, however, they are not regarded by their advocates as the most forceful that can be advanced against discount rates *per se*. Those rest with arguments about intergenerational justice.

Future generations

The higher the rate of discount, the greater will be the discrimination against future generations. First, projects with social costs that occur well into the future and net social benefits that occur in the near term will be likely to pass the standard cost–benefit test, the higher the discount rate is. Thus future generations may bear a disproportionate share of the costs of the project. Second, projects with social benefits well into the future are less likely to be favoured by the cost–benefit rule if discount rates are high. Thus future generations are denied a higher share of project benefits. Third, the higher the discount rate, the lower the overall level of investment will be, depending on the availability of capital, and hence the lower the capital stock 'inherited' by future generations. The expectation must be, then, that future generations will suffer from rates of discount determined in the market place since such rates are based on current generation preferences and/or capital productivity.

It might be thought, however, that existing preferences do take account of future generations' interests. The way in which this might occur is through 'overlapping utility functions'. What this means is that my welfare (utility) today includes, as one of the factors determining it, the welfare of my children and perhaps my grandchildren. Thus, if *i* is the current generation, *j* the next generation and *k* the third generation, we may have:

$$U_i = U_i\,(C_i,\; U_j,\; U_k)$$

where *U* is utility and *C* is consumption. In this way, we could argue that the 'future generations problem' is automatically taken account of in current preferences. Notice that what is being evaluated in this process is the current generation's judgment about what the future generation will think is important. It is not therefore a discount rate reflecting some broader principle of the rights of future generations. The essential distinction is between generation *i* judging what generation *j* wants (the overlapping utility function argument) and generation *i* engaging in resource investment so as to leave generation *j* with the maximum scope for choosing what it wants.

The issue is whether such an argument can be used to substantiate the idea that current interest rates reflect future generations' interests. The basic reason for supposing that the argument does not hold is that current rates are determined by the behaviour of many individuals behaving in their own interest. If future generations enter into the calculus, they do so in contexts when the individual behaves in his or her 'public role'. The idea here is that we all make decisions in two contexts – 'private' decisions reflecting our own interests, and 'public' decisions in which we act with responsibility for our fellow beings and for future generations. Market discount rates reflect the former context, whereas social discount rates should reflect the public context. It is similar to an 'assurance' argument, namely that people will behave differently if they can be assured that their own action will be accompanied by similar actions by others. Thus, we might each be willing to make transfers to future generations, but only if we are individually assured that others will do the same. If we cannot be so assured, our transfers will be less. The 'assured' discount rate arising from collective action is lower than the 'unassured' discount rate.

There are other arguments that are used to justify the idea that market rates will be 'too high' in the context of future generations' interests. The first is the 'super-responsibility' argument. Market discount rates arise from the behaviour of individuals, but the state is a separate entity with the responsibility of guarding collective welfare and the welfare of future generations as well. Thus, the rate of discount relevant to state investments will not be the same as the market discount rate and, since high rates discriminate against future generations, we would expect the state discount rate to be lower than the market discount rate.

The final argument used to justify the inequality of the market and social rate of discount is the 'isolation paradox'. This is often confused with the assurance problem (see above), but the isolation paradox says that individuals will not make transfers even if assurance exists.

14.4 PROBLEMS WITH THE CRITIQUE OF DISCOUNTING

The criticisms of discounting may not individually be very powerful, but it is difficult to conclude that, taken together, they leave little concern about discounting. It surely does seem a little odd that we 'allow' impatience to determine discount rates where in most other contexts we would universally regard impatience as reprehensible. Discounting because of the utility-of-consumption argument seems to beg the question, making a discount rate depend on expected growth which itself may depend on environmental quality which in turn may depend on the choice of discount rate. The opportunity cost argument is less readily criticised but the effects of positive rates based on capital productivity certainly appear to be inconsistent with sustainability, conservation and some form of intergenerational fairness.

The implication of the criticisms is that we should lower discount rates from whatever they are when determined by the STPR and SOC arguments. If we accept this we have an immediate problem in that the criticisms do not tell us by how much we should lower discount rates. We are left with an indeterminate theory of discount rate selection. Some environmentalists argue that the only proper discount rate is a zero discount rate. Then we would treat £1 or $1 as being the same whenever it occurred. It is also possible to find mention of 'negative' discount rates, but such views reflect a misunderstanding. Negative discount rates would mean that we should always postpone consumption now in favour of consumption tomorrow, leaving current generations perhaps at some subsistence level of consumption. Worse, the postponement is perpetual – we would never actually have a generation above subsistence, each one being willing to postpone 'the big feast' to the next generation. So, the choice seems to be between zero rates and rates that are positive and determined by STPR and SOC considerations. In practice, the latter considerations tend to set a limit of between 5 and 12 per cent, although much higher rates are sometimes advocated in poor countries where the potential for high rates of return on investment is large (the reader is left to ponder the resulting paradox in terms of the effects on the environment and hence poverty!).

But there is a problem with the idea of lowering discount rates because of environmental considerations, for in fact, there is no unique relationship between high discount rates and environmental deterioration as is often supposed. Thus, high rates may shift cost burdens forward to later generations, but, if the discount rate is allowed to determine the level of investment, they will also slow down the general pace of development through the depressing effect on investment. Since natural resources are required for investment, a relationship established through the mass balance principle, the demand for natural resources is generally less with high discount rates than with low ones. High rates will also discourage development projects that compete with existing environmentally benign land uses, e.g. watershed development as opposed to an existing wilderness use. Exactly how the choice of discount rate impacts on the overall profile of natural resource and environment use in any country is thus ambiguous. This point is important since it reduces considerably the force of arguments to the effect that conventionally determined discount rates should be lowered (or raised, depending on the view taken) to accommodate environmental considerations.

14.5 AN ALTERNATIVE TO ADJUSTING DISCOUNT RATES

The previous sections have extended some sympathy towards the environmental criticisms of discounting. Nonetheless, the implication of these criticisms, that discount rates should be lowered, at least when decisions are thought to involve environmental effects in a significant way, could be counter-productive. An alternative to integrating environmental concerns into decision-making would thus be desirable.

One important way of achieving environmentally sensitive

decision-making is to ensure that the environmental effects are better understood and that they are valued in economic terms. That means understanding the various interlinkages in the environmental and economic systems (see Chapters 2 and 3) and valuing the environmental effects on the basis of the total economic value concept (Chapter 9). But we need more than this if we are to account for the environment concerns voiced by the critics of positive discount rates. One approach, which we stress is not fully worked out here, is to integrate a *sustainability requirement* into the analysis. We observed that conserving the stock of natural capital is consistent with the idea of sustainability. Therefore, when evaluating decisions, we can integrate sustainability into the decision by imposing the constraint that, whatever the other benefits and costs associated with the decision, the stock of environmental capital should be constant. Essentially, the normal decision-aiding rules would be applied, for example that benefits must exceed costs for the decision to favour a given development. But it would be a requirement that whatever environmental damage is done by the development should be compensated for by restoration and rehabilitation.

As stated, this would be a very strict rule and almost certainly an unworkable one. As each development takes place we cannot require each tree, each piece of lost soil, each fine view to be restored. But what could be done is to consider a whole range of decisions about developments and impose the sustainability requirement on them. Now what would happen is that the sum of the environmental damage done by a whole sequence of projects would have to be offset by separate projects within the 'portfolio' of decisions being made. These separate projects would aim at compensating for the reduced capital stock by deliberate creation and augmentation of environmental capital. Moreover, they would not be required to pass any test relating costs to benefits. Their justification would lie in the fact that they are honouring the sustainability requirement.

Clearly, the sustainability requirement idea is controversial; but then so is making decisions with conventionally determined discount rates. Adjusting discount rates is likely to be inefficient and clumsy. The sustainability requirement may be less so.

15 · ENVIRONMENTAL ETHICS

15.1 ENVIRONMENTAL VALUES

Environmental concerns all share a common characteristic in that they are interrelated and therefore should not be analysed in isolation. So questions about the economic value of species preservation or the costs and benefits of pollution control, for example, soon spill over into deeper questions about values, ethics, equality and individual rights.

The term value has been variously interpreted but *three environmental value relationships* seem to underlie the policy and ethics adopted in industrialised societies: values expressed via individual preferences; public preference value; and functional physical ecosystem value (see Figure 15.1). We have stressed in earlier chapters that economic systems are dependent on ecological foundations and ultimately on the maintenance of the global life-support system. The long-run survival of human society also depends on certain functional requirements that are met by a set of social norms (i.e. principles of behaviour that ought to be followed). Over time, such norms must be consistent with the natural laws governing ecosystem maintenance if sustainability is the accepted policy goal.

Neither market-based economies nor planned economies are systems with in-built features that would guarantee sustainability. Individuals are assumed to operate according to their own preferences within the context of physical requirements and social norms. Some analysts claim that there is a need to emphasise the intuitive environmental ethic which is present in society but which has remained largely dormant. Currently, however, there is no consensus among philosophers as to whether this intuitive ethic can be given rational and theoretical support, or what the content of the ethic should be.

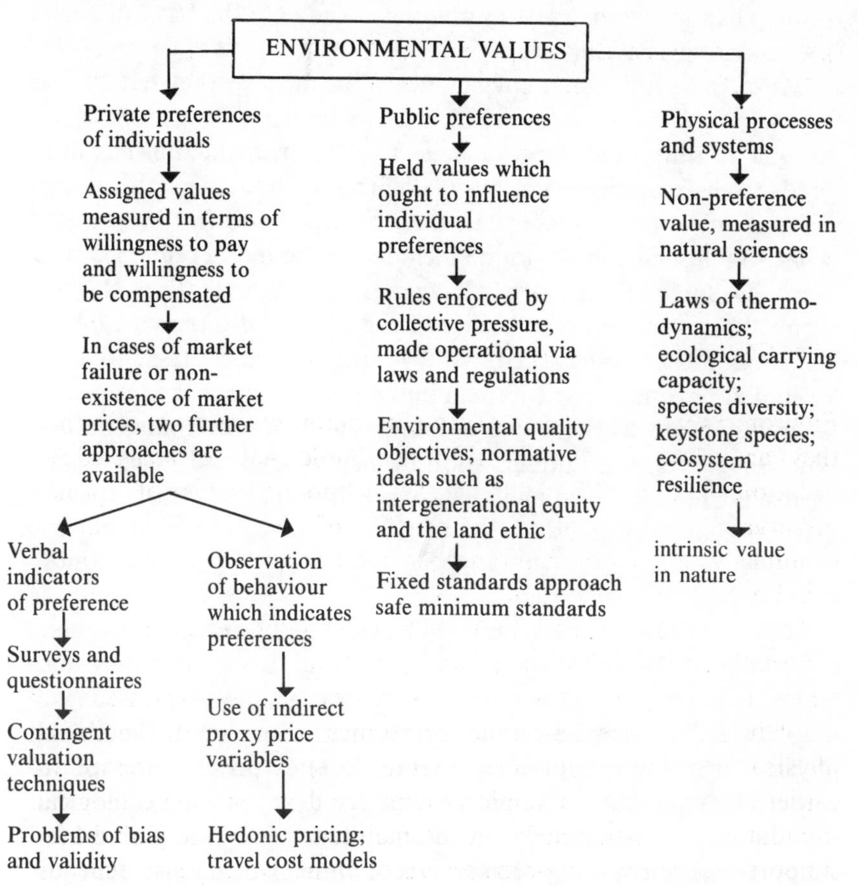

Figure 15.1 Environmental value relationships.

The traditional explanation of how value occurs, is based on the interaction between a human subject (the valuer) and objects (things to be valued). Any one individual person has a number of so-called *held values* which are the basis of individual *preferences* and which in turn result in objects being given various *assigned values*. The total economic value of a resource (use + option + existence value) is therefore a form of assigned value (see Chapter 10).

Debates about the need for and content of a *new environmental ethic* highlighted supplementary and alternative measures of value. It is argued that individuals have *public preferences* as well as private preferences and related assigned values. Public preferences are said to involve opinions and beliefs about what *ought* to be the case rather than individual desires or wants. They are the basis of social norms and legislation.

More radically, some environmentalists believe that nature has *intrinsic* or *inherent* value which exists whether or not humans are around to sense and experience it. This latter distinction has been used to help define two basic environmental ethical positions. Simplifying matters somewhat, these viewpoints differ in terms of what should and should not be included in the moral reference class (i.e. what sort of things should have moral rights and interests). They also differ over the acceptability of the notion of intrinsic value in nature. It has been suggested that the moral reference class should be extended beyond current human individuals to cover the rights and interests of future generations of humans and/or non-human nature. The issue of species preservation (which is analysed in economic terms in Chapter 17) will be used to distinguish more clearly the two basic ethical positions.

15.2 ENVIRONMENTAL ETHICS AND SPECIES PRESERVATION

Figure 15.2 summarises some of the main elements in the ethical debate over the question of whether species preservation policy should be strengthened given the escalating rate of species loss that the planet is experiencing. Some analysts have argued that ethical rules are required to prohibit current human behaviour that has effects upon the long-range future as well as the present. Humans must accordingly become generally more aware of nature's regulatory mechanisms and must achieve some sort of homeostasis (i.e. position of balance) in the utilisation of natural resources. The exploitationist (see Chapter 1) world view constrained only by an adequate rate of savings (to benefit future generations) must be rejected in favour of a conservationist or eco-preservationist world view. According to the exploitative view, current obligations to future generations involve only a guarantee that the productive potential of the economy that is passed on has been expanded. The conservationist view encompasses the notion of an economic growth process constrained by the criterion (applied to particular productive

Figure 15.2 The ethics of preserving species[1]

Consequentialist (teleological) positions (i.e. the results or consequences of actions determine their rightness or wrongness).

A defence of species protection or rejection of it on the grounds of the consequences of the act of protection.

↓

Anthropocentric (human-based) utilitarianism
It is the case that many species yield direct and indirect USE VALUE, as well as option and existence value (see Chapter 17). So from a utilitarian position, species protection could be justified in order to *conserve* these benefits which humans value.

↓

Expanded rights view (future human generations)
Rights and interests of future generations of humans should be considered. Species loss may mean harm to future human prospects and need provision.

↓

Non-anthropocentric utilitarianism
Mere *sentience* (i.e. the capability of sensing pleasure or pain) should be the criterion for inclusion in the moral reference class.
Animal rights view: species loss is wrong only to the extent that the interests of individual animals are unjustly harmed.
Loss of a species would be judged wrong only to the extent that plants and inanimate nature cannot be harmed because they are non-sentient; their value is only as a resource for sentient beings.
Entire ecosystem protection is only right if such policies yield the highest net level of utility.

Non-consequentialist (deontological) positions
(i.e. based on duty or obligation to perform or refrain from actions that are right or wrong, quite apart from considerations of consequences).

A defence of species protection on the grounds that this protects what is valuable in itself (intrinsic value in nature), or on grounds of fairness and justice, rather than making claims that good consequences will result from protection.

↓

Intergenerational equity principle
Species loss represents an unfair/unjust *restriction on the options* available for future human generations. Support for this justice principle has a number of sources – *stewardship arguments; justice as opportunity arguments* i.e. that the resource conservation task should encompass the 'Lockean Standard', in that the present generation has done enough morally if it leaves enough and as good for the next generation. As wide a range of resource use opportunities as possible should be passed on from one generation to the next.

↓

Ecocentric preservationist position
Preservation policy applied to *all* the constituent parts of entire ecosystems (all elements are assumed to have a right to exist and flourish).
Additional constraints on resource exploitation are required, probably in the form of pre-emptive environmentally-based constraints (global preservation zones, stringent environmental quality standards).
Leopold's 'Land Ethic': whole ecosystems have rights and interests and should be protected.

Figure 15.2 *cont.*

Gaian arguments: protection of habitats, species and individuals is important, although this need not entail any absolute principle to the effect that no species ever be allowed to vanish. Particular environments, especially coastal margins, may require special protection; but the Gaian self-regulatory mechanism carries on regardless of all except the extreme forms of human interference.

Note: 1. The ethical positions are arranged in such a way that as we go down the columns increasingly 'radical' departures from traditional philosophy are encountered.

---

systems) of sustainability of resources productivity. Hence our obligation to future generations is now two-fold – save and expand the production potential of the economy, while at the same time ensuring that such progress is sustainable indefinitely.

Preservationist arguments applied to entire ecosystems, and geographical zones require the imposition of pre-emptive environmentally-based standards on economic activity. Only after these constraints, necessary to protect biological diversity and biotic functions, have been satisfied is it permissible to use economic cost-benefit analysis and discounting to evaluate projects or policy. Non-preservationist positions accept the discounting procedure without such qualification although the rate of discount chosen is a matter of some controversy (see Chapter 14).

It seems to us that the sustainability notion which we developed in Chapter 3 could satisfy the majority of the ethical concerns and objectives outlined in Figure 15.2. Recall that the basic rules for the maintenance of sustainable economic development implicitly contain the idea that the resource stock K_N should be held constant over time. Maintaining the natural capital stock lends necessarily to the protection of environmental functions and services useful to humans as well as actual habits and ecosystems required by flora and fauna. Habitat and ecosystem protection will need to be quite extensive in order to avoid 'island effects' which lead to species decline. Our sustainability rules also clearly satisfy the intergeneration equity objectives required for 'fair' treatment of future generations.

Figure 15.2 has simplified and aggregated a number of ethical arguments and positions. It does, however, highlight the fact that there is little consensus among philosophers as to whether a new environmental ethic is required, or alternatively that traditional forms of ethical reasoning need supplementing. Three ethical issues in particular lie at the heart of this controversy:

1. Should the world view adopted be essentially anthropocentric or ecocentric?
2. Should distributional issues be decided on utilitarian, libertarian or contractarian-egalitarian value systems?
3. Should individualism or collectivism be the locus of environmental values?

Each of these issues will be briefly analysed in the next section.

15.3 THE SEARCH FOR A NEW ENVIRONMENTAL ETHIC

Anthropocentrism versus ecocentrism

While a number of different types of philosopher are agreed that there must be some extension of the moral reference class, in order to support ethical principles necessary for a more stringent environmental protection policy, the limits of this process are hotly disputed. An *anthropocentric environmental ethic* (referring to a comprehensive, coherent set of principles, obligations and values) would recognise the full range of instrumental value (use, option and existence value) in nature. It could also be extended to encompass the notion of intergenerational equity and perhaps even the existence of moral interests held by non-human but potentially conscious beings (known as the 'expanded-rights view').

Supporters of the 'expanded-rights view' by and large believe that an ethic which restricts moral standing to humans and/or all consciously sentient beings would still be sufficient to safeguard the basic integrity of the global environment. The basis for this argument rests on the fact that nature is a storehouse of instrumental value. Provided that this value is generally recognised, then humans will make arrangements so as to protect adequately their environmental assets. The number and extent of nature conservation/preservation policies that would be deemed justifiable could be quite large. If some animals merit moral consideration, for example, then preservation of their habitats will also, more often than not, be justified.

This anthropocentric line of ethical arguments has proceeded incrementally, gradually moving away from received moral judgements about familiar issues and their consequences and on to analogous cases involving the future for humans. It comes to a halt part way into the non-human realm and refuses to discard the traditional concern for individual beings in favour of any irreducible concern for biotic systems as a whole. The outer limits of the anthropocentric environmental ethic debate is well illustrated by the so-called *extended stewardship ideology*. Such a position can be described and defended in terms of human interests (i.e. humans should act as nature's steward and practice careful husbandry of natural resources, for their own sakes and to protect the interests of other creatures). Thus even if, as seems likely, the moral significance of non-sentient living creatures is slight, it is as well to be aware of it. This would be especially the case if large numbers of present or future organisms are in question.

Ecocentric ethics require that non-human nature (both conscious and non-conscious) be capable of being inherently valuable (i.e. possess intrinsic value not dependent on human sensations) and possess moral rights. The class of morally considerable beings should be extended to *systems* and not just individuals. The anchor point for these positions is *Leopold's 'Land Ethic'*. According to the Land Ethic, human life, however worthwhile, is of value only insofar as it preserves the sustainability of the global environment system (the biosphere). Individuals cannot possess absolute intrinsic value, only relative value determined by their relative contribution to biosphere integrity. The *summum bonum* of the Land Ethic is the integrity, stability and beauty of the biotic community. This ethic seems to support the ecological principle that the characteristic structure of the ecosystem (its objective beauty) is vitally important for the preservation of its stability. However, the scientific literature should not be interpreted as saying that the more diverse an ecosystem is the more robust it is in dealing with outside interference. It is much more plausible to claim that the stability of an ecosystem is a function of its characteristic diversity.

The Leopold maxim and its implied moral code of conduct carries some controversial, and in our view, unacceptable consequences. Food aid to some developing country peoples (e.g. in the Sahel region of Africa) could be deemed wrong insofar as the continued existence of such human communities leads to further deterioration of their local ecosystem because of their wood gathering, dung burning and other practices. Even moderately intensive agricultural systems anywhere on the globe could be morally suspect because of their threat to the integrity and stability of ecosystems.

Alternatively, if the Land Ethic is merely used as a supplementary ethic alongside conventional ethics, then ecosystem good would have to be weighted along with human good in order to distinguish moral action. The range of 'critical' interest conflicts could be narrowed if a version of the *Gaian* notion proves acceptable.

The Gaia hypothesis is a scientific theory which implies that life (all the biota) profoundly modifies the environment and that this modification acts to stabilise the environment. The sum total of the living and non-living components of this system is Gaia. So according to the strong version of the hypothesis, the earth is a single huge organism intentionally creating an optimum environment for itself. More credible scientifically is a weak version of the hypothesis which holds that life has controlled its environment within limits narrow enough that life continued. This is a homeostatic Gaia devoid of purposefulness but still powerful. Many scientists reject even this version, arguing that mindless chemistry and physics is enough to explain the mechanisms that affect the composition of the global oceans and atmosphere.

The Gaia debate has stimulated much interesting science on the links between the living and non-living world. There may be 'core' regions of Gaia which are especially important in terms of biospherical stability (e.g. the original hypothesis held that regions between latitudes 45° North and 45° South, particularly continental shelves, wetlands, tropical forests and scrublands, represented 'core' zones, which also contained key species). Human intervention in the 'core' regions should be especially circumspect.

Gaia has been taken as the basis for a holistic ethical system similar in many ways to the Leopold Land Ethic. In this system it is argued that humans want and need a biosphere that can sustain itself without constant human effort. The Gaian notion fills that need and, although far from fragile, its mechanisms are not invulnerable.

Humans are both part of, and apart from, nature. They should therefore not seek futile change but endeavour to intervene as constructively as possible. The creation of the contemporary English countryside in the wake of the eighteenth century Enclosure Acts is often cited as an example of constructive intervention and worthwhile conservation. Overall, protection of the global environment need not be regarded as a non-negotiable constraint; not every particular eco-system is necessary for global system survival.

For the *deep ecologists* only 'biotic egalitarianism in principle' (equal rights for all species must be recognised at least in theory) offers adequate environmental protection. Non-human nature (conscious and non-conscious) is thought to possess *intrinsic value* in the widest sense. Two major problems are raised by this extreme preservationist position. If inherent/intrinsic value and not just instrumental natural value exists, what is it and how do we discover it? It seems reasonable to conclude that we either justify our acceptance of intrinsic value in nature at an intuitive level only, or we look for support via appeals to 'expert judgement'. Both of these forms of justification seem problematic.

A second major difficulty with the preservationist case involves the adjudication of cases where human and non-human moral interests come into conflict. Even among humans, who are conventionally treated as equal moral persons, there are often extremely difficult controversies over rights. But acceptance of the preservationist position would mean that interspecies and animal–natural habitat conflicts, as well as human interest conflicts, would need to be equitably and practicably resolved. Principles or rules for the settlement of such moral interests conflicts have not so far been worked out on anything like a satisfactory basis.

Notions of compensatory and distributive justice

Another central concern of the environmental ethicists is the need for a new theory of justice. Neither *utilitarianism* (i.e. that an act or rule is to be judged most desirable if it maximises either average or total utility, even if some individuals, the minority, lose out) nor *libertarianism* (i.e. an action is desirable or right as long as the individual involved has legitimate entitlement or claim to support the action) can support an ethical basis for principles of intra- and inter-generational equity (fairness) requiring equal treatment for all persons.

These traditional systems of ethics do not in general support the proposition that current generations have moral obligations to future people similar to those that are owed, say, to contemporary strangers. Less consideration is owed to the future, it is argued, because of the following:

1. The very temporal location of future people.
2. Ignorance of future people's wants and needs, and the supplementary argument that in a contemporary growth society, investments in science and technology will almost certainly make people better off in material terms in the future (inherited stock of capital and knowledge).
3. The contingency of future people.

A counter argument to (1) and (2) has it that as long as some people will exist, and will be in no relevant way unlike current rights holders, they are worthy of equal consideration. Further, whatever the uncertainty about the extent of future preferences, it is clear that basic needs will exist and will not be substantially different from contemporary ones. The satisfaction of these basic needs will be a prerequisite of the satisfaction of most of the other interests of future people regardless of their uncertainty.

Future people may also inherit more than an enhanced stock of capital and knowledge. Technologically advanced industrial economies seem to be generating an increasing number of 'environmental risk' situations (e.g. the generation of hazardous wastes with long-term pollution potential and genetic engineering). If global life-support systems become seriously impaired, future generations may have little opportunity to ameliorate or to adapt to their grossly polluted world.

The contingency of future people argument highlights the fact that such individuals may not exist at all (possible people) and that their actual number depends in large measure upon current actions and decisions. It is therefore argued that current policy cannot be governed by reference to harm to the interests of future individuals, because those policies determine who those individuals will be and what interests they will have. Utilitarians, for example, may conclude then that the present generation has no obligations to future persons. They would reach this conclusion because of their acceptance of what is known as the *Person-Affecting Principle*. This principle lays down that a person has not been wronged by another unless he has been made worse off by the other's act, i.e. he is the other's victim. Therefore nothing which we (the current generation) can do, can wrong future actual people (brought about by our actions), unless what we do is so bad that these actual future people turn out to wish they had never been born at all.

The *contractarian approach* (i.e. based on actual or hypothetical negotiations which are said to be capable of yielding mutually agreeable principles of conduct, which are also binding upon all parties) and in particular an amended form of John Rawls' idealised decision model (designed to produce procedural rules for a just society) have been cited as possible sources of *intergenerational equity criteria*. Efforts based on the Rawlsian framework are not, however, totally convincing. In the original formulation, Rawls has rational individual representatives from contemporary society deciding on the rules for a fair society in what he calls the 'original position' (the negotiations) and operating from behind 'a veil of ignorance'. In other words, the representatives make their choices without knowing the strata of society they themselves belong to. Various principles, it is argued, would be rationally chosen, e.g. equal opportunity for all individuals and the 'maximin' criterion (i.e. an acceptable standard of living for the least well off in society).

The dual nature of Rawls' theory – justice as rational cooperation and justice as hypothetical universal assent – produces conflict as soon as the analysis moves away from the self-contained society of contemporaries. In the intergenerational context, if all generations are represented in the 'original position' then the representatives themselves could work out how many actual generations there will be. However, this is one of the issues which in some measure is supposed to depend upon their deliberations. Perhaps the representative should only know that some (but not all) generations are represented. Both the rights of possible people (if it is accepted that they have any) and non-human species are not catered for in any set of Rawlsian rules. We are then left with the proposition that the rational representative should behave as if he were a 'possible person'.

Individualism versus collectivism

Most egalitarians do not restrict their conception of human nature to that exemplified by the 'rational economic person' model. They also emphasise the social and communal aspects of human nature and the possibility of some sort of social contract. Adapting a more collectivist approach allows us to recognise so-called *'generalised obligations'*. In the intergenerational equity context, such obligations could be interpreted to mean obligations on the current generation to maintain a stable flow of resources into the future (an inheritance of environmental and conventional economic goods and services), in order to ensure on-going human life rather than just meeting individual requirements.

From this collectivist viewpoint, all value is still found in human loci, but it is not restricted to satisfactions of felt preferences of human individuals. Individuals may also hold *considered preferences*, expressed after careful deliberation, including a judgement that the desire or need is consistent with the argument that individuals hold both private (self-interested) and public (in the public interest or in the interest of a group or community) preferences.

It has been argued that private and public preferences belong to different logical categories, because the latter do not involve desires or wants but opinions or beliefs. Some environmentalists have concluded that since public preferences are not reflected in market-type situations (real or hypothetical) they are therefore not amenable to quantification via monetary cost–benefit analysis. These community-regarding values are nevertheless reflected in legislation which has passed through the political process. Policy priorities are set on the basis of non-economic grounds, i.e. scientific, cultural, historical, ethical, as well as economic grounds.

The fixed-standards approach to environmental protection would put primary emphasis on the non-economic factors and utilise economic analysis in a cost-effectiveness role. But the distinction between cost-effectiveness and full cost–benefit analysis in the environmental policy context is not clear-cut. Economic analysis need not be restricted to the determination of the least costly methods of achieving social goals. It may also play an important role in the process of goal determination, separating the practicable from the impracticable.

Given a collectivist position the principle of intergenerational equity can now be interpreted in terms of a 'justice as opportunity' argument and the 'Lockean Standard'. Each generation should leave 'enough and as good for others that follow on'. Our sustainability principle set out in Chapter 3, with its constant natural capital stock K_N can fulfil this equity objective. In practical terms this means, among others, policies directed at the conservation of renewable resources, enhanced technology and recycling innovation. What is proposed is that future generations are owed compensation for any reduction (due to the activities of the current generation) in their access to easily extracted and conveniently located natural resources. So the future's loss of productive potential must be compensated for if justice is to prevail. The current generation pays the compensation via improved technology and increased capital investment designed to offset the impacts of depletion.

In conclusion, we have seen that the passing-on of the resource base 'intact', i.e. constant natural capital stock K_N, over the next few generations is central to the concept of sustainable economic development (see Chapter 3). Such a managed growth policy, although directed primarily toward the satisfaction of human needs, would also necessarily ensure the survival of the majority of non-human nature and its natural habitats. Adequate environmental safeguards are available therefore without the need to adopt any of the radical 'deep ecology' arguments and ethics. In particular it is not necessary to have to accept the notions of intrinsic value in its widest sense, or of equal rights for all species. Our sustainability principle is general enough to encompass the environmental ethical concerns of consequentialist philosophy, as well as meeting the intergenerational equity objective.

PART IV

THE ECONOMICS OF NATURAL RESOURCES

16 · RENEWABLE RESOURCES

16.1 INTRODUCTION

The essential feature of a renewable resource is that its stock is not fixed and can be increased as well as decreased. It will increase if the stock is allowed to regenerate. An obvious example is a single species of fish or a forest. Nonetheless, there is a maximum stock – no renewable resource can regenerate to levels above the carrying capacity of the ecosystem in which it exists. Thus, left alone, blue whales could increase in number, but they would not carry on increasing *ad infinitum*. This potential for increase is important because man can 'cream off', or harvest, the increase in the size of the stock and, provided certain conditions are met, the stock will grow again, be harvested, grow again, and so on. Other things being equal (e.g. the conditions within the relevant ecosystem), there is reason to suppose that this process of harvesting can carry on for very long periods indeed. But, as we shall see, the potential for over-harvesting a renewable resource is significant: it is quite easy to make a renewable resource disappear. This will obviously happen if the rate of harvest exceeds the rate of natural growth of the resource persistently. It can also happen if the resource population falls below some critical level, perhaps because of over-harvesting or for some other reason unconnected with the direct use made of the resource (e.g. habitat destruction).

This chapter investigates some of the theorems which have been derived with regard to the optimal use of a renewable resource. Two major caveats are in order at the outset. First, conventional usage has it that resources which exhibit continuous flows through time are also called 'renewable'. Energy from the sun or the waves or the tides are examples of continuous flow resources. The analytics of this chapter are not relevant to these resources. Second, while even the analytics might appear complex, at first sight anyway, the fact is that we concentrate on the use of a single species without paying attention to the fact that species are *interdependent*. One highly relevant example of interdependence in ecosystems is the predator–prey relationship: one species requires another as its food source. The prey, in turn, is a predator on another species, and so on. What the optimal use of a set of interdependent species might look like is exceedingly complex. While the results presented here are important, they do not approximate the complex world of ecological interdependence. It is this context that conservationists have in mind when they ask us to have due regard to the 'fragility' of ecosystems and to adopt cautious policies of use.

16.2 GROWTH CURVES

Consider a single fish species; then its stock (or biomass) may exhibit growth through time as shown in Figure 16.1. The curve shown is a

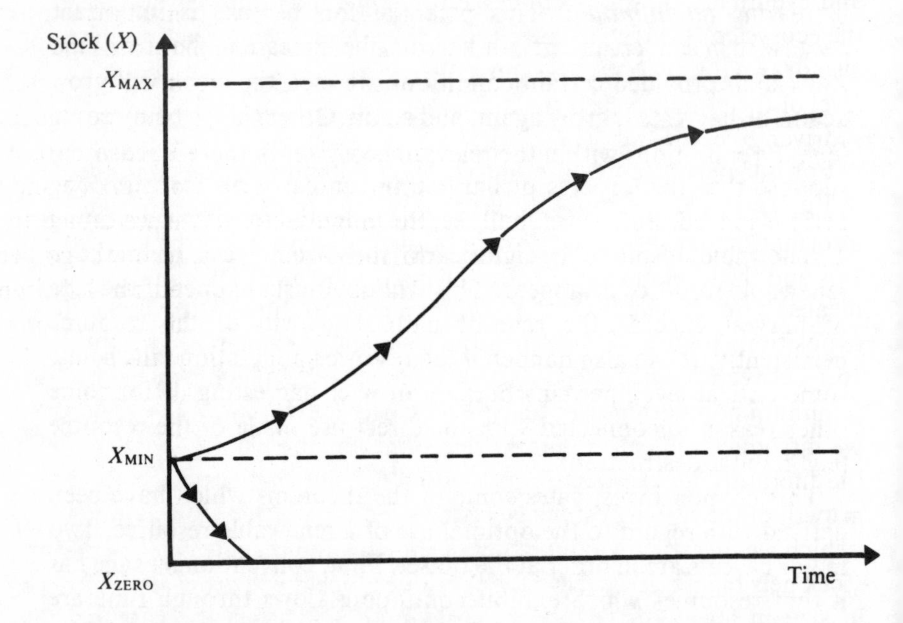

Figure 16.1 Logistic growth curve of a renewable resource.

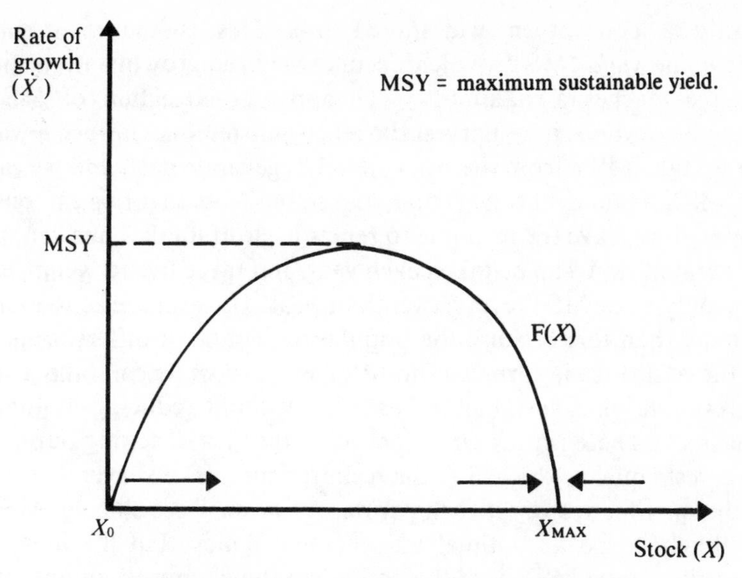

Figure 16.2 Pure compensation growth curve.

logistic function: at low levels of stock the fish multiply, but as they begin to compete for food supplies their rate of growth slows down and eventually the stock converges on some maximum level X_{MAX}, the ecosystem's carrying capacity for that species. Note that we have also drawn the curve as beginning at X_{MIN}: this is a critical minimum level of the population. If the numbers go below this level the species is driven to extinction (X_{ZERO}).

For our purposes it is useful to look at the information contained in Figure 16.1 in a slightly different way. Figure 16.2 plots the same information but shows the *growth* of the resource, $\dot{X}$, on the vertical axis and the level of stock, X, on the horizontal axis. The notation $\dot{X}$ means 'rate of change in X with respect to time'. For ease of exposition, Figure 16.2 ignores the segment of the curve in Figure 16.1 between X_{MIN} and X_{ZERO}. i.e we now assume there is no critical minimum population size. The analysis becomes far more complex if we introduce critical minimum sizes – producing what is known in the literature as 'depensation'. Figure 16.2 tells us that the rate of growth of the resource stock is positive at first, reaching a maximum (which we investigate in more detail shortly) and then declines as the stock gets bigger. If we leave the resource alone it will grow and grow in size in terms of its total biomass until it reaches the carrying capacity of its environment at X_{MAX}. Figure 16.2 permits us to identify a concept in widespread use. This is the *maximum sustainable yield* (MSY) which occurs when the growth rate of the resource reaches a maximum. The apparent attraction of MSY should be obvious: if we harvest the renewable resource in such a way that we take MSY from the stock, it will regenerate itself and we can take MSY again next time round, and so on. Note that this can only happen if we leave the resource to renew itself. If it takes one year to regenerate, MSY can be taken each year. If it takes twenty years, we must only take MSY every twentieth year. (In practice it is more complex than this because the population will be of different ages, but the basic idea is correct.) But MSY is the *most* we can take from the resource on a sustainable basis, i.e. without reducing its long-term stock. There is thus an attraction in the idea of setting our rate of harvest equal to the MSY: the resource survives 'for ever' and we get the maximum from it each period. As we shall see shortly, MSY is unlikely to be an optimal management policy, but the idea of harvesting at the MSY is still a commonly-held view about optimal resources use.

16.3 THE RATE OF EXPLOITATION

We can now introduce the level of exploitation or 'harvest' (or 'yield') of the resource. The simplest hypothesis to use is that the *effort*, E, expended in harvesting is equal to the ratio of the actual harvest, H, to the stock, X. That is:

$$E = \frac{H}{X} \tag{16.1}$$

the bigger the effort, the greater the proportion of the stock that would be harvested. We can rewrite equation (16.1) as

$$H = EX \tag{16.2}$$

Then, the rate of harvest can be shown on our basic diagram. This is done in Figure 16.3 which shows how the choice of the effort level will determine the harvest and the stock level, i.e. where EX is just equal to the rate of growth of the resource. This gives the harvest H^* and the stock X^*. Any harvest level along the line EX to the right of X^* will mean that the harvest is greater than the sustainable yield X and the stock will fall. A harvest level along the line EX to the left of

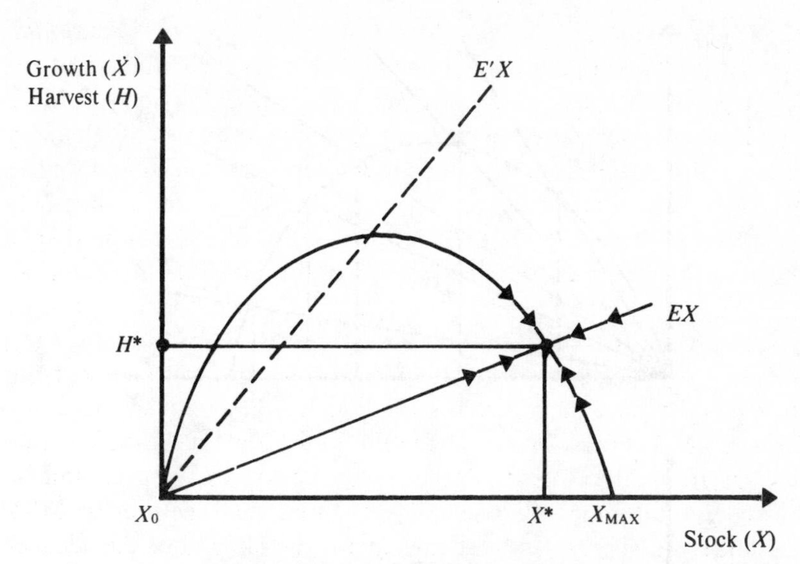

Figure 16.3 Effort–growth equilibria.

X^* is less than the yield through natural regeneration, and the stock will grow.

Note that H^* is *not* the maximum sustainable yield, but we could easily introduce a management policy which says that effort should be changed so as to take the MSY. In this case, E becomes the instrument of management and the harvest rate is set equal to $E'X$ as shown in Figure 16.3. Thus, introducing the effort level helps us to determine the harvest and stock level but tells us nothing as yet about the desirable level of exploitation. For that we shall need some concepts of cost and revenue.

16.4 COSTS AND REVENUES

In order to introduce costs and revenues we need to transform Figure 16.3 into one showing the relationship between the harvest or yield, and the level of effort. Figure 16.4 shows how this is done. It shows various equilibria for various degrees of effort where $E^4 > E^3 > E^2 > E^1 > E^0$ (E is the slope of the line EX). The harvest or yields for each equilibrium effort level are shown as h^0, h^1, etc. It is then a simple matter to plot the levels of effort in the lower half of the diagram and read off the asociated harvest levels. (*N.B*: if you are drawing this, a

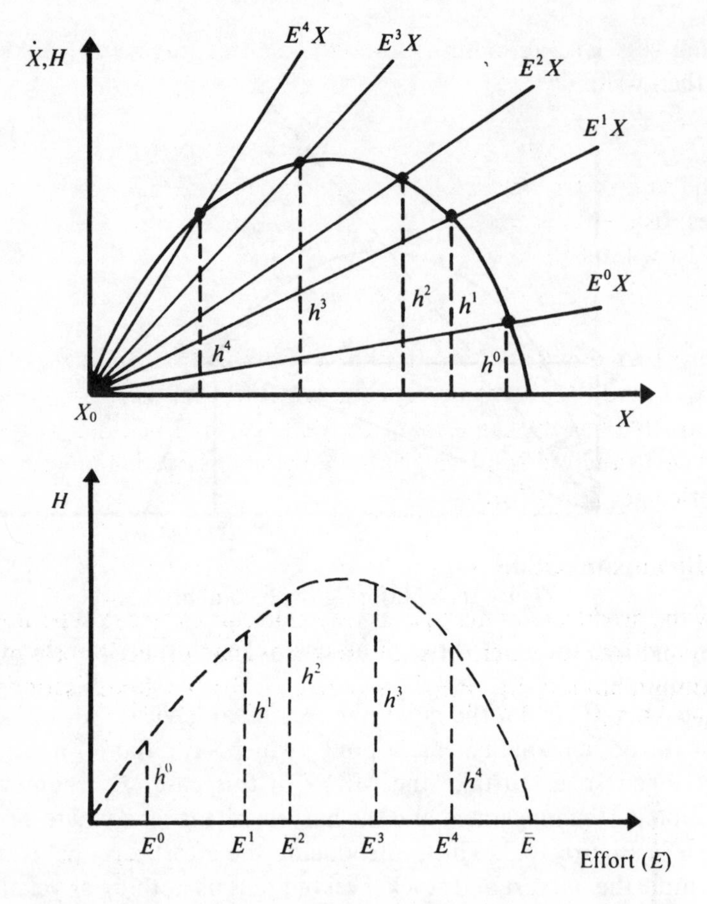

Figure 16.4 From growth–effort to effort–harvest functions.

doubling of the slope EX in the upper part of the diagram is the same as a doubling of E on the horizontal axis in the lower half of the diagram.)

The effort–harvest (or effort–yield) curve thus looks very much like the growth–harvest curve, but observe that the yields are read off the upper part of Figure 16.4 in such a way that they appear as a mirror-image in the lower half. Thus, X_{MAX} corresponds to *zero effort,* and X_0 to $\bar{E}$ in Figure 16.4.

Now, the effort–harvest curve in Figure 16.4 is easily translated into costs and revenues. If we assume that effort is the only factor of production involved, then total cost, TC, will be equal to the level of effort multiplied by the price of effort, say the ***ruling wage rate,*** W (in practice, effort will involve capital as well, e.g. lumber equipment, fishing vessels). For simplicity we assume this wage is constant. We can then write:

$$TC = WE \tag{16.3}$$

In the same vein, we assume that the price of the harvested product (trees, fish, etc.) is constant at P. Hence total revenue, TR, from the harvest will be;

$$TR = PH \tag{16.4}$$

Figure 16.5 shows the resulting total revenue and total cost curves. Since P and W are assumed constant, the total revenue curve in Figure 16.5 has the same shape as the lower half of Figure 6.4. The total cost function is linear and its (constant) slope is the wage rate, or 'price per unit effort'.

Profit maximisation

Now the producer has his cost and revenue curves he can superimpose them and find at which level of effort revenue exceeds costs by the maximum amount i.e. R−C = MAX → Profit Maximisation (see figure 16.5a).

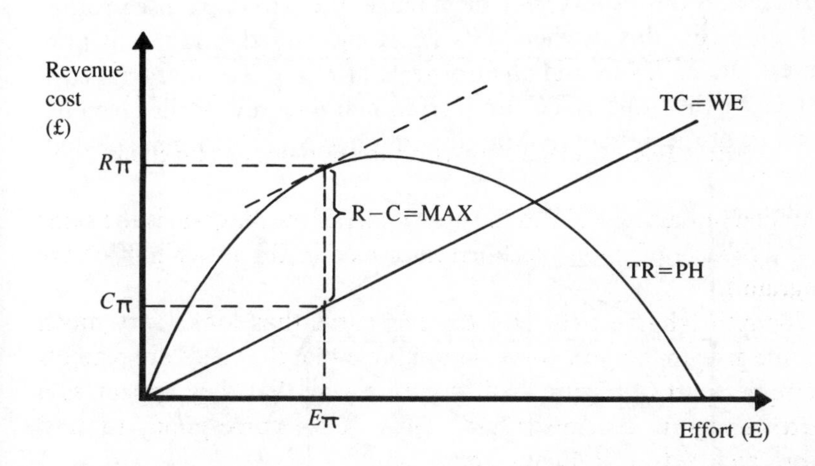

Figure 16.5a Profit Maximisation.

Another way of looking at this is to remember that profit is also maximised when marginal revenue = marginal cost.

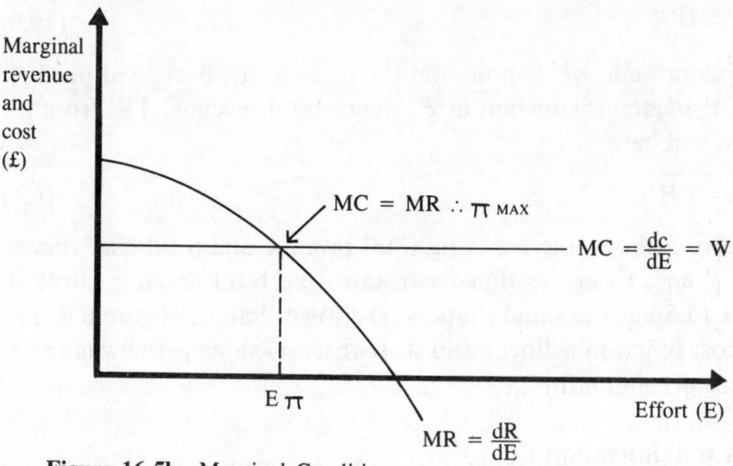

Figure 16.5b Marginal Conditions.

Two possible equilibria are shown in Figure 16.5. If the renewable resource can be placed under single ownership, or joint ownership in such a way that the owners act collectively, it is reasonable to suppose that the resource will be managed so as to maximise profits. In Figure 16.5 this is where TR–TC is maximised, and this is at a harvest rate of H_{PROF} and an effort rate of E_{PROF}. The marginal cost is given by the slope of TC, or W. The marginal revenue is given by the slope of the TR curve. Marginal revenue equals marginal cost at

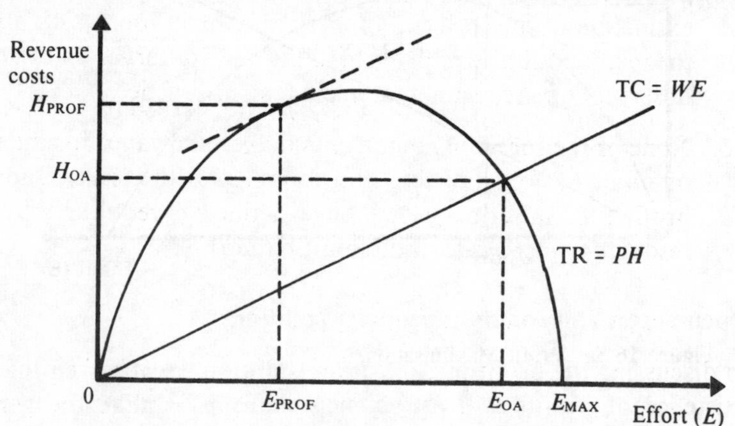

Figure 16.6 Profit maximisation and open access equilibria.

H_{PROF}, E_{PROF}, and in turn, this is the condition for profit maximisation. There are several observations to be made about this equilibrium:

1. Unless the owners of the resource can keep new entrants out of the resource industry, the profit will be dissipated as new entrants come in. As we shall see, this will happen if ownership rights are not well defined.
2. The profit-maximising equilibrium does *not* coincide with MSY – the highest sustainable harvest. Indeed, in our diagram the stock is actually larger than that corresponding to MSY since E_{PROF} lies to the left of the level of effort corresponding to MSY (remember that in the effort–harvest diagram, maximum stock is at zero effort). In the absence of information about externalities, profits and social benefits may be assumed to be the same thing. Hence MSY does not appear to be a socially desirable management practice.
3. Clearly, the price of effort (wages in our example) could be so high as to produce a profit-maximising solution nearer still to maximum stock (i.e. nearer to the origin in Figure 16.5), or so high that no exploitation takes place at all (TC might lie above TR at all points). At the other extreme, if effort is costless, $W =$ O and the TC curve will be coincident with the horizontal axis and the MSY will coincide with profit maximisation.
4. Profit maximisation does *not*, in this example, lead to the extinction of the species, contrary to casual observation about the incompatibility of profit-making and conservation by some observers of environmental management practice.
5. None of the foregoing comments makes allowance for the role of *time*. As we shall see, this complicates the issue. (Indeed, profit maximisation as we have defined it requires that the resource owner's rate of discount be zero.)

Open-access and common-property solutions

In discussing the profit maximisation solution we observed that the existence of profits will attract new entrants. Unlike the general microeconomic theory of new entry, however, we can observe that new entrants cannot come about if the resource is wholly owned by a single owner. The obvious examples are again forests and fisheries. However, while it is common to think of privately-owned forests and privately-owned fishing rights in rivers, private ownership is not typical of major forests or sea fisheries. Instead, we may have government territorial ownership – as with territorial waters – with free access by the relevant country's nationals. And for a great many species of interest – for example, whales – there may be no territorial rights either: the sea may simply be internationally common property. What happens in these cases?

In either case we may get the 'open-access solution'. If less than normal profits are being made (TR less than TC) some resource exploiters will go out of business. If abnormal profits are being earned (TR greater than TC) new entrants will come in and the solution will be at E_{OA} ('OA' stands for 'open access') with a harvest rate H_{OA} in Figure 16.5. The equilibrium is where profits are dissipated and each resource exploiter secures normal profits only.

The observations about the open-access solution are fairly obvious:

1. The stock is less than that associated with profit maximisation, as we would expect, and the harvest rate is lower.
2. Open access does not coincide with MSY, unless by chance, TC cuts TR at the latter's maximum.
3. Open access generally does *not* lead to the extinction of the species, contrary to the widespread arguments by some environmentalists that open-access resources are inevitably doomed. The conditions under which extinction will occur are (a) that effort is costless – effort is at E_{MAX} in Figure 16.5 and the stock goes to zero – or (b) harvesting takes place at levels persistently above the natural rate of regeneration, i.e. the harvest is *non-sustainable*.

It is important to distinguish between open-access equilibria and *common property* equilibria. We have seen that open access means that no one owns the resource and access is open to all. There are no limits on new entrants. A common-property resource, however, is one that is owned by some defined group of people – a community, a nation. It is possible that within this group of people there will be open access, that is, each individual member of the group will be permitted to make whatever use they wish of the resource. But it is more likely that the group will develop rules of use, limiting the use that any one individual is allowed to make of the resource. These 'common property rules' are widespread where common property exists, e.g. tribal control of rangelands in large parts of Africa, national parks and so on. The reason that such rules emerge is that unconstrained use by each individual is *more likely* to lead to resource extinction, damaging the welfare of everyone and perhaps imposing an irreversible cost on future generations. In terms of Figure 16.5 we might expect a common-property solution to be somewhere between the profit-maximising solution and the open-access solution. Since the profit-maximising solution involves least effort and hence larger stocks, the risks of extinction appear to be low. The open-access solution is much more likely to risk extinction, although it is not inevitable, as we saw. This is because effort is much greater and stocks are consequently lower. The common-property solution lies somewhere in between.

As it happens, the open-access solution will carry high risks of resource extinction if there is a critical minimum size to the population (see Figure 16.1). Similarly, common-property solutions can break down if the defined group gets larger and larger because of population growth. It may then pay any one individual to 'break ranks' and maximise individual utility at the expense of the community's interests overall.

16.5 PRESERVATION VALUE

The arguments so far are fairly reassuring for the conservationist: essentially, if the resource is exploited in an *equilibrium* manner (i.e. harvest rates do not exceed regeneration rates on a persistent basis) then either profit maximisation or open access will generally be consistent with species preservation. But single ownership or control of the resource will be preferred to open-access solutions on conservation criteria because the latter has the potential for depleting the stock substantially, especially if cost of harvesting is low. These remarks suggest that conservationists will prefer a larger stock of the resource to a smaller one. However, nothing in the analysis so far has allowed for the preferences of conservationists: it has all been carried out in terms of the costs and benefits to those taking the resource for some specific use which involves the death of the actual quantity harvested (fish for food, trees for wood-processing or pulp etc.).

It follows that the profit-maximisation solution can only be optimal in the *social* sense if the preferences of conservationists imply zero 'preservation value' for the resource. The way in which the spill-over effects of one group of individuals on another group can be accomodated in economics is via the introduction of externalities into the analysis. For example, if conservationists prefer larger to smaller stocks, their utility loss will be related to the difference between the maximum possible stock (the natural equilibrium, or carrying capacity stock) and the actual stock that results from the resource using activity. We could write:

$$-U_C = f(X_{MAX} - X_E) \qquad (16.5)$$

where $-U_C$ is the loss of utility to the conservationists and X_E refers to the various equilibrium levels of stock. Figure 16.6 shows a possible valuation function that is consistent with this hypothesis, but there are other possibilities. Much depends on how conservationists see the 'reference point'. We have assumed they regard X_{MAX} as the most desired position. Hence deviations from it are what matter. This is consistent with the importance some conservationists attach to 'homeostasis', the steady state that comes about by not using the resource for human purposes. It is also consistent with the 'ordinary person's' viewpoint that certain species should be left alone. But others may prefer to see the resource operating so that growth is maximised – i.e. X_{MSY} becomes the reference point. Still others will have some idea of a 'permissible' use

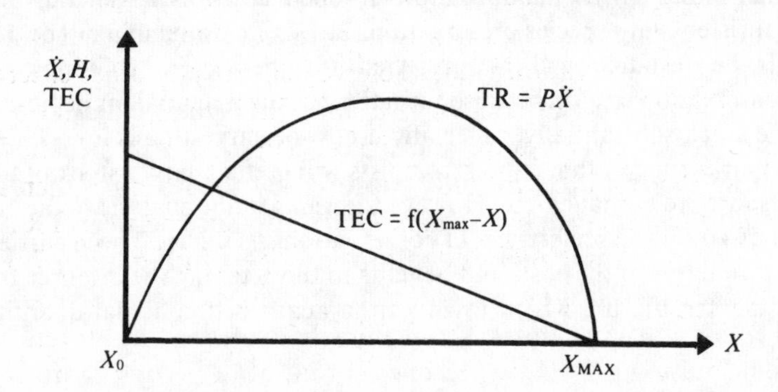

Figure 16.7 Preservation values.

of the resource, perhaps comprising some small harvest. Our preservation value function is also likely to be discontinuous when very low stocks are reached: the last few members of the species will have very high values.

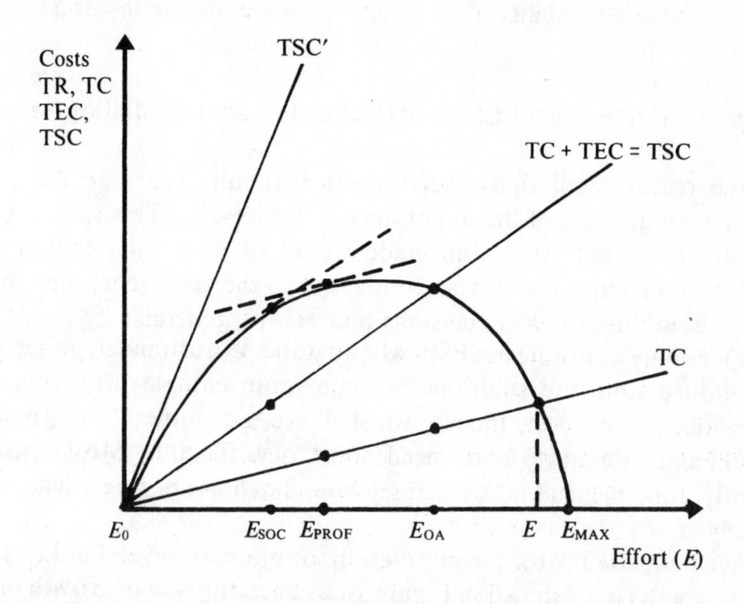

Figure 16.8 Possible social optima for renewable resource exploitation.

Now, Figure 16.6 shows stock on the horizontal axis. It will be recalled that when transferred to the effort–harvest diagram we get mirror-images of the curves. So, Figure 16.7 shows the total cost/total revenue diagram with preservation value added. It will be seen that the preservation values behave just like any other external cost functions. We can add the TC and TEC (total external cost) curves together to get total social cost (TSC). If the benefits of exploiting the resource are fully captured by the revenue, then the *social optimum* is given by the maximum difference between TR and TSC. As shown, this could be at effort level E_{SOC} in Figure 16.7. For illustration, we also show the case where TSC lies everywhere above the TR curve: the resource should not be exploited at all and the optimal solution is at E_0, which corresponds to X_{MAX} in Figure 16.6. The addition of externalities to the analysis therefore suggests the following:

1. The optimal stock will be higher when the aim is net benefit maximisation compared to simple profit maximisation.
2. If external costs are very large, the resource will be optimally 'managed' if it is left alone to reach its natural equilibrium.
3. Introducing social costs still confers no particular attraction on the social desirability of stock levels corresponding to MSY.

16.6 A MORE COMPLETE MODEL: INTRODUCING TIME

Some readers will find what follows difficult. We urge them to persevere because of the importance of the results. This said, a full treatment would reveal the inadequacies of even this amount of 'dynamics'. Thus, those who do make it to the end are warned that they should not draw conclusions that are too general.

Our analysis to date has been wholly static. Unfortunately, when we introduce time the solutions become more complex still, but of considerable interest. Indeed, we shall see that some of the general statements we have made need some qualification. Most importantly, time means that we can see how the *discount rate* affects our analysis.

We have used X for the population, or biomass, level. Let $F(X)$ be the growth curve shown in Figure 16.2. Then the rate of growth of X over time is dX/dt (t is time), and is equal to

$$\frac{dX}{dt} = F(X) - H(t) \tag{16.6}$$

where $H(t)$ is the rate of harvesting. That is, the rate of growth of the population is equal to its natural growth rate $F(X)$ minus its rate of harvest.

We now introduce a production function:

$$H = Q(E,X) \tag{16.7}$$

which says that the harvest rate is a function of the level of effort and the population level. More specifically, we give (16.7) a Cobb–Douglas form as

$$H = AE^{a}X^{b} \quad \text{where } a + b > 1 \tag{16.8}$$

For simplicity, we set $a = 1$ and we write $AX^{b} = G(X)$ so that

$$H = EG(X) \tag{16.9}$$

or

$$E = \frac{H}{G(X)} \qquad (16.10)$$

Effort, E, produces a profit π equal to

$$\pi = TR - TC$$

or

$$\pi = PH - CE \qquad (16.11)$$

Putting (16.10) into (16.11) gives us

$$\pi = PH - \frac{CH}{G(X)} \qquad (16.12)$$

Now let $C/G(X) = C(X)$, and we have

$$\pi = PH - C(X)H \qquad (16.13)$$

Given (16.13) we are now in a position to re-introduce the objective of maximising profits for a single owner. We shall *assume*, along with most of the literature, that the aim of a single owner, or a social agency controlling the resource, is to *maximise the present (discounted) value of profits.*

We thus write the profit maximisation objective as:

$$\text{Maximise} \quad PV(\pi) = \int_0^\infty [P - C(X)]He^{-st}dt \qquad (16.14)$$

where s is the relevant discount rate. From (16.6) we know that $H(t) = F(X) - dX/dt$. Writing dX/dt as $\dot{X}$, we can substitute $H(t)$ in (16.14), to give:

$$PV(\pi) = \int_0^\infty [P - C(X)][F(X) - \dot{X}]e^{-st}\,dt \qquad (16.15)$$

Now, deriving the solution to this equation involves some maths which we do not elaborate on here. We simply assert that the solution is:

$$\frac{dF}{dX} - \frac{dC/dX\,F(X)}{P - C(X)} = s \qquad (16.16)$$

or

$$F'(X) - \frac{C'(X)F(X)}{P - CX)} = s \tag{16.17}$$

16.7 INTERPRETING THE FUNDAMENTAL RULE OF RENEWABLE RESOURCE EXPLOITATION

Equation (16.17) tells us the conditions needed to maximise the present value of profits from the resource *whilst simultaneously taking a harvest equal to the rate of reproduction of the resource.* It is in this sense that the solution also provides for a 'steady' or 'stationary' state. (Note that, from our previous discussion, there is no unique rate of reproduction: it depends on the stock.) We shall now suggest several approaches to an intuitive understanding of equation (16.17). *Note that (16.17) assumes the resource price is constant.* In some of our interpretations we shall relax this assumption.

Marginal rules

In our first attempt at an intuitive understanding of equation (16.17) we rearrange it as follows:

$$F'(X)\,[P - C(X)] - C'(X)\,F(X) = s\,[P - C(X)] \tag{16.18}$$

Now if we differentiate the expression

$$[P - C(X)]\,F(X)$$

with respect to X we obtain

$$\frac{d}{dX}\left\{[P{-}C(X)]F(X)\right\} = [P{-}C(X)]\,F'(X) - F(X)\,C'(X)$$

the right-hand side of which is the left-hand side of equation (16.18). (The differentiation above uses the product rule.) Hence (16.18) can be rewritten as:

$$\frac{d}{dX}\left\{[P{-}C(X)]F(X)\right\} = s\,[P{-}C(X)] \tag{16.19}$$

In a stationary state the rate of harvest H(t) equals the rate of reproducton F(X), so that F(X) in (16.9) can be replaced with H(t). The expression $\{P - C(X)\}\,H(t)$ can then be interpreted as the level of *rent* or *profit* that would be sustained at population level X. If we write this sustainable rent as R, (16.19) can be rewritten as

$$\frac{d}{dX}[R(X)] = s\,[P - C(X)] \tag{16.20}$$

or

$$\frac{1}{s} \cdot \frac{dR}{dX} = P - C(X) \tag{16.21}$$

Consider what happens if we reduce the stock of the resource by harvesting a small amount. There will be an immediate gain of $P - C(X)$, the right-hand side of (16.21). but we shall incur a loss of future sustainable rent of dR/dX, the present value of which (over an infinite time horizon) is the left-hand side of (16.21). This is because we lose the offspring of the marginal stock. Thus (16.21) restates the fundamental rule of the optimal use of renewable resources to say that *the immediate marginal gain from an increase in the current harvest of the resource must be equal to the present value of the future losses in rental brought about by that change.*

An alternative look at the marginal rules

Look at the equation (16.17) again. $C'(X)F(X)$ measures the increase in future costs of harvesting due to a reduction in the stock now brought about by an increase in current period harvesting. (And vice versa – if we relax the current harvest and increase the stock, this will reduce future costs.) Hence $C'(X)F(X)$ is the 'marginal stock effect' or 'direct welfare effect' of the stock. We denote this direct welfare effect of the stock as $U'(X) = dU/dX$, where U refers to utility. But $P - C(X)$ is the net (utility) gain of consuming now. (We write this as $U'(H) = dU/dH$ where H is harvest, or consumption.) Hence (16.17) becomes:

$$F'(X) + \frac{U'(X)}{U'(H)} = s \tag{16.22}$$

Note the plus sign in (16.22). $C'(X) < 0$ because a stock reduction now increases future costs. Hence the middle expression in (16.17) is negative.

Equation (16.22) is the 'Ramsey rule' which states that the *net* rate of return on an asset should be equal to the rate of discount. The net return is greater than the 'own' marginal product $F'(X)$ if $U'(X) > 0$ and less if $U'(X) < 0$. $U'(X)$ will be positive for renewable resources and negative for the stock of 'bads' such as pollution.

The renewable resource rule when prices change

Equation (16.17) assumes the price of the harvested resource is 'given' (it is 'parametric'). If we now let prices be a function of time, we shall have $P = \mathrm{P}(t)$. (16.17) is altered when this is the case to:

$$\mathrm{F}'(X) - \mathrm{C}'(X)\mathrm{F}(X) = s - \frac{\dot{P}}{P - \mathrm{C}(X)} \tag{16.23}$$

The most obvious reason for prices changing with time is that demand goes up or down, other things remaining the same.

The intuitive meaning of (16.23) is best captured by assuming that harvesting is costless, i.e. let $\mathrm{C}(X) = 0$, and hence $\mathrm{C}'(X) = 0$. Then (16.23) becomes

$$\mathrm{F}'(X) = s - \frac{\dot{P}}{P}$$

or

$$\mathrm{F}'(X) + \frac{\dot{P}}{P} = s \tag{16.24}$$

This equation is important and will be used several times. Even if you have found these mathematical sections difficult you are urged to understand this result as best as you can.

Equation (16.24) offers the clearest intuitive understanding of the renewable resource extraction rule. Basically, $\mathrm{F}'(X)$ is equivalent to the marginal productivity of the resource – i.e. the natural rate of growth of the stock is $\mathrm{F}(X)$, which is the marginal productivity of the resource. The second expression is the rate of increase in the (real) price of the harvested resource. Thus by consuming the resource later rather than sooner, we know that the resource owner can reap a capital gain through the price increases that will occur. *Equation (16.24) then says that the marginal productivity of the resource plus the growth of the marginal capital gain should equal the discount rate.* As long as the value of the asset is growing faster than the discount rate, it pays to leave the asset 'in the ground' (or 'in the sea' depending on the type of asset under consideration). This 'leaving alone' of the asset is investment. It seems odd to think of it this way, but it should be possible to see the sense of it when people talk about 'investing in trees' – they mean that the rate of appreciation of their asset (trees) exceeds their discount rate. Thus we have the sense in which investment in resources means leaving the resource alone and waiting for its capital gains. In the renewable resource case this means waiting for the gains both from price rises and the natural rate of growth of the asset.

We have two general rules for renewable resources. The value of X that solves equation (16.24) is the optimal stock. If we write that as X_{OPT} our basic rules are as follows:

1. If the initial stock $X_{INIT} < X_{OPT}$, then invest in the resource by letting it grow and secure capital gains as well.
2. If the initial stock $X_{INIT} > X_{OPT}$, then disinvest in the resource, i.e. extract it until the stock approaches X_{OPT}.

The speed with which one should adjust to these two disequilibrium situations is the subject of some of the natural resource literature. We do not dwell on it here.

16.8 IMPLICATIONS OF THE MORE COMPLETE MODEL

What lessons can be learned from the previous section? We list below some of the interesting results. Again, it must be borne in mind that even the model we have used severely understates true life complexity.

1. From equation (16.21) we can see that if $s = 0$, $\mathrm{d}R/\mathrm{d}X = 0$, so that sustainable rent is maximised. If the discount rate is zero, then, we secure maximum sustainable rent: very simply, any future gain from a current reduction in harvest lasts forever and, since the resource owner is indifferent between present and future ($s = 0$), a current sacrifice is always worthwhile.
2. In the same vein, if $s \rightarrow \infty$, rent tends to zero. Operating with an infinite discount rate is thus analogous to the open-access solution where rent is dissipated.
3. What happens if s is positive but not infinite? If $\mathrm{d}R/\mathrm{d}X$ can be assumed to decline as X increases, then Figure 16.8 shows how the optimal population size X^* is determined. Since $C(X)$ will *decrease* with X (the bigger the stock the 'easier', and hence cheaper, it is to harvest the resource), the function $s[P - C(X)]$ will *increase* with X. From equation (16.21) we see that the intersection of the two functions in Figure 16.8 will determine X^*. Moreover, we can see how X^* varies with the various parameters. The results are, other things being equal:

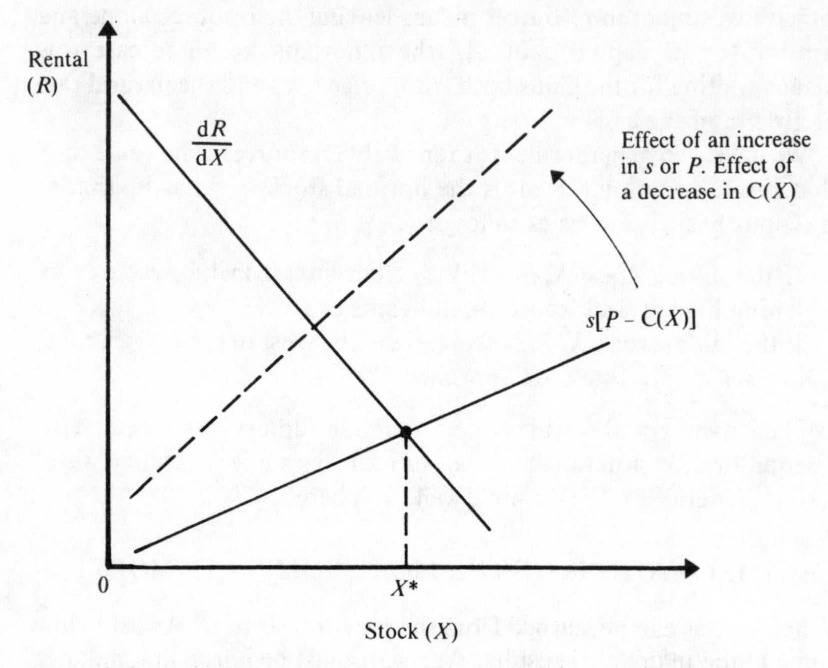

Figure 16.9 Optimal resource stocks: discount rates, costs and prices.

- the optimal stock is lower, the higher the discount rate
- the optimal stock is lower, the lower is cost per unit harvest
- the optimal stock is lower, the higher the unit price

4. Equation (16.17) states that

$$F'(X) - \frac{C'(X)F(X)}{P - C(X)} = s$$

Hence if $C'(X) = 0$, i.e. costs are unrelated to stock size, we have $F'(X) = s$. What is $F'(X)$? It is in fact the rate of change of $F(X)$ where $F(X)$ is as depicted in Figure 16.2. For any given stock, X_0 say, $F(X_0)$ tells us the addition to the stock in the next period. (See Figure 16.9.) If we consider a slightly larger stock, X_1 say, $F(X_1)$ gives us the addition to the stock given that the initial stock is X_1.

It follows that, if X_1 is just slightly bigger than X_0, the increase in the stock from X_0 to X_1 is associated with an increase in the addition to the stock of $F(X_1) - F(X_0)$, or $F'(X)$. $F'(X)$ is thus the percentage change in the population size per period. It is the resource's *own* rate of return, or its marginal product, as we have already seen.

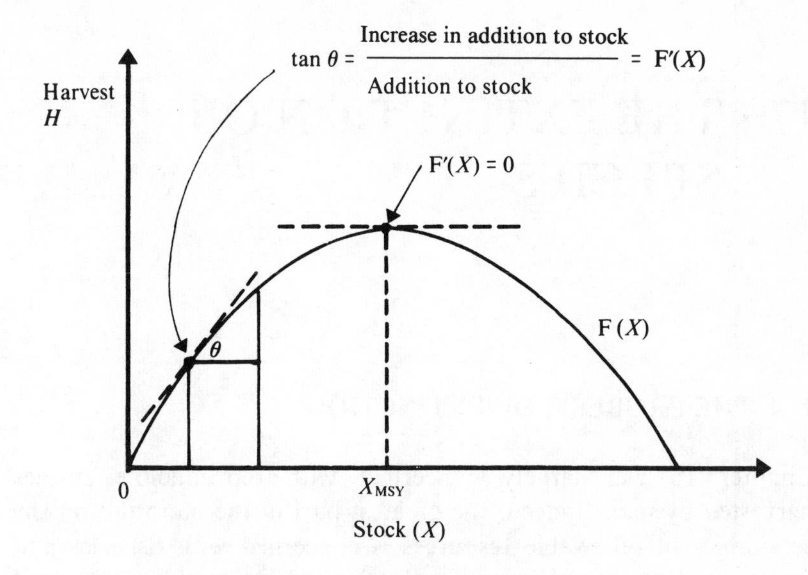

Figure 16.10 Interpreting F′(X).

Thus, equation (16.17) tells us that, *if costs are unrelated to stock size, the optimal stock is reached when the own rate of return equals the discount rate.* This immediately permits some further observations. First $F'(X) = 0$ when maximum sustainable yield is achieved (see Figure 16.9), so that MSY turns out to be an optimal policy if (a) costs are unrelated to stock size and (b) the discount rate is zero. Second, if costs *are* related to stock size, the own rate of return must be *less* than the discount rate (since $C'(X)$ is negative). Note that when prices are *not* constant, this rule about the rate of return is still preserved except that we add capital gains to $F'(X)$.

5. Returning to the case where $C'(X) = 0$, we observed that optimality requires $s = F'(X)$ for the case when prices are constant. What happens if the discount rate is held *above* the marginal productivity level? Then, like any investment, it pays to secure, as quickly as possible, the revenues from the resource and transfer them to some other investment yielding s, as the resource stock declines its marginal productivity should rise. But if s persistently lies above $F'(X)$ then the 'switching' could continue and the resource could be eliminated. *Hence the use of discount rates in excess of own rates of return on a renewable resource will have the tendency of eliminating species.*

17 · THE EXTINCTION OF SPECIES

17.1 THE PROBLEM OF EXTINCTION

Chapter 16 was largely concerned with reproducible species harvested by man. Indeed, the greatest part of the literature on the economics of renewable resources is concerned with fisheries and forests. The danger of species extinction applies to these harvested resources: we saw that any resource with a significant minimum critical size faces a real prospect of extinction, particularly if the resource is subject to 'open-access' harvesting. Second, any resource will risk extinction if the 'steady-state' condition is not met – i.e. if the rate of harvest exceeds the rate of reproduction. The fact that such an operational rule damages long-run profits is not a guarantee against myopia or plain ignorance of a resource's reproductive capacity. Third, high discount rates threaten resources, particularly slow-reproducing species such as whales.

This chapter investigates these phenomena in a little more detail. But we also need to widen the concern to species other than fish, whales and commercial forests. Some experts estimate that the world as a whole loses perhaps 1,000 species every year, and conjecture that this loss rate could rise to 10,000 per year by 1990. Such estimates are disputed, but not to the point where anyone would argue that species extinction is proceeding at an unprecedented rate. If the figures are correct, perhaps 1 million species will disappear by the year 2,000, out of a global total of 5–10 million – i.e. a staggering loss of 10–20 per cent. Moreover, there is no conceivable way that we can suggest such losses are consistent with evolutionary trends. Species extinction occurs because of (a) non-steady state exploitation and (b) habitat destruction or modification. Of the two causes the latter is by far the more important and includes drainage of wetlands, destruction of tropical moist forests, flooding of wilderness areas, the effects of pollution on wildlife and the introduction of 'exotic' species into hitherto ecologically stable environments.

Table 17.1 shows some illustrative statistics of endangered species. Table 17.2 shows what has happened to whale catches since 1920 and reveals the obvious truth that man as predator has been directly responsible for the dramatic decline in stocks of the blue, humpback, fin and sei whale.

Table 17.1 World population of selected, endangered and threatened species, 1947–83.

Year	African elephant	Black rhino[a]	Bengal tiger[b]	Kemp's (Atlantic) ridley sea turtle[c]	Mountain gorilla	Golden lion marmoset	Mauritius kestrel
1947				40,000			
1960					450		
1968						600	
1969		90,000					
1970	5,000,000					400	
1972			1,827				
1973							6
1975						250	
1977			2,484				13
1978					200		19
1979	1,300,000	15,000		750			
1982	1,300,000	15,000	2,000		300	200	15
1983	1,300,000	15,000	2,000	1,500–1,600	300	200	

Source: (US) Council on Environmental Quality, *Environmental Quality 1983*, Washington DC.
[a]Includes those in Kenya and Tanzania only.
[b]Includes those in India only.
[c]Refers to nesting females.

We have conclusive evidence that species are being made extinct and that mankind is responsible. But so what? We could argue that mankind is, in some sense, the superior being on earth and that other species are there to serve man's objectives. If the fulfilment of those objectives involves species loss this is not a cause for concern. Many people do think this way and if the 'horizon of concern' is short, there

Table 17.2 Whale abundance and catch, by species, 1920–80. (Total is greater than the sum of species listed individually. Brydes and others were omitted.)

Whale abundance	Blue	Humpback	Fin	Sei	Sperm	Gray	Minke	Total
Virgin stock (thousands)	215	50	448	200	922	11	361	220
1970s stock (thousands)	13	7	101	76	641	11	325	1,174
Percent of virgin stock remaining	6	14	22	38	69	100	90	53
World catch, by year								
1920	2,274	545	4,946	1,120	749	NA	NA	9,634
1925	7,548	3,342	9,121	1,093	1,439	NA	NA	22,543
1930	19,079	1,919	14,281	841	1,126	NA	NA	37,246
1935	16,834	4,088	14,078	962	2,238	NA	NA	38,200
1940	11,559	528	19,924	541	4,091	NA	NA	36,643
1945	1,111	303	2,653	218	1,661	NA	NA	5,946
1950	6,313	5,063	22,902	2,471	8,183	NA	NA	44,932
1955	2,495	2,713	32,185	1,940	15,594	NA	NA	54,927
1960	1,465	3,576	31,064	7,035	20,344	NA	NA	63,484
1965	613	452	12,351	25,454	25,548	NA	NA	64,418
1970	0	0	5,057	11,195	25,842	NA	4,539	46,633
1975	0	17	1,634	4,975	21,045	NA	11,221	38,892
1976	0	11	785	1,866	17,134	NA	10,176	29,972
1977	0	14	155	2,179	12,279	NA	12,398	27,025
1978	0	32	650	634	10,274	NA	9,018	20,608
1979	0	19	743	150	8,536	NA	9,900	20,449
1980	0	16	472	102	2,091	NA	11,709	15,129

Source: (US) Council on Environmental Quality, *Environmental Quality 1983*, Washington, DC.
NA, Not available.

may be little that can be done to convince them that species extinction does matter.

Yet there are good reasons for concern even in terms of a man-centred ethic that confers no rights on other species. We list below just some of the relevant arguments.

1. For many people other species generate direct welfare benefits. This is true of birds, marine life, insects and so on. The enormous membership of ornithological societies, nature trusts and so on testifies to this. Even more telling is the popularity of television programmes on wildlife. Whereas the ornithologist gains direct experience, watching television programmes is 'indirect' but indicative of *option value* (see Chapter 9). While nature societies can express market preferences in favour of preservation, e.g. by buying land and retaining it as wilderness, there is no market in option values. Hence these values go unrecorded and extinction can readily occur because of this.

2. Many present-day drugs are derived directly from wild plants. Even those made synthetically are often based on 'blueprints' supplied by plants. Reserpine from the plant *Rauwolfia serpentina* is used as a tranquilliser. Digitalis, used for heart disorders, comes from *Digitalis purpurea*. Another heart drug is digoxin from *Digitalis lanata*. Two species of yam from forests in Mexico and Guatemala gave us the basic materials for cortisone and the first oral contraceptives. Prescriptions based on just these five species run into hundreds of millions every year.

It seems likely that hardly any of the major plant-based drugs used in Western medicine are at risk because of threats to the relevant habitats. The issue at stake is what value has yet to accrue from plants which *are* endangered but which have yet to be 'screened' for medicinal properties. Brief appraisals exist for only one plant in ten. Detailed assessments exist for only one in a hundred. Yet once lost they cannot be recreated: the loss is irreversible. Extinction means the irreversible loss of probable medicinal benefits.

3. Wild plants are critical in terms of *genetic diversity*. In turn, such diversity is critical for plant-based foods. Disease resistance in wheat, for example, depends on the repeated introduction of new varieties combining disease-resistant genetic types. Stripe rust in American wheat was defeated in the 1960s using germ plasm from a wild wheat found in Turkey. An Ethiopian barley gene was used to

protect some American barley crops against the yellow dwarf virus. And the same story can be told for rice, potatoes, coffee, tomatoes and other plants. But not only do wild varieties assist in disease resistance, they are in continuous use to improve yields regardless of disease reduction. The moral is that reduced species mean reduced diversity and this in turn means a lower probability of improved crop productivity and a higher probability of vulnerability to disease.

4. Living species also serve 'life-support' functions for mankind. This is especially true of forests which affect climate and ground water supplies and which clean the air through natural sulphur 'fixing' activities. Some of these functions are understood and some, no doubt, can be substituted. Airborne sulphur, for example, can be removed at source, at a cost. But many of the life-support functions are not understood. It may be possible to eliminate many species and their habitats with no life-support threat to mankind. The problem is that we do not know how far this process can go. In a world of uncertainty it is not rational to behave as if we do know and as if there is no limit to the process.

5. Living species also serve scientific purposes, giving us clues to evolutionary processes, the links between habitats and life-forms and so on.

None of the foregoing reasons require us to protect species for other than purely utilitarian reasons. Each of the reasons provides us with the evidence for supposing that, if markets existed in the option to preserve species, the price of such options would be positive. Moreover, there is an urgency about the problem of extinction, for the losses being incurred are irreversible: it is not a matter of regretting a loss and restoring it. The regret is perpetual. The manner in which this combination of irreversibility and uncertainty can be handled is the subject of the rest of this chapter.

17.2 OPEN ACCESS AND SPECIES EXTINCTION

We saw in Chapter 16 that open access significantly increases the risk of species extinction by resulting in lower stocks than would occur under sole ownership.

One widely-based equation to describe population growth $F(X)$ is the logistic equation or Verhulst model. This takes the form

$$\frac{dX}{dt} = F(X) = rX\,(1 - \frac{X}{K}) \tag{17.1}$$

where K is the carrying capacity and r is the net proportional growth rate of population. Deducting the harvest rate and making the harvest rate equal to the population growth (the steady state) gives

$$\frac{dX}{dt} = rX(1 - \frac{X}{K}) - EX = 0 \tag{17.2}$$

This logistic equation describes the *biological* component of a model of reproduction resource use. If we postulate open-access then Chapter 16 suggests that the *economic* component of the model is given by

$$\text{TR} - \text{TC} = \text{PEX} - \text{CE} = 0 \tag{17.3}$$

From (17.3) we see immediately that

$$X^* = \frac{\text{C}}{\text{P}} \tag{17.4}$$

where X^* is the equilibrium population. Substituting (17.4) in equation (17.2) gives

$$E = r\,(1 - \frac{C}{PK}) \tag{17.5}$$

From (17.5) we see that if $C > PK$, then $E < 0$, i.e. the resource is not exploited at all. High costs of exploitation (fairly obviously) preserve the resource. But, conversely, low costs are likely to make the resources extinct. In the limit, for example, equation (17.4) tells us that if $C = 0$, then $X = 0$, the population is zero. In general, *the lower is the ratio of costs to price* (C/P) *the lower is the population size under open access conditions.* Moreover, if price exceeds cost at very low levels of population, extinction is a distinct possibility in open access conditions.

Thus, while it is correct to say that open-access conditions are consistent with positive stocks (the result obtained in Chapter 16), it is easy to see how those same conditions expose the population to a greater risk of extinction than would otherwise be the case. The 'tragedy of the commons' is not at all a necessary result of common ownership or common access, as Chapter 16 showed, but 'commonality', is a potentially dangerous condition for the living species of commons.

17.3 PROFIT MAXIMISATION AND EXTINCTION

Now consider the case where a sole owner seeks to maximise the present value of profits. Chapter 16 gave the solution as:

$$\mathrm{F}'(X) - \frac{\mathrm{C}'(X)\,\mathrm{F}(X)}{P - \mathrm{C}(X)} = s \tag{17.6}$$

If $\mathrm{C}'(X) = 0$, i.e. unit harvesting costs do not depend on stock size, the solution becomes

$$\mathrm{F}'(X) = s \tag{17.7}$$

i.e. the growth rate of the stock should equal the discount rate. As we saw in Chapter 16, if $s > \mathrm{F}'(X)$ the owner of the resource has the incentive to reduce the stock to zero since the 'own' rate of return on the asset is less than the opportunity cost of capital. In this sense of profit maximisation, extinction is 'optimal'. In the absence of any externalities (e.g. non-market valuations of the stock which are ignored by resource users) extinction also appears to be *socially* optimal. That such a result tends to be greeted with disbelief by non-economists (and many economists) suggests that externalities *are* important.

But Chapter 16 also showed that where costs are dependent on stock size, equation (17.6) can be written

$$\mathrm{F}'(X) + \frac{\mathrm{U}'(X)}{\mathrm{U}'(Y)} = s \tag{17.8}$$

where the left hand side of the equation is now the *net* rate of return on the asset. For a replenishable resource $\mathrm{U}'(X) > 0$, i.e. an increase in the stock now reduces future cost. Equation (17.8) thus effectively incorporates *one* form of externality – an externality through time. If the future cost-reductions benefit later generations we have an example of an *intergenerational externality*. Notice that for extinction to be optimal in these circumstances, the discount rate must exceed *the net rate of return* (and, as it happens, price must be above cost as well).

While Chapter 16 suggested that sole ownership and profit maximisation would preserve species more than is the case under open access, we see that the effect of introducing positive discount rates is to raise the (distinct) possibility that species can be driven to extinction simply because they are construed as one more capital asset.

17.4 WHY DOES EXTINCTION OCCUR?

The preceding discussion provides us with the basic clues as to why species are disappearing at a highly significant rate.

First, many species can be 'harvested' at extremely low cost. The elephant is an obvious example. Unless closely protected unlawful poaching takes place. Indeed some 80 per cent of the world's supply of ivory comes from illegal poaching. Price remains positive as long as there is a demand for ivory and, indeed, prices can increase in real terms through time as incomes increase and 'supplies' fall. The condition that $P > C(X)$ when X is very low is fulfilled.

Second, the discount rates of hunters and poachers tend to be high. The condition that $s > F'(X)$ therefore tends also to be met: hunters and poachers have no incentive to curb their killing rate to preserve species for future culling. The first and second reasons combined are, as we have seen, powerful incentives for extinction.

Third, common-property and open-access conditions increase the probability of extinction and it is no accident that experts suggest that the majority of endangered species are in tropical moist forests where open access *de facto* describes the situation. Three further features of these conditions are important:

1. The 'harvesting' of one species can lead to the extinction of a *separate* species that happens to be the incidental victim of the deliberate harvesting policy. In effect, there is 'joint production' as far as harvesting effort is concerned.
2. For a very large number of species, market, or 'perceived' *price* is zero or on the basis of the rules derived previously, near zero. While zero prices *should* dictate conservation (no one should be bothered to kill them) what happens is that their *habitat* has positive value. A tropical forest may therefore be readily cut down because of the value of wood or the value of land for development. With the habitat gone the species goes too. The (apparent) absence of a species preservation price means that

habitat preservation also has a zero price. The development option 'wins'. As suggested previously, this outcome may be due to the absence of markets in preservation values.

3. As some 'species are eliminated' so the predators which require these prey as food are eliminated.

There is an interesting variant on the 'zero price – habitat destruction' example; the Californian condor is very much at a critical minimum population size (some forty pairs in 1980). The social costs of *preservation* largely comprise oil and phosphates that cannot be developed because the condor is protected. Yet the oil tends to have a rising real value through time: its rate of real price increase could even exceed the relevant discount rate. If so, the condor is doing the energy consumer a favour by keeping the oil in the ground. To be fair, one suspects that oil company discount rates exceed real oil price increases and that, anyway, conservationists will have longer time horizons than oil companies – the 'oil conservation' argument only works for fairly short time horizons: after that, oil companies lose patience!

The fourth point has been dealt with in Chapter 16. The 'prices' referred to in the preceding paragraphs are the prices to the exploiter of the resource. They neglect the price of the resource to anyone wishing to *preserve* it. We cannot *add* to the two prices – one is a price for harvesting the resource, the other is a price for *not* harvesting it. Essentially we have a problem of *externality*: the price received by the whaler, for example, fails to reflect the negative price, the cost, to those who wish to preserve whales. What most of the case studies of extinct and endangered species show us is that we have a *conflict of values*. In essence, it is a conflict between '*development' and preservation*. We can encompass many things by the term 'development' – it can be the destruction or modification of habitats for agriculture, industrial development, energy (especially hydroelectricity), commercial forestry and social infrastructure (roads, housing, etc.) or the exploitation of a resource for commercial purposes (whale oil, whale meat, fisheries, forests).

18 · EXHAUSTIBLE RESOURCES

18.1 INTRODUCTION

Chapter 16 derived the basic conditions of the optimal use of a renewable resource. We noted there that if the harvest was always constrained to be less than, or equal to, the sustainable yield, the stock of the resource would not be depleted. Thus we spoke of the optimal rate of *use* of the resource. With exhaustible resources, on the other hand, we cannot speak of sustainable yield. The resource will be depleted so long as the use rate, the 'harvest', is positive. We therefore need to rephrase the question in this context: we need to know the optimal rate at which to *deplete* the resource.

18.2 THE FUNDAMENTAL PRINCIPLE OF EXHAUSTIBLE RESOURCE USE

Chapter 16 showed that the basic rule for optimally utilising a renewable resource with costless extraction was:

$$F'(X) + \frac{\dot{P}}{P} = s \tag{18.1}$$

For an exhaustible resource there is no growth function, the resource has a fixed size. Hence $F'(X) = 0$. Modifying equation (18.1) accordingly, we get:

$$\frac{\dot{P}}{P} = s \tag{18.2}$$

Equation (18.2) is the fundamental principle of the economics of exhaustible resource use. It says that the resource should be depleted in such a way that the rate of growth of price of the extracted resource should equal the rate of discount. Note that this rule applies in only the most simple of cases: for example, we have assumed zero costs of extraction so that the price of the extracted resource is the same as the price of the resource 'in the ground'. Some of the other simplifications will become evident as we proceed.

The fundamental equation (18.2) is known as the *Hotelling rule* after an important demonstration by Harold Hotelling in 1931. An alternative variant, which says the same thing, is:

$$P_t = P_0 e^{st} \tag{18.3}$$

That is, the price of the resource in any period t is equal to the price in some initial period (0) compounded at the rate s, the discount rate. The owner of the resource should be indifferent between a unit of resource at P_0 now and the same unit at P_0e^{st} in t years time. This underlines one of the features observed in Chapter 16: natural resource economics treats resources 'in the ground' as capital assets. By leaving resources in the ground (preserving them) the resource owner can expect capital gains as the resource price rises through time. The owner will be indifferent between holding the resource in the ground and extracting it if the rate of capital gain ($\dot{P} / P$) equals the rate of interest (s) on alternative assets.

Now, so long as we maintain the unrealistic assumption of costless extraction, the price in the ground is the same as the price of the extracted resource (the 'wellhead' price as it is often known). Once extraction costs are positive, however, the two prices differ. The wellhead price is P, but the price in the ground is $P - \mathrm{C}(X)$ since it costs $\mathrm{C}(X)$ to extract it. The price in the ground is better referred to as the *royalty* (a term that derives from sovereigns' rights to property in the ground).

We need to see what happens to the Hotelling rule when extraction costs are positive. For convenience, let these costs $\mathrm{C}(X) = C$, i.e. extraction costs are constant. Recall the full version of the basic equation for optimal renewable resource use given in Chapter 16:

$$\mathrm{F}'(X) - \frac{\mathrm{C}'(X)\mathrm{F}(X)}{P - \mathrm{C}(X)} = s - \frac{\dot{P}}{P - \mathrm{C}(X)} \tag{18.4}$$

Setting $\mathrm{F}(X)$ and $\mathrm{F}'(X) = 0$ and $\mathrm{C}(X) = C$ gives us:

$$\frac{\dot{P}}{P - C} = s \tag{18.5}$$

But we have just seen that $P - C$ is the royalty. Writing this as R, we have

$$\frac{\dot{P}}{R} = s \tag{18.6}$$

which is one formulation of the Hotelling rule when costs are positive (and constant). Equation (18.6) can also be written

$$\frac{\dot{R}}{R} = s \tag{18.7}$$

because the royalty in period $t + 1$ minus the royalty in period t is $\dot{R}$. That is $R_{t+1} - R_t = \dot{R}$ but, the left-hand side of this equivalence can be written $(P_{t+1} - C) - (P_t - C) = \dot{P}/\dot{R}$. Hence $\dot{P} = \dot{R}$. A further important relationship derives from the observation that $P = C + R$. Since C is assumed constant, it is both the average and marginal cost of extraction. In any period, t, we can therefore write:

$$P_t = C_t + R_t \tag{18.8}$$

The price will be equal to marginal extraction cost plus the royalty on the marginal unit of the resource. In the literature R is known as the *royalty* (the term we have been using in this chapter), the *resource rent* (or *rental*), which is the term we sometimes used in Chapter 16, the *depletion premium*, and the *marginal user cost*. The different terms are unfortunate since they add to the potential for confusion. The reader simply has to watch out for the context in which they are used. Using one of the new terms, we can therefore say:

optimal price = marginal extraction cost + marginal user cost

We make use of this equivalence later.

18.3 A DIAGRAMMATIC EXPOSITION OF OPTIMAL RESOURCE USE

We are now in a position to show the Hotelling rules in diagrammatic form. Return to the costless extraction case. We derived the result that

$$P_t = P_{0e}{}^{st}$$

But we did not say how P_0 was determined. P_0 has itself to be an optimal price. Moreover, we need to investigate how long it takes to exhaust the resource. Call the time period in which the resource is exhausted T. We wish to know how to determine P_0 and T. To do this we make use of a diagrammatic approach. Figure 18.1 shows four quadrants. The top right-hand quadrant shows the price path of the

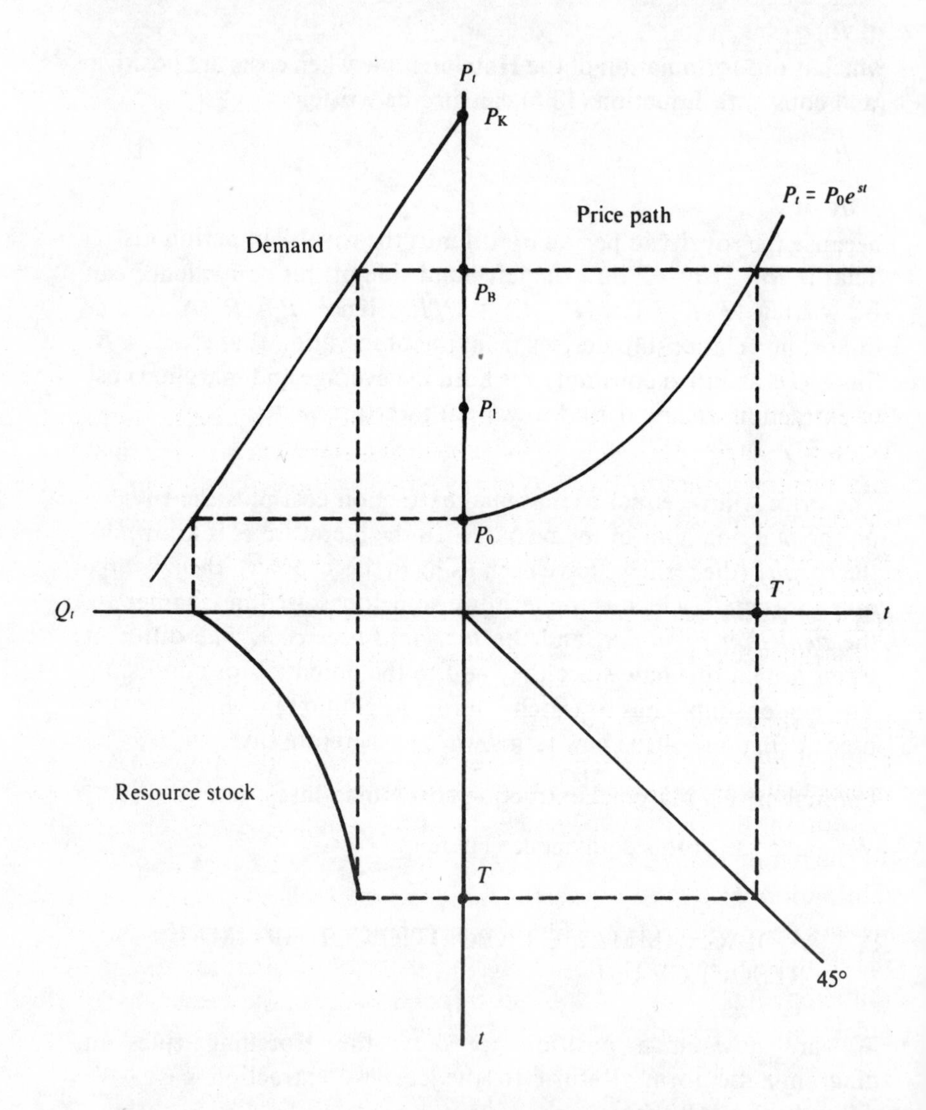

Figure 18.1 The Hotelling price path.

resource against time. The path shown is actually the optimal path and exhaustion occurs at T. The shape of the time path of price is determined by the Hotelling rule, i.e. price rises at the rate of discount (interest). In the top left-hand corner we show the demand for the resource 'back to front', i.e. it is a conventional demand curve but shown in reverse. The higher the price of the resource, the lower is the quantity demanded. The bottom right-hand quadrant is a 'dummy' quadrant: the 45° line simply permits us to transfer a measure of the time period on the downwards vertical axis on to the horizontal rightwards axis. The bottom left-hand quadrant shows the relationship between quantity demanded (and hence extracted), time, and the *cumulative* amount of extraction. The *area* under the curve in this quadrant shows cumulative extraction. Since the stock is finite, we fix the area such that the cumulative production levels equal the stock of the resource. To see that P_0 and T are the 'right' values, consider P_1 and assume it is arbitrarily set as the initial price. We wish to demonstrate that P_1 is not an optimal initial price. Given P_1 as the initial price, price in the next period will be P_1e^s, and for any period t will be P_te^{st}. Thus the path of prices with P_1 as the initial price will be everywhere above the path shown in the top right-hand quadrant. Demand will be less in each period than shown by the curve in the top right-hand quadrant. Cumulative extraction will therefore be less and hence the stock will last longer than T. That P_1 is not an optimal initial price depends on the price P_B which we have not so far considered. P_B is the price of the *backstop technology*. What this means is that as the price of the finite resource rises there will be some substitute for the resource. For example, oil from conventional sources is much less expensive than oil from tar sands or shale. As the price of oil rises it must eventually 'hit' the cost of extracting oil from these more expensive sources. P_B thus serves to 'anchor' the demand, price and production system in Figure 18.1.

The reason that P_1 is not an optimal initial price can now be seen. The path of prices with P_1 as the initial price will 'hit' P_B before the path of price with P_0 as the initial price. This will mean that demand is less in each period with P_1 as the initial price than with P_0 as the initial price. But if demand is less, cumulative production is less and stock will not be exhausted at the time when P_B is reached. Effectively, we shall have run the stock down in such a way that some is left over just at the time when it will become cheaper to switch to an alternative source – the 'backstop' source. Exactly the same reasoning can be applied to any initial price below P_0. This will result in a price path that 'hits' P_B later than T, but it will also result in rates of demand which will exhaust the resource *before* T, and yet the price will not have reached P_B at time T. We will have exhausted the resource only to find that the backstop resource is either not available or is available but at a significant 'jump' in the price. Thus P_0 is the optimal initial price since it permits a price path that will deplete the resource at a rate which 'smoothly' permits the transition from the existing resource to a backstop resource.

The assumption of a backstop technology is not essential to the determination of the price path. So long as the demand curve intercepts the price axis at some finite price (e.g. P_K in Figure 18.1) the analysis can be repeated such that the price path just exhausts the resource and the price in period T equals P_K, the price at which demand falls to zero. In empirical studies, however, it is usual to see the idea of a backstop technology invoked.

18.4 RESOURCE PRICES AND BACKSTOP TECHNOLOGY

Our previous analysis showed us that:

$$R_t = R_0 e^{st} \tag{18.9}$$

This result is easily obtained by re-arranging equation (18.7). It follows then that:

$$R_0 = \frac{R_T}{(1+s)^T} \tag{18.10}$$

where, it will be recalled, T is the time period at which either the backstop technology comes into play, or where demand falls to zero. But R_T is the royalty (rental, user cost) in period T and equation (18.8) shows us that this must be equal to price in period T less cost. Keeping our assumption that costs of extraction are constant, this means

$$R_T = P_{B,T} - C \tag{18.11}$$

i.e. the royalty equals the price in period T, which is equal to the cost of the backstop technology, less cost. Substituting (18.11) in (18.10) gives:

$$R_0 = \frac{P_{B,T} - C}{(1 + s)^T} \qquad (18.12)$$

That is, the royalty, or marginal user cost, in the initial period is equal to the price of the backstop technology minus the cost of extracting the resource, all discounted back to the present. We have thus established a link between the backstop technology price and the royalty in the initial period. More generally, the royalty in any period t will be equal to:

$$R_t = \frac{P_{B,T} - C}{(1 + s)^{T-t}} \qquad (18.13)$$

The time path of price will be as shown in Figure 18.2 and that for quantity is as shown in Figure 18.3.

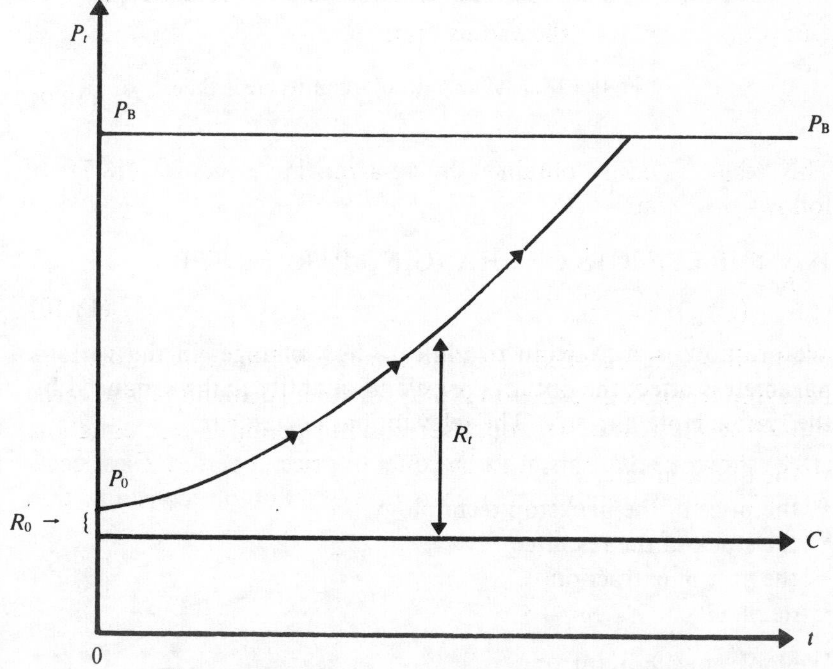

Figure 18.2 The path of resource price over time.

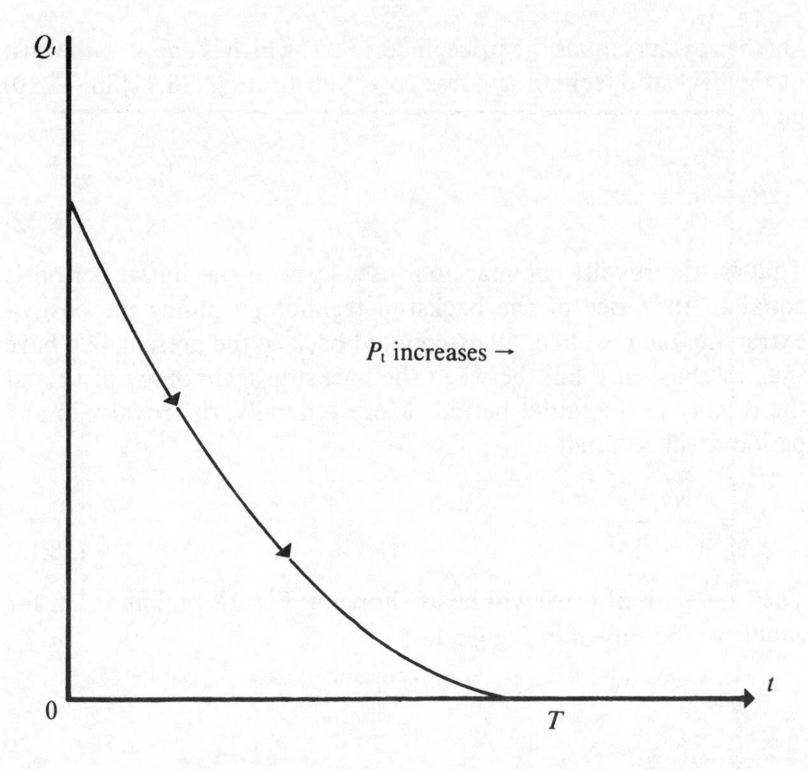

Figure 18.3 The path of quantity over time.

18.5 THE EFFECTS OF CHANGING PARAMETERS

We are now in a position to indicate how changes in the various parameters affect the optimal price and quantity paths produced by the simple Hotelling rule. The relevant parameters are:

- the discount rate, s
- the price of the backstop technology, P_B
- the stock of the resource
- the cost of extraction, C
- demand.

We look at each in turn.

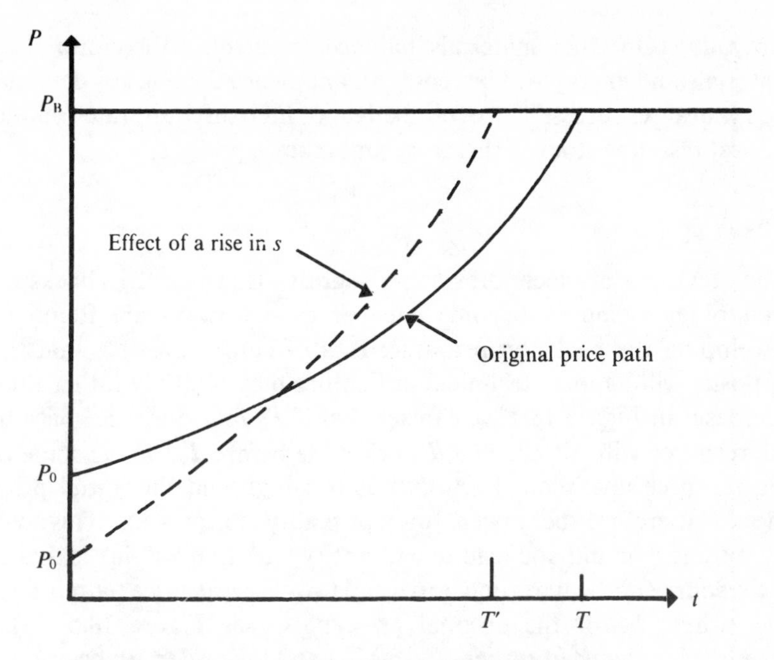

Figure 18.4 The effects of increasing the discount rate

Changing *s*

Since the royalty must grow at the discount rate, an increase in *s* will change the price path. In terms of Figure 18.1, if P_0 remains the same then P_t will be everywhere above the path shown and the limit, P_B will be reached with some of the resource left unexploited. Thus, if the discount rate increases, P_0 cannot be the starting point. That point must lie *below* P_0. The effect is shown in Figure 18.4: initial price is lowered, later prices are raised, and T is reduced to T'. This shows that *higher discount rates tend to mean more rapid exhaustion of the exhaustible resource*. The lower prices in the early periods encourage demand and the higher prices in later periods discourage demand. High discount rates mean that the resource owner wishes to secure the benefits of extraction now rather than later. We have now formalised one of the concerns of the conservationist: positive discounting of the future encourages more rapid exploitation of natural resources.

As we noted in Chapter 14, however, even this result is uncertain. Higher discount rates will tend to discourage capital investment generally. On the materials–balance principle, therefore, less materials and energy will be required, and hence there is less demand for natural resources. The overall effect of high discount rates on the natural resource stock is therefore ambiguous.

Changing P_B

One feature of technological progress is that the backstop technologies tend to become cheaper as new ways are found of developing the backstop resource. It is of course possible that the opposite will happen: technical difficulties may multiply rather than decrease. In Figure 18.1 we can see that if P_B *falls*, the price path of the resource will 'hit' the new P_B price line before T, leaving some of the resource unexploited. P_0 is thus too high and the initial price should therefore be lower for optimality to prevail. This will encourage demand and lead to a quantity path that will just exhaust the resource at the new time horizon. However, the price path is now everywhere *below* the original price path (see Figure 18.5). The resource is exhausted earlier, at time T′, and the price path beginning at P_0' must still rise at the discount rate, s, which is unchanged.

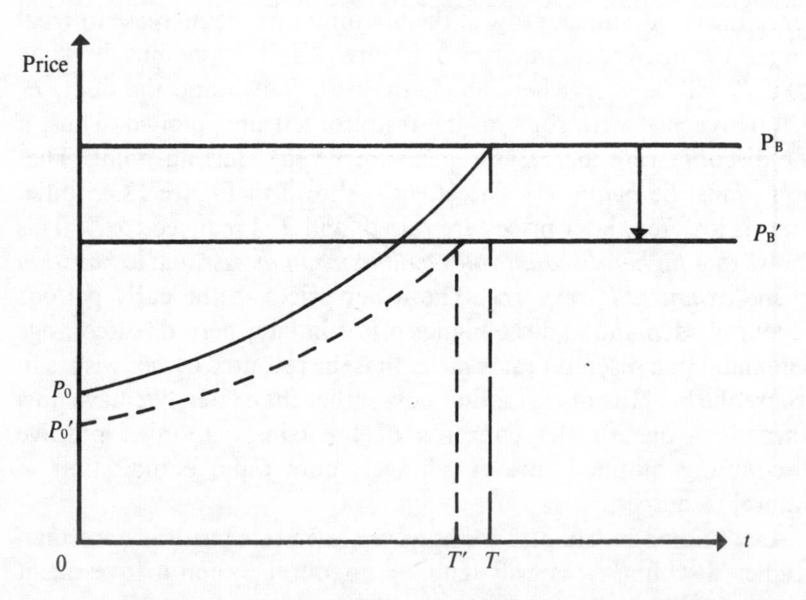

Figure 18.5 The effects of a decline in backstop price.

Changing the resource stock

It may seem odd to consider a change in the size of the stock of an exhaustible resource. In practice, however, estimates of the size of reserves of exhaustible resources such as coal and oil are under constant revision. New seismic technologies mean that discovery rates are improved. New recovery techniques mean that what were once considered known but unrecoverable reserves often become recoverable stocks. Another ingredient in this process is the role which resource prices play. As prices rise, what were hitherto uneconomic stocks become economic to recover. A conspicuous example is the exploration for and exploitation of crude oil in areas such as the North Sea after the OPEC-stimulated price rises of 1973–1974. Geologists often operate with a gradation of reserves: 'proven', 'probable' and 'possible'. *Proven reserves* are what are known to exist and what can be recovered at current prices and costs. *Probable reserves* usually relate to known fields with proven reserves and are a best guess at the additional amounts that could be recovered in those fields. *Possible reserves* usually define reserves available in geological structures adjacent to proven fields. The sum of proven, probable and possible reserves, however, is less than more speculative estimates of what is 'ultimately recoverable'. This larger figure will be geologically determined and tends to take no account of the costs of recovery.

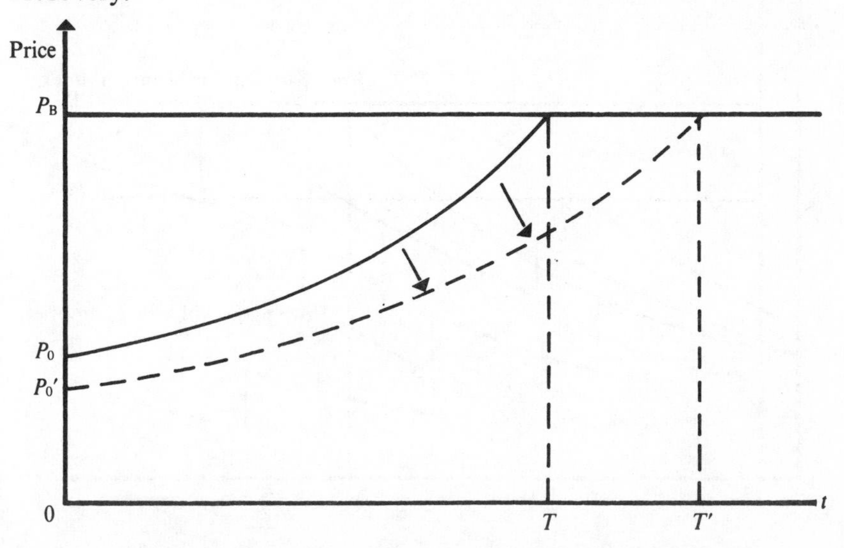

Figure 18.6 Effects of an increase in resource stocks.

Obviously, then, the concept of a 'stock' of an exhaustible resource is rather elastic. How does an increase in resource stocks affect the Hotelling price path? As long as the demand curve itself does not change, the available stock will last longer. In Figure 18.1 the price path beginning at P_0 will leave reserves unutilised. Hence the price must be lower than P_0 in the initial period and the time to exhaustion is extended. The situation is shown in Figure 18.6.

Figure 18.6 offers a useful insight into the path of *actual* resource prices. Chapter 19 provides the empirical evidence which shows that, until recently anyway, resource prices have tended to *decline* over time. This appears to be inconsistent with the Hotelling rule. Figure 18.6 shows that if resource discoveries are made on a frequent basis, we might expect a sequence of connected paths as shown in Figure 18.7, each discontinuity coinciding with a resource discovery. The general *trend* of prices could therefore be downwards, as the lowest curve in Figure 18.7 shows.

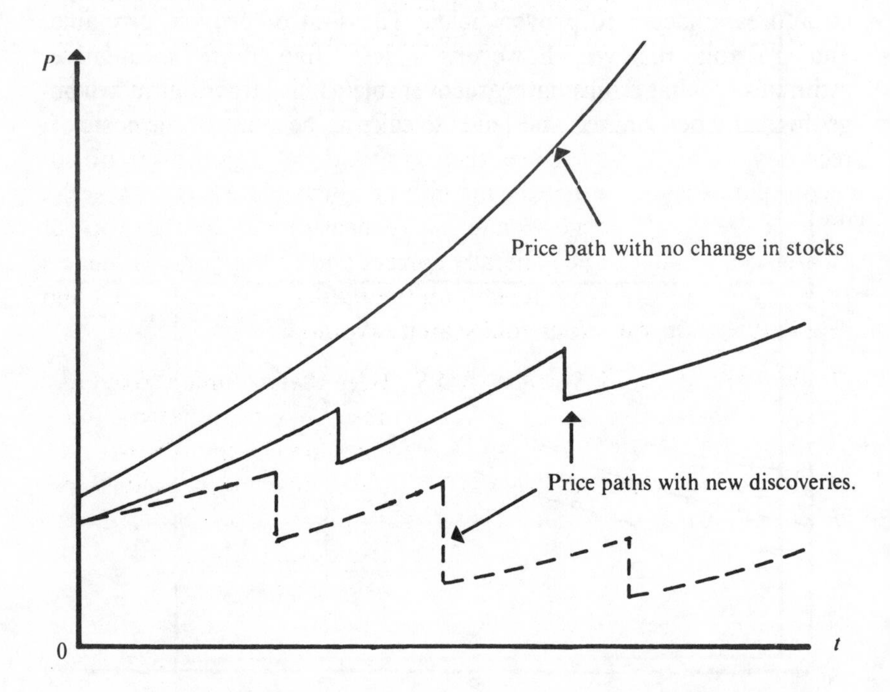

Figure 18.7 Price paths with frequent resource discoveries.

Changing extraction cost, *C*

Figure 18.8 shows what happens when extraction cost falls from C to C'. If C falls and P_0 stays the same, $P_0 - C$ would be the initial royalty and this would grow at the discount rate. This would cause P_B to be reached before the resource is exhausted. So P_0 cannot be the starting point for the new price path. P_0 must fall to P_0', causing more of the resource to be extracted in the earlier periods. The new price path must therefore be steeper than the old one, and the resource is exhausted earlier, at T'. The steeper path may seem odd: the discount rate has not changed. But it is the royalty that rises at the discount rate and the royalty in the initial period has increased – the fall in P_0 is not as great as the fall in C.

Changing demand

In the analysis so far, the demand curve has been static: its position is the same in each period and the changes in quantities demanded in

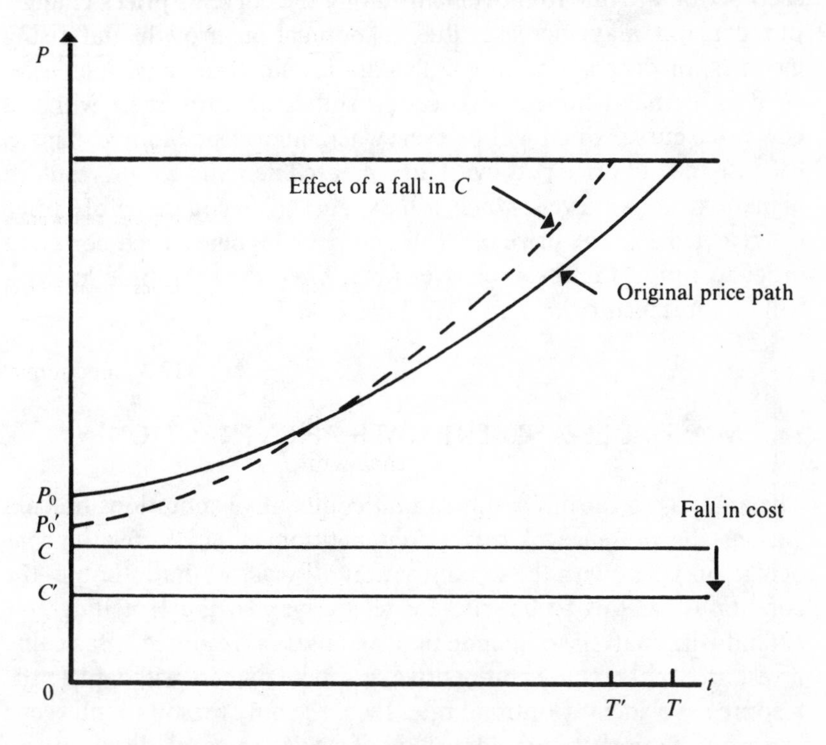

Figure 18.8 The effects of a fall in extraction costs.

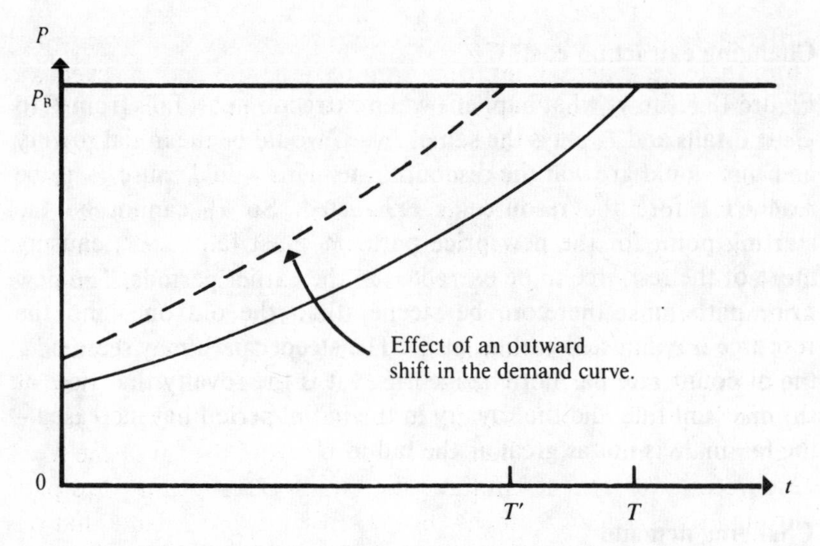

Figure 18.9 The effects of an increase in demand

each period are due to movements along the curve as prices change. But demand may increase, due to population growth and rising incomes, or decrease as substitutes are found. If demand increases, i.e. if the demand curve in Figure 18.1 shifts outwards, there will be a new price curve which will lie everywhere above the old one. This is because the old price path with an increased demand would result in higher extraction levels which will exhaust the resource before price has reached P_B. The price must therefore be higher in each period in order to ration the resource over time. Even then the time horizon will be shortened from T to T' in Figure 18.9.

18.6 MONOPOLY AND THE RATE OF EXTRACTION

The analysis so far has assumed that competitive conditions prevail. Just as the presence of perfect competition is widely regarded as being consistent with the maximisation of social welfare (though the conditions for this to be true are restrictive), so the Hotelling rule extends the analysis to include time and natural resources. Basically, if resource markets are competitive, resource owners will deplete the resource at a socially optimal rate. If so, the only reason to intervene in resource markets would be: (a) if social rates of discount are different to the private rate of discount used by the resource owner; (b) if there are externalities from resource use; and (c) if markets are not competitive. The effects of altering the discount rate have already been discussed: if social rates are lower than private rates, we would expect the socially optimal price path to encourage more conservation, i.e. a less rapid exploitation of the resource. The presence of external effects would also imply that social rates of depletion are lower than private rates. The effects of relaxing the competition assumption now need to be discussed.

The general presumption is that monopolists restrict output and raise prices compared to those that would prevail under perfect competition. This leads us to surmise that P_0, the initial price, will be higher under monopoly than under competition. But given the fixed resource stock, a higher initial price entails a less steep price path through time. The outcome is shown in Figure 18.10. Note that the effect of monopoly is to increase the 'life' of the resource: *monopoly turns out to be an ally of the conservationist.*

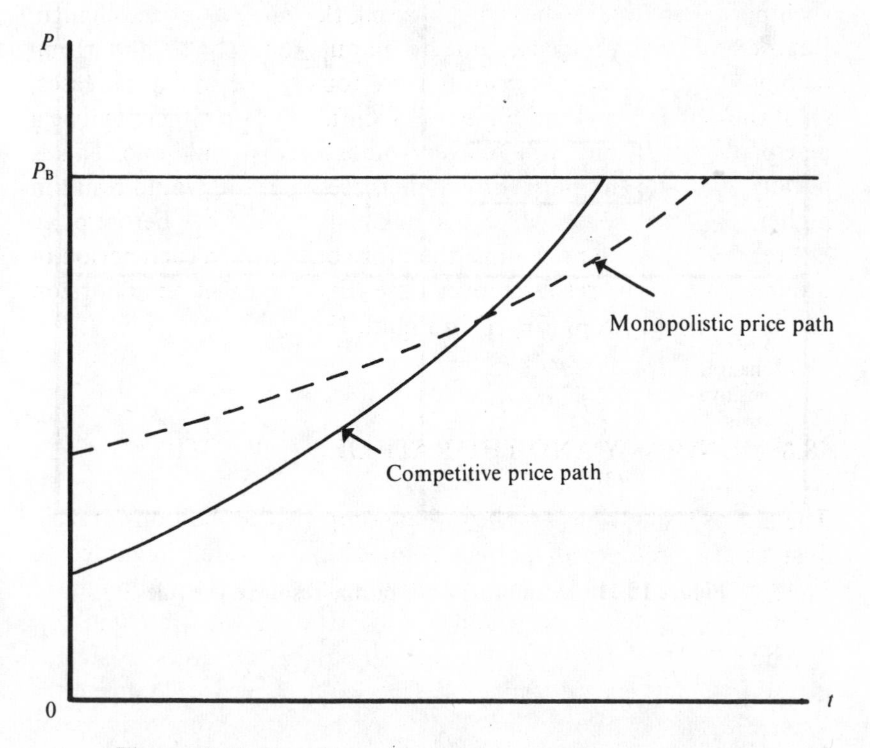

Figure 18.10 The effects of monopoly on resource extraction

This general proposition accords with intuition, but in practice it depends on particular values of the relevant paramaters, e.g. the elasticity of the demand curve. Overall, we might say that monopoly *tends* to lead to resource conservation.

	Renewable resources $F(X) > 0$	Exhaustible resources $F(X) = 0$
Prices 'given'	$\frac{1}{s}\frac{dR}{dX} = P - C(X)$ (1)	Costless extraction $s = \frac{\dot{P}}{P}$ (4)
Prices change, zero costs	$F'(X) = s - \frac{\dot{P}}{P}$ or $s = F'(X) + \frac{\dot{P}}{P}$ (2)	Constant costs $s = \frac{\dot{P}}{R}$ (5) or $s = \frac{\dot{R}}{R}$
Prices change, positive costs	$F'(X) - C'(X)\,F(X) = s - \frac{\dot{P}}{P - C(X)}$ (3)	

Figure 18.11 A summary of optimal resource use rules.

18.7 SUMMARISING THE OPTIMAL USE AND DEPLETION RULES

Figure 18.11 brings together the various rules derived in Chapter 16 and in this chapter. In all cases the rules derive from the assumption that the aim is to maximise the *present value* of the resource to society. The really important rules are highlighted within boxes, and it is advisable to keep these rules in mind even if it may not have been easy to understand how they were derived. Equation (1) in Figure 18.11 says that the gain (P – $C(X)$) from reducing the renewable resource stock by a small amount must equal the foregone rent (in present value terms) from the loss of biomass growth that would have occurred if the stock was left alone. Equation (2) says that, for a renewable resource, the discount rate must equal the natural marginal product *plus* the capital gain from leaving the resource *in situ*. Equation (3) modifies this rule for the case of positive costs of harvesting. Equation (4) is the simplest form of the Hotelling rule for exhaustible resources: extract the resource in such a way that the (real) price over time rises at a percentage rate equal to the discount rate. Equation (5) gives a more realistic version of the Hotelling rule, but even this has costs that are *independent* of the size of the resource stock ($C(X) = C$). Once costs of extraction depend on the size of the stock the analysis becomes far more complex.

19 · MEASURING AND MITIGATING NATURAL RESOURCE SCARCITY

19.1 INTRODUCTION: MALTHUSIAN AND RICARDIAN SCARCITY RECOGNITION

While concern over the adequacy of natural resources is certainly not new (see Chapter 1), there is still no consensus on what in fact has turned out to be a far from straightforward issue. The appraisal of natural resource availability and scarcity involves a combination of physical science, materials science/engineering and economic considerations. In simple economic terms, scarcity will be reflected in costs and relative prices. However, scarcity is not the only influence on prices and prices often do not fully reflect scarcity especially if natural resources encompass the full range of environmental functions, amenity provision, waste assimilation and life-support and raw materials supply.

From the *Malthusian perspective*, it is the absolute physical limit to non-renewable resources which is important and which is predicted to become binding in the near/medium term future. Forecasts in this tradition are based on static stock index calculations. They utilise supply data covering only proved reserves and assume an exponential trend rate of growth in resource demand. This approach is epitomised by the publication *Limits to Growth* (Meadows *et al.*, 1972). A related *neo-Malthusian position* emphasises the importance of 'environmental' limits to resource exploitation (e.g. mineralogical thresholds in terms of required energy inputs for extraction, thermal and other pollution costs).

According to the *Ricardian perspective*, the 'depletion effect' of resource exploitation is felt in terms of rising costs and materials prices over time, as the 'quality' of available resources declines.

Resource adequacy forecasts in this tradition are much more optimistic. They typically assume non-homogeneous resources, an expansion of the recoverable resource stock (which is greater than proved reserves) via increased exploration effort and success, and rapid technological progress (in particular the establishment of a backstop energy technology). Thus advances in processing technologies such as, for example, *in situ* solution mining (direct leaching technique) and biotechnology, enable extractors to bypass costly mineral beneficiation processes. This has led to the recovery of precious metals and copper from relatively low-grade ores. Such Ricardian analyses indicate no scarcity dilemma within at least the next hundred years or so. The 'depletion effect' is further mitigated by compensatory processes stimulated by the market. These processes include material and function substitutions, industrial process changes and scrap recycling.

The *cornucopian view* of technical progress has forecast that rather than adapting existing natural resources to economic end uses, technological advances may soon permit the creation of entirely new synthetic materials. This will be achieved via the reconstruction of molecular structures, tailored to meet a specific need. It is then suggested that progress in material science and engineering will supplant resource scarcities as the ultimate constraint on the rate at which key sectors (utilising *advanced materials*, i.e. materials that exhibit greater strength, higher strength/density ratios, greater hardness, and/or one or more superior thermal, electrical, optical, or chemical properties when compared with traditional materials) of the economy can grow. Advanced ceramics, metals and polymers, including composites of these, offer the promise of decreased energy consumption and better performance at lower cost. In the context of these advanced materials, material processing and fabrication considerations may override issues of resource cost and availability.

For the industrialised economies the advent of advanced materials may also lessen their dependence on imports of *strategic materials*. At various times during the twentieth century fears about long-term shortages of resources have been compounded by anxieties, in times of war or international political tension, about short-term physical scarcities of raw materials. Influenced by this possibility of *transitional Malthusian scarcity*, both governments and mineral markets have come to recognise the existence of a category of *strategic materials*. A strategic or critical material is one for which the quantity required for essential civilian and military uses exceeds the reasonably secure domestic and foreign supplies, and for which acceptable substitutes are not available within a reasonable period of time. Implicit in this definition is the justification of special measures, such as conservation and stockpiling, in order to optimise supplies of these 'special' materials for essential applications.

19.2 EMPIRICAL EVIDENCE ON RESOURCE PRICE PATHS

Physical indicators

Physical measures of scarcity rely on looking at geological estimates of reserves, and relating this in some way to the level of demand. We noted in Chapter 18 that estimates of the size of reserves of exhaustible resources are under constant revision. The McKelvey Box diagram (Figure 19.1) illustrates the reserves gradation and the concept of the 'utimately recoverable resource'. Feasible natural resource exploitation potential is at its highest in cases which fit the conditions as laid down in the top left-hand section of the diagram, i.e. fully demonstrated and measured resource deposits which are economic to work. The feasibility of exploitation then deteriorates by stages, through to the lower right-hand section, i.e. in the context of submarginal deposits found in currently uneconomic concentrations, in districts yet to be explored.

A very simple indicator, for example, is to divide an estimate of resources by the current level of consumption to give a period of years to exhaustion. If demand is expected to rise, this can be built into the indicator. For example, if consumption now is C_0 (in physical units such as tonnes) and annual growth of demand is at k per cent and the stock of the resource is S, then the number of years to total exhaustion (T_E) is given by the solution of:

$$\int_0^{T_E} C_0 e^{kt} = S$$

T_E is something known as 'the exponential index' and it was made particularly famous in the publication *Limits to Growth* (Meadows *et al.*, 1972). We can illustrate how to calculate T_E.

If the current extraction rate is 10 units (C_0 = 10), the demand is

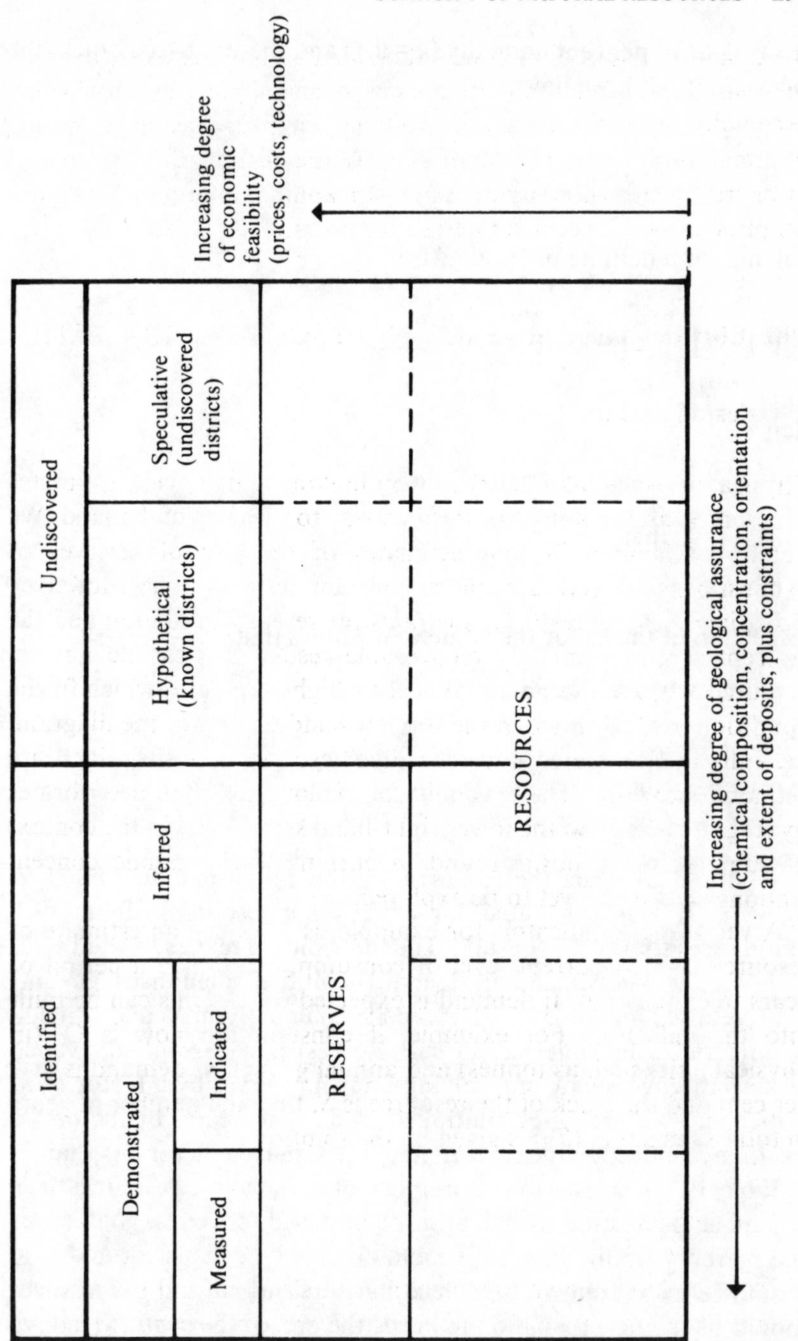

Figure 19.1 McKelvey box-type diagram: resources and reserves.

growing at 10 per cent annually ($k = 0.1$) and there is a stock of 1,000 units ($S = 1,000$), we have:

$$10\int_0^{T_E} e^{0.1t} = 1,000$$

Solving for a definite integral gives;

$$[0.1e^{0.1t}]_0^{T_E} = 1,000/10 = 100$$

Hence

$$e^{0.1TE} - e^0 = 100 \times 0.1 = 10$$

$e^0 = 1$, so that

$$e^{0.1TE} = 11$$

Inspection of tables for the value of e^x shows that

$$e^x = 11 \text{ when } x = 2.398$$

Therefore,

$$0.1T_E = 2.398$$
$$T_E = 23.98$$

In less than 24 years the resource will be exhausted. In *Limits to Growth*, T_E was calculated for exhaustible resources. Table 19.1 summarises some of the results. The first column relates to what was, in 1972, the known global reserve (identified, demonstrated and measured, see Figure 19.2). The second column (estimated identified, demonstrated and inferred global reserves) makes some allowance for new discoveries by multiplying known reserves by a factor of 5. Notice that with reserves multiplied by 5, T_E increases by factors of less than 5, indeed in the case of molybdenum by a factor less than 2.

Table 19.1 suggests that a number of exhaustible resources face imminent exhaustion. In the absence of new discoveries, gold, silver and mercury 'in the ground' should already be exhausted, and zinc should be mined out by 1990, lead by 1993 and natural gas reserves should be facing exhaustion towards the end of the century. Yet we hear of no such problems. Why?

One reason is that the reserve figures in *Limits to Growth* are

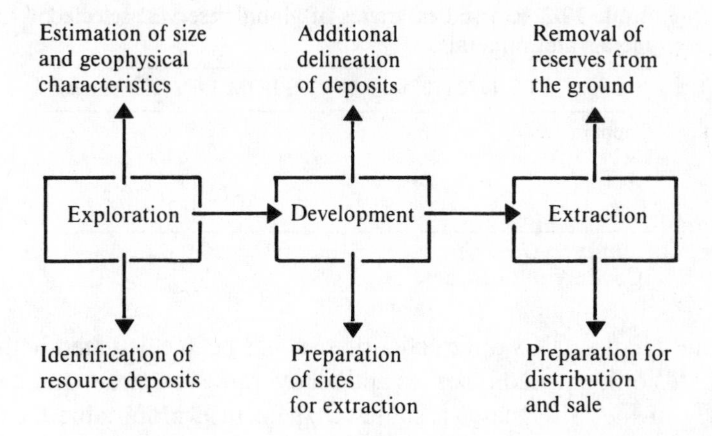

Figure 19.2 Non-renewable resource supply process.

Table 19.1 The 1972 exponential exhausation indices (T_E) in years (S = known global reserves).

	S	$5{\times}S$		S	$5{\times}S$
Aluminium	31	55	Molybdenum	34	45
Chromium	95	154	Natural Gas	22	49
Coal	111	150	Nickel	53	96
Cobalt	60	148	Petroleum	20	50
Copper	21	48	Platinums	47	85
Gold	9	29	Silver	13	42
Iron	93	173	Tin	15	61
Lead	21	64	Tungsten	28	72
Manganese	46	94	Zinc	18	50

Source: Meadows *et al.* (1972), pp. 56–60.

themselves already out of date. Table 19.2 shows some selected data ten years on: nickel reserves are roughly as stated in *Limits to Growth* but copper, lead and zinc reserves have increased. A second reason is that exponential indices fail to account for: (a) variations in forecast demand; (b) the effect of rising real prices on demand via reduced demand, the substitution of other materials for the scarce ore and conservation in terms of reduced resource requirements per unit of output; and (c) the effect of rising real prices on supplies in respect of increased recycling and new exploration/discovery.

Table 19.3 suggests that over the long run, physical scarcity is unlikely to be a significant problem for most of the materials

Table 19.2 Revised estimates of global reserves: selected metals and minerals.

	LTG 1972 (10^6 tonnes)[a]	US BOM 1980 (10^6 tonnes)[b]
Copper	308	505
Nickel	66	64
Lead	91	127
Zinc	123	162

[a] *Limits to Growth.*
[b] US Bureau of Mines.

currently in use. This optimistic forecast has been calculated both on estimates of identified reserves and recoverable resources and particularly in the latter case, assumes that no insurmountable technological, energy or environmental constraints present themselves over the next one hundred years or so. Even in the case of advanced materials, supplies appear to be adequate for the foreseeable future. Of forty-seven elements known to have advanced materials applications, only eleven seem to have potentially insufficient measured reserves. If demonstrated reserves are used as the basis for the calculation then the potentially scarce category contains just eight elements – indium, tin, tantalum, mercury, silver, thallium, gold and bismuth (see Table 19.4).

Table 19.3 Depletion prospects based on recoverable resource estimates: selected metals and minerals.

Material	Demand growth per annum (per cent)	Estimated identified reserves (tonnes)	Depletion of estimated reserves by year 2100 (per cent)	Depletion of estimated recoverable resources by year 2100 (per cent)
Chromium	3.3	1.0×10^{10}	12	–
Cobalt	2.8	5.4×10^{6}	150	36
Manganese	2.7 – 3.3	2.8×10^{9}	120	18
Molybdenum	4.5	2.1×10^{7}	249	5
Nickel	4.0	2.1×10^{8}	152	35
Titanium	3.8	7.1×10^{8}	102	38
Tungsten	3.4	6.8×10^{6}	236	11
Zinc	2.0	3.3×10^{8}	581	37

Source: Goeller and Zucker (1984).

Table 19.4 Advanced material elements: reserve adequacy.

Elements	(1) Annual growth in demand	(2) World cumulative demand 1983–2000	(3) Measured reserves	(4) Measured reserve/demand ratio	(5) Demonstrated reserves	(6) Demonstrated reserves/demand ratio
Tin	1.0	3,930,000	3,000,000	0.78	3,000,000	0.78
Mercury	1.4	127,000	138,000	1.1	250,000	2.00
Silver	2.1	168,000	244,000	1.5	336,000	2.00
Indium[a]	1.6	1,000	1,700	1.4	3,000	0.13
Thallium[a]	1.2	200	380	1.9	650	2.10
Gold	1.8	21,000	40,000	1.9	45,000	2.20
Bismuth[a]	1.3	77,000	90,000	1.2	204,000	2.20
Barium[a]	2.2	132,000	16,800	1.3	500,000	3.40
Tantalum[a]	3.2	19,500	27,200	1.4	41,000	1.40
Cadmium	1.8	345,000	555,000	1.6	970,000	2.80
Arsenic	1.5	590,000	100,000	1.7	1,500,000	2.50

[a]Demonstrated reserves/demand ratio based on higher cumulative demand than that assumed in columns (1) and (2).
Source: Adapted from Fraser, Barsotti and Rogich (1988).

19.3 A TYPOLOGY OF SCARCITY AND PRICE/COST SCARCITY INDICES

In Chapter 18 we used some economic theory to derive some optimal resource use rules, under a set of simplifying assumptions. The models of non-renewable resource supply underlying this analysis are based on profit-motivated firms operating in competitive markets and also on the monopoly supplier case. In Chapter 18 our concern was to demonstrate some fundamental economic principles of resource use, using as simple an economic analysis as possible. In this chapter we have briefly surveyed some of the complexities that surround the real-world supply situation. We are now in a position to outline a more realistic non-renewable resource supply model, although formal proofs are beyond the scope of this text.

The supply of non-renewable resources can be viewed as a three-stage process with complicated dynamic interrelationships (see Figure 19.2). Current decisions are related to past decisions, as well as to expectations of the future prices and costs. It is assumed that producers take these intertemporal cost decisions into account in planning their decisions at each stage of the process. Thus the current rate of extraction affects the amount that may be extracted in future periods. Consequently, the cost of extracting a unit of, say, a mineral today depends not only on the current level of mineral usage and necessary inputs and their prices, but also on the level of input usage in the past as well as on the impact of current extraction on the profitability of the deposit. For example, for a given inventory of discovered sites, a decision to develop relatively low-cost sites today leaves only higher cost sites for the future. Reductions in the total stock of resource-bearing sites and the tendency for earlier discovery of large and more accessible sites lead to increased exploration costs over time.

Activity at each stage of the supply process satisfies a derived demand for inputs to subsequent stages, e.g. new discoveries are an input to development and new reserves are an input to extraction. Changes in the price of final output and in supply costs influence decisions at all three supply stages. Thus final output price changes influence exploration and development decisions, as well as extraction decisions. Development cost changes clearly affect development activities, but also the derived demands for new discoveries and extraction decisions through changes in the cost of replacing reserves.

The decision process is subject to a significant degree of uncertainty. Because producers lack full knowledge of future prices and costs, they must predict both future prices and the uncertain consequences of current decisions on future prices.

Finally, the supply process is affected by the impact of numerous governmental regulatory changes (taxation and environmental protection in particular), as well as by varying institutional and market structures.

Fundamentals of an economic theory of individual supply decisions

The following are assumed:

1. The cost of any particular rate of extraction *increases* over time as the reserve stock falls or as cumulative production increases.
2. Development costs *rise* as cumulative developed reserves increase and development effort increases.
3. Discovery (exploration) costs *rise* as cumulative discoveries increase and exploration effort increases (earlier discovery of larger and more accessible sites).

As Chapter 18 showed, producers seek to maximise the sum of discounted net revenues (present value of gains minus present value of costs) over time, rather than current profits, and given their expectations of future prices and costs, they attempt to balance the gains from increased extraction (revenues) dependent on price, with the costs of increased extraction. The latter are made up of increased operating costs plus the user cost of extraction. This is the fall in expected net present value from increased operating costs in the future. So current user cost is a function of future planned rate of extraction, which is itself a function of expected future prices. Current operating costs for extraction are a function of cumulative past production. So current extraction rates are also a function of cumulative past production. Current extraction costs are therefore related to cumulative past production and current extraction rates, as well as to expectations about future prices and costs. At the margin, marginal extraction operating costs plus user cost is compared with ore price, in order to determine optimal (economic efficiency) output over time, i.e. output Y_1 (see Figure 19.3).

A decline over time in developed reserves as a result of depletion increases the user cost of extraction and will stimulate an acceleration of development activity, as well as a fall in extraction (current). A fall over time in the stock of known prospects capable of development increases the user cost of development. This retards further development of reserves, increases the user cost of extraction and slows current production. Geological/physical limits on the ultimate size of reserves and the volume of discoveries do not enter directly into producers' decision calculus, but enter indirectly through the effect of depletion on costs over time. But this analysis is still not complete; it does not fully capture the minimum economic value possessed by the resource (the marginal social opportunity cost). Thus as well as the operating costs, the user cost concept must be expanded to encompass all the external costs of resource extraction (pollution and landscape and amenity losses).

Our analysis in Chapters 2 and 3 highlighted the need for a better understanding of what we have called an existence theorem. We require more comprehensive analysis that relates the scale and configuration of an economy to the set of environment–economy interrelationships underlying that economy (i.e. the general life-support function encompassing the resource supply, waste assimilation and aesthetic commodity). The important point here is

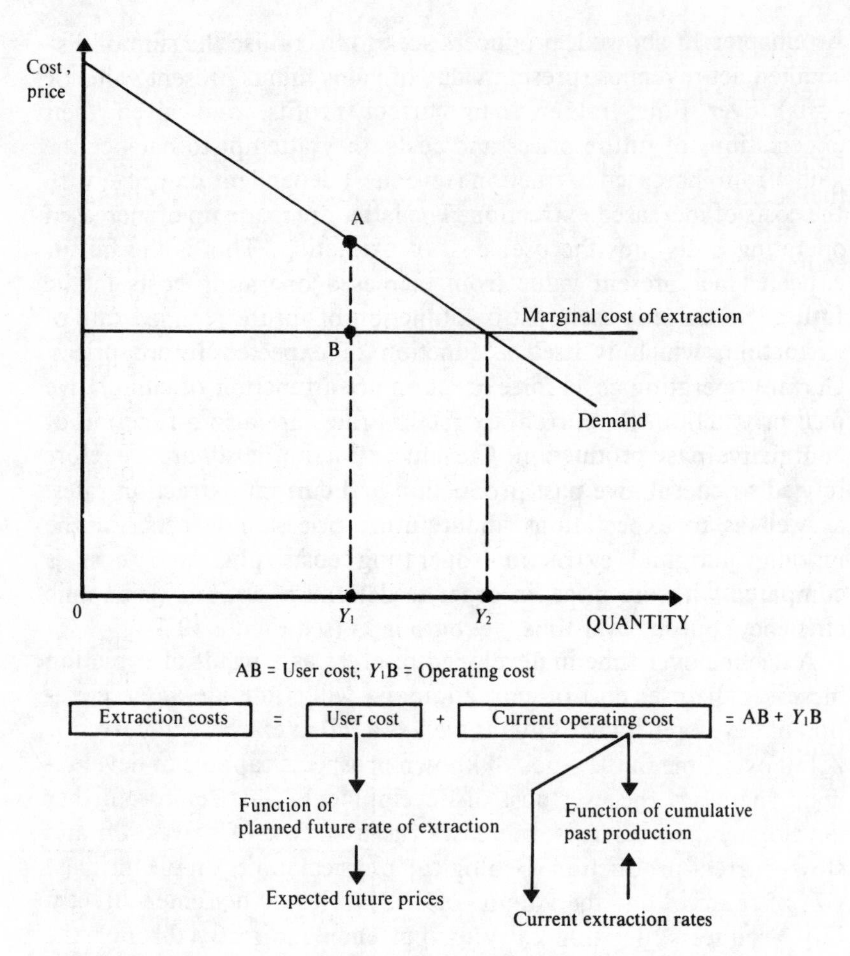

Figure 19.3 Economic optimum level of output.

that while there is obvious direct dependence between human well-being and natural resources, the functioning of the economic system itself is interrelated with, and threatened by, natural-resource depletion. The difficulty is that many of these 'external costs' will be displaced in both space and time. They are intertemporal and lateral external costs. Looking only at the direct and immediate costs of resource use will almost certainly understate the true costs of resource depletion.

Economic scarcity indices

There are a number of studies which attempt to measure scarcity by some indicator of price or cost. In principle, a scarcity index ought to be measured with just one essential property: it should encompass the sacrifices, direct and indirect, made to obtain a unit of the resource. So far, three measures have been used generally in the economics literature:

1. Real unit costs of production of the resource – the real quantity of homogeneous inputs required to provide a unit of constant quality extractive output.
2. Real price of resource-intensive goods.
3. Rental rate of the resource (i.e. true user cost) – this is unobservable, and so it has been suggested that for some non-renewable resources the user cost is equal to the shadow price of a unit of its stock. The shadow price reflects the present value of the benefits of shifting the production of a unit of the resource to the future. This value is the value of an additional unit of proven reserves, which is equal to the cost of producing that additional unit (known as the marginal replacement cost or discovery cost).

Unfortunately, increasing physical scarcity may be associated with either increases or decreases over time in any or all of the proposed scarcity measures. Because they respond to different technico-economic factors, neither the magnitude nor the direction of scarcity change need be the same for alternative measures. No one measure of scarcity can claim to be generally correct and so the appropriateness of a given indicator will depend on the *nature of scarcity*. Hall and Hall (1984) suggest a four-fold scarcity typology:

1. *Malthusian Stock Scarcity* (*MSS*) Here the resource is fixed in absolute size and extraction costs are constant.
2. *Malthusian Flow Scarcity* (*MFS*) Here the resource is fixed in absolute size but extraction costs rise with the rate of extraction.
3. *Ricardian Stock Scarcity* (*RSS*) No absolute size constraint exists but extraction costs rise with the rate of extraction *and* how much of

	the resource has been extracted to date.
4. *Ricardian Flow Scarcity* (*RFS*)	No absolute size constraint exists, but extraction costs rise with the rate of extraction.

To illustrate, the *Limits to Growth* analysis was essentially 'stock Malthusian' – absolute limits to resource stocks were assumed. On the other hand, our discussion of 'backstop' technology was 'stock Ricardian' because, although we were concerned to find a price path which did exhaust the stock at a point when the resource price equalled the backstop price, the implication is that some of the resource *could* be left over when a backstop technology takes over.

Hall and Hall (1984) show that the price of the resource output (P) is related to cost for the four types of scarcity as follows:

(1) *MSS*: $P = \mathrm{AC} + \mathrm{UC}$
(2) *MFS*: $P = \mathrm{AC} + \mathrm{PV(FC)} + \mathrm{UC}$
(3) *RSS*: $P = \mathrm{AC} + \mathrm{PV(FC)}$
(4) *RFS*: $P = \mathrm{AC}$

Where AC = average cost, UC = user cost and PV(FC) is the present value of future cost increases due to extraction now. Notice that only for the RFS case will a unit cost measure alone be appropriate. For all other cases, it is price itself which is the appropriate measure since price should capture the user cost element and the factor reflecting the impact of current extraction on future costs. Yet some of the most influential literature (e.g. Barnett and Morse, 1963; Barnett, 1979) has used unit cost measures only. These are only appropriate if Ricardian Flow Scarcity is the problem. Hall and Hall (1984) analyse data for the USA for 1960 to 1980. They test the hypothesis that scarcity (measured by unit costs and relative prices respectively) decreased in the 1960s and increased in the 1970s. We summarise some of their results in Table 19.5. The main observation is that for oil, gas and electricity, scarcity increases in the 1970s. Hall and Hall get the same result for timber. These results extend those of Barnett and Morse who used unit costs only and found no evidence of scarcity. In later observations Barnett concluded that apparent increases in scarcity in the 1970s were mainly due to OPEC's oil cartel.

Table 19.5 USA resource scarcity 1960–80*

Resource	Unit cost test	Relative price test
Coal	60s down, 70s up	Not significant
Oil	60s down 70s up	60s down, 70s up
Gas		60s?, 70s up
Electric power	60s down, 70s up(?)	60s down, 70s up
Non-ferrous metals	60s up, 70s down	60s, 70s?

Source: Hall and Hall (1984), pp. 369–70.
Notes: 'Down' means scarcity *decreases*, 'up' means scarcity *increases*. ? means the result is statistically not significant or the direction of change is indeterminate.

19.4 GEOCHEMICAL AND STOCK POLLUTION CONSTRAINTS ON RESOURCE EXPLOITATION

Geochemical theory

Some geological research work suggests that as the chemical features of any particular material affect the physical occurrence of that material in the environment, these same geochemical properties will affect the costs of mining, concentrating and smelting. But more troublesome may be the finding that impacts may not be smooth, because the transformations involved change with the level of concentration and form of the deposit. These same chemical characteristics are clearly also important to the materials' end uses. It may well be that the supply function for resources with the 'right' attributes is upward-sloping and potentially discontinuous (stepped function). In the case of copper, for example, the energy costs of extracting and processing increasingly lower-grade ores do not increase smoothly. A major discontinuity is likely to be experienced when ore grades of 0.1 per cent are exploited. Breaching this *mineralogical threshold* will require a significant extra energy input and consequent thermal pollution risk.

This geochemical complexity argument is important because it qualifies one of the main resource scarcity mitigation factors implicitly assumed in the optimistic Ricardian-type forecasts of future resource adequacy. This is that there is a general principle which relates to the exploitation of all lower-grade ores. The principle is that the quantity of recoverable reserves increases geometrically as ore grade declines arithmetically. But the newer geological theory suggests that geochemically less abundant metals (e.g. cadmium, cobalt, copper, zinc, silver, gold, tin, chromium, nickel, tungsten and others) do not fit the generally expected exploitation profile. For these metals the prospect is that, as their grade diminishes so too does the total available reserve base, at least until very low-grade deposits are encountered. The resource discovery effort will therefore become very much more demanding.

Stock pollution and other environmental damage

Resource exploration and mining activities can potentially impose significant (and sometimes irreversible) damage impacts on the environment. Wilderness areas or semi-wilderness (national park) areas are especially at risk from such activities. Because of the multifunctionality of such environmental assets, resource extraction can, for example, mean the direct loss of landscape quality and both direct and indirect ecological damage via pollution.

We have argued throughout this book that the services provided by the whole range of open-access environmental resources ought to be encompassed within the concept of the natural resource base and its economic exploitation. Thus any analysis of resource scarcity must include the effects of resource extraction-induced pollution and amenity loss on society's overall level of welfare. Stock pollutants accumulate in the environment and therefore represent a type of depletion, i.e. the depleted ability of the ambient environment to assimilate wastes. Landscape and ecosystem losses may well turn to be irreversible and again represent stock depletions.

19.5 RESOURCE SCARCITY MITIGATION: RECYCLING AND SUBSTITUTION

The Ricardian analysis of scarcity emphasises that increasing scarcity leads to increases in the rate of resource price appreciation, which then induces greater substitution of other materials and increased recycling. In this section we examine in more detail the complexities surrounding the issue of increased materials recycling and also the nature of real-world substitution processes.

The materials-balance and resource recovery sub-system

The materials-balance model examined in Chapter 2 emphasised that the economist's terms, production and consumption, are not precise terms and they should not be taken to imply that matter simply disappears through use. The term waste is also a misnomer since in nature, nothing is wasted, everything is part of a continuous cycle. More properly wastes should be re-termed 'residuals' and placed in an economic context. What are non-productive outputs in any given context may be re-used or become a useful by-product in other or future contexts.

Millions of tonnes of materials currently discarded by modern societies could be physically recovered but are not recycled because given current market conditions the financial costs of recovery exceed the value of the recovered materials. The term 'resource recovery' (reclamation) has evolved into a general concept referring to any productive use of what would otherwise be a residual requiring disposal. It is instructive to view resource recovery as a sub-system within the larger materials balance generated by an economic system. The sub-system could encompass the following:

1. The re-use of products in the same form and for the same purpose – typified by the returnable bottle.
2. The processing of residuals in order to recover and re-use the materials in the same production activity – examples of such direct recycling (closed-loop recycling) are the recovery and re-use of old paper fibre in the 'furnish' make-up for new paper and board products or the use of used glass (cullet) in new glass products.
3. The re-use of residuals as inputs into another type of production activity – this indirect recycling (open-loop recycling) is typified by the use of ground-up glass in road-surfacing material and the conversion of the organic functions of municipal refuse to compost.
4. The processing of residuals in order to recover their energy potential – for example the incineration of municipal solid waste with heat recovery.

The outputs of the resource recovery (reclamation) sub-system are what can be labelled secondary materials in order to distinguish them from primary (virgin) material (the original natural resources) and from end-use products.

The extent to which materials and/or energy recycling is practised is a function of the cost of recovered residuals as a raw material input into production relative to virgin (primary) raw material inputs. These relative costs will be determined by a multitude of factors including the following: production process technologies (primary and secondary); virgin material extraction technologies; secondary and primary materials market conditions and structure; end-product output specifications; and the extent of government intervention in the form of tariffs, discriminating freight rates, severance taxes, depletion allowances or tax credits for secondary materials usage.

The important characteristics which will determine the value of a secondary raw material and therefore the path taken by a residual are: quantity (mass); location (dispersion of mass); level of contamination; and quality (homogeneity). Thus a large mass of high-quality materials concentrated near to production or market facilities will have a high value regardless of whether they are secondary or primary raw-material inputs. Residuals must be generated in sufficient quantity to be worth recovering and must not be too dispersed in terms of the location of their generation sites. The location of the residuals collection area with respect to the nearest processing plant equipped to utilise secondary materials is also important. Contamination, the extent to which the desired material occurs in combination with other materials and the complexity of separation, may or may not pose problems. Homogeneity, the material's consistency in terms of form and quality, affects the 'production risks' involved in the use of secondary material inputs and therefore the material's recycling potential.

The characteristics outlined above will vary according to the stage in the economic process at which the residual is generated. The residuals generated by processing and basic manufacturing activities (home scrap) are generally excellent substitutes for the material input and output and are recovered as a matter of course. Converting/fabrication residuals (prompt scrap) tend to exhibit the desirable characteristics of large concentrated mass, homogeneous quality and known contamination problems. Private reclamation industries have developed specifically to collect and process the large volumes of converting residuals that are generated in industrialised economies. The value of post-consumer residuals (old scrap) is, on the other hand, often unfavourably affected by such factors as location, mass and quality. These residuals arise in dispersed locations (individual households and small-trader premises) in fairly small amounts and often in a contaminated state mixed in a general solid waste flux. Post-consumer residuals thus present a much more difficult and financially costly recycling problem.

The role of recycling in extending the resource base

Because of the Second Law of Thermodynamics, recycling of secondary materials cannot take place such that 100 per cent of waste is recovered. Economic models of recycling have indicated that both the growth rate of the overall economic system and the average product lifetimes of the eventually scrapped good play an important role in determining the extent of recycling's contribution to resource conservation.

If we let d years be an average product lifetime and α a fixed proportion of scrap recycled and supplied to a free market (i.e. price of scrap adjusts continously to equate scrap demand and supply), and if total material demand is growing at a rate of g per cent annually, the total demand for material input at time t (M_t) can be written as:

$$M_t = M^p_t + M^s_t = M_0 e^{gt}$$

where M^p_t = demand for primary material, M^s_t supply of scrap, and M_0 = initial demand.

The supply of scrap M^s_t can be written as:

$$M^s_t = \alpha M_{t-d} = \alpha M_0 e^{g(t-d)}$$

Hence the steady-state share (β) of scrap in total input demand is:

$$\beta = M^s_t / M_t = \alpha e^{-gd}$$

So the higher the growth rate of the economy and the longer the average product lifetime, the smaller is recycling's relative contribution to resource conservation. Both the total stock of scrap potentially available at any given time and the increasing marginal costs of scrap supply (increasing costs of collection) serve to limit recycling activity in a growth economy. This sort of model contains a number of simplifying assumptions, e.g. perfect substitutability between scrap inputs and primary material inputs, but nevertheless the control message is clear – the role of recycling is limited.

19.6 SUBSTITUTION PROCESSES

Technological alternatives exist which are capable of mitigating potential resource scarcity and supply vulnerability problems. Increasing emphasis may have to be placed on materials conservation strategies (i.e. a combination of more effective materials utilisation measures, including recycling and materials substitution) if real economic resource scarcity has begun to increase, as additional energy requirements and environmental damage costs become more evident. On the other hand, the *intensity-of-use hypothesis* suggests that some leeway may be present because of a general and persistent decline in the materials intensity of national output in industrialised economies.

Material substitution may result from the introduction of new technology, from shifts in the composition or quality of final goods, and from changes in the mix of factor inputs used in producing these goods. It would appear that inter-material competition has become more intense and substitution more prevalent, particularly since the early 1960s. The number of materials has increased and the properties of existing materials have been enhanced, in order to penetrate new markets. The initiating causes and the process of substitution itself are under-researched topics. Empirical evidence is limited, but studies examining the use of tin in various applications suggest that three factors – relative material prices, technological change, and government regulations – are important.

Economic models often assume that material prices are the primary driving force behind substitution. The conventional theory assumes that the price variable and material demand are linked together in a lagged and reversible relationship. The nature of the lagged demand response most often assumed is one relevant to situations in which existing technology and equipment permit the use of one material for another in the production process. The absence of a materials balance constraint (i.e. physical constraints) means that substitution can respond immediately to changes in material prices above a certain threshold level. Further, if and when price falls again, the substitution process is reversible.

Thus in Figure 19.4 a desired level of output, given by the isoquant Q_1 can be produced by a range of combinations of substitutable primary and secondary (scrap) material inputs. Giving the existing prices of the primary and scrap inputs the slope of isocost line C_1

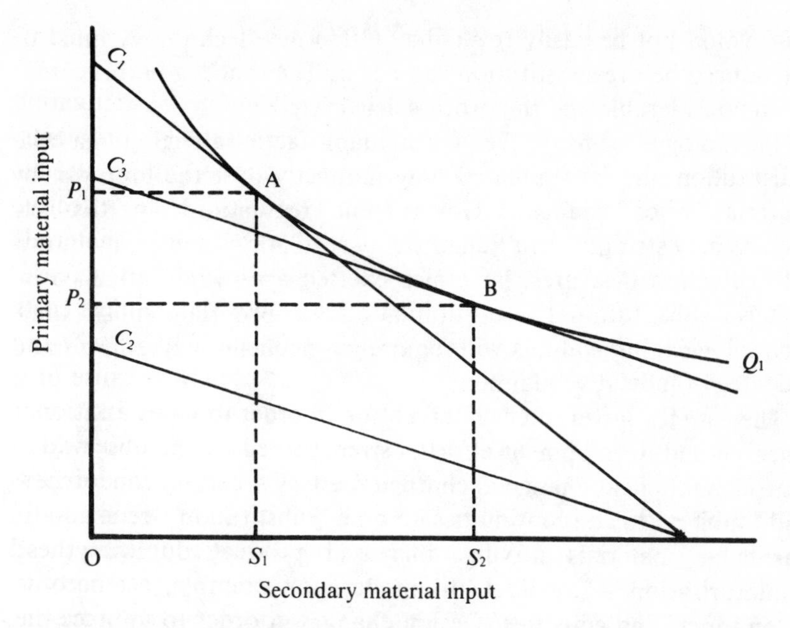

Q_1 isoquant, measuring the amount of feasible output given different combinations of inputs $P + S$;

C = isocosts, indicating various quantities of inputs a firm can purchase for a given expenditure

Figure 19.4 Material substitution.

indicates the ratio of the scrap input price to the primary input price. Faced with such circumstances a firm would minimise its costs of production for an input level Q_1 by operating at point A, using S_1 scrap inputs and P_1 primary inputs. Now if the price of the primary input increased because of supply shortage, this would be reflected in a change in the slope of the isocost line. Let the isocost line rotate from C_1 to C_2. In order to continue producing output Q_1 at minimum cost, the firm substitutes, in the short run, increased inputs of scrap, S_1 to S_2, for primary inputs, P_1 to P_2.

It remains to be demonstrated, however, that in the real world, situations which are not subject to implementation lags are the norm rather than somewhat special cases. It seems feasible that the more numerous situations are likely to be those in which equipment must be modified and/or replaced before significant levels of substitution can take place. Substitution would be subject to a lengthy time lag and would not be easily reversible. Often new technology must be introduced before substitution can occur. The time lag in this case is often considerable and the process is irreversible.

Technology seems to be a dominant factor affecting material substitution and is stimulated only indirectly over the long run by material price changes. Government regulations, particularly increasingly stringent health and safety regulations and environmental protection measures, have also exerted a growing influence on material substitution. Direct stimulation of appropriate 'substituting' technological innovations will require government interventions and a degree of indicative planning.

The case for government intervention in order to focus and direct research and development effort is strengthened by the observation that technological change is characterised by a certain randomness and subject to discontinuities. Some substitution trends – in particular materials mixing, increased product durability and miniaturisation – are likely to hinder, for example, attempts to extend recycling activities. Design changes in order to improve the recyclability of eventually scrapped products have not so far been particularly evident. Materials-mixing can produce potentially incompatible combinations as far as recycling is concerned, while at the same time producing technically and/or aesthetically superior products. Miniaturisation reduces the amount of recoverable material per unit of scrap and often requires an intricate recovery operation.

PART V

DEVELOPMENT AND ENVIRONMENT

20 · DEVELOPMENT, PRESERVATION AND CONSERVATION

20.1 CONSERVATION AND PRESERVATION

The debate between conservationists and those who wish to destroy or modify natural environments for purposes of industrial or commercial development is frequently a debate about 'zero-one' or 'discrete' choices. What this means is that a given habitat, say, can either be developed or preserved in its natural state: there is no compromise. It seems reasonable to maintain the word 'preservation' to describe the non-development option in this case. 'Conservation' might be better used to describe options in which the essential features of the natural habitat are maintained but some of the habitat area or some of its features are traded off for development benefits. Alternatively, the natural habitat is maintained but the resource itself is used ('harvested') for commercial purposes. Typical examples of conservation might therefore be national parks in which visitors are encouraged but efforts are made to keep the natural features that attract visitors. Similarly, wildlife may be maintained for the benefit of tourists or hunters, as with wildlife areas in many African countries, wildfowl and game shooting areas in the developed nations, and so on. Conservation might also take less obvious forms: wildlife may be 'ranched' to supply hides, meat and other products, reducing the pressure on wild populations which are less subject to the quality control that ranched animals may provide. This latter example shows how preservation and conservation might be mixed: some areas are earmarked for non-use and other areas for direct commercial use.

Some of the fiercest debates in the environmental literature are between preservationists and conservationists. It is easy to see why.

If, for example, there is a belief in the rights of wildlife or the rights of living things in general, it is difficult to square these, superficially anyway, with the idea of managing a natural resource for human benefit. Similarly, many preservationists feel that conservation as a compromise between development and preservation gives too much ground – it is the 'thin end of the wedge' in the fight to protect natural environments. In many cases, the conservation option does not really arise: either a given habitat is preserved because it is the minimum critical area need for species survival, or it is destroyed for development. These cases of discrete choice present some of the most challenging problems in environmental economics. It seems fair to say that economists have not 'solved' the issue: there must be unease about the prescriptions so far advanced in the literature. Nonetheless, very little science advances in giant leaps and bounds. The methods that have been proposed for dealing with discrete preservation – development options have been fruitful. This chapter reviews these procedures.

20.2 DEVELOPMENT AND TOTAL ECONOMIC VALUE

The first observation we need to make is that it is important to capture the nature of the economic benefits that arise from preservation. Chapter 9 introduced the idea of *total economic value*. Consider the choice between developing, say, a wetlands area or preserving it: a discrete choice problem because any commercial use of the wetlands would destroy its preservation benefits. Then, writing $PV(B_D)$ as the present value of development benefits, $PV(B_P)$ as the present value of preservation benefits, $PV(C_D)$ as the development costs and $PV(C_P)$ as the direct costs of preservation (e.g. policing, maintenance and monitoring costs), the rule we have been using would indicate that we should develop the wetland if:

$$\{PV(B_D) - PV(C_D)\} > \{PV(B_P) - PV(C_P)\} \tag{20.1}$$

or

$$\{PV(B_D) - PV(C_D) + PV(C_P)\} > 0 \tag{20.2}$$

Dropping the PV notation for convenience, we know that we can also write:

$$B_P = \text{TEV} = \text{OP} + \text{EXV}$$
$$= \text{E(CS)} + \text{OV} + \text{EXV} \quad (20.3)$$

where TEV is total economic value, OP is option price, EXV is existence value, E (CS) is the expected consumer surplus from use of the wetlands in its preserved form, and OV is the option value from preservation. Recall that OV could be negative but that considerations of supply uncertainty are likely to make it positive (see Chapter 9).

Thus our rule for deciding on development becomes:

$$\{B_D - C_D - C_P\} > \{\text{OP} + \text{EXV}\} \quad (20.4)$$

In terms of *measurability*, it will be clear that development benefits, development costs and preservation costs are likely to be the subject of well-defined monetary estimates. This raises two immediate cautions. First, since OP and EXV are difficult to measure (difficult, not impossible), there is a danger of 'misplaced concreteness'. The things that can be measured might appear to be somehow more important than those which cannot be measured. This is a false deduction, for the economic values embodied in non-market preferences are just as important as those embodied in market preferences. Second, because something is easy to measure it does not mean that the estimate is correct. It always pays to scrutinise the alleged development benefits. *Ex post* evaluations of development projects frequently show that development benefits are exaggerated at the time of the proposal: there is an in-built 'benefit optimism' on part of planners and developers. This bias, for example, has been present in energy planning with respect to nuclear power, and in the building of large hydroelectric dams. Another reason for bias is the underestimation of technological progress: as technology advances it tends to displace the technology that is generating the development benefit.

We can now investigate two approaches to the discrete choice problem, one based on the work of John Krutilla and Anthony Fisher, the other on the work of S. V. Ciriacy-Wantrup and Richard Bishop.

20.3 IRREVERSIBILITY AND THE KRUTILLA-FISHER ALGORITHM

In the discrete choice case, if development goes ahead, the preservation benefits are probably lost forever. There is an *irreversibility*. Consider a development project costing only \$1. Let the development benefits from this project be \$$D$ per annum for ever. Then we can immediately write the present value of this project as

$$PV(D) = -1 + \int_0^{\infty} De^{-rt}dt \qquad (20.5)$$

where r is the discount rate. Equation (20.5) reduces to

$$PV(D) = -1 + \frac{D}{r} \qquad (20.6)$$

(The procedure here is that we are working with *perpetuities* so that the present value of \$$D$ for ever is simply D/r. This simplifies the analysis considerably). Now, we need to compare the development benefits in (20.6) with the costs of the development. But bearing in mind the definition of cost as opportunity cost, it will be evident that the cost of the project is not simply the \$1 expended on capital and operating costs of the development. It must also include the foregone benefits of the destruction of the environment as a natural asset. Let us call these foregone preservation benefits P per annum. Then the present value (PV) of these foregone preservation benefits will be

$$PV(P) = \frac{P}{r} \qquad (20.7)$$

and we shall have to write the net present value (NPV) of the development project as

$$NPV(D) = -1 + \frac{D}{r} - \frac{P}{r} \qquad (20.8)$$

For the development project to be admissible we require that NPV(D) be greater than zero, i.e. on rearranging (20.8) we will have

$$\frac{D-P}{r} > 1$$

or

$$(D-P) > r \qquad (20.9)$$

Now we can investigate the nature of the preservation benefits in a little more detail. First, we can observe that, over time, the *relative price* of P is likely to rise as the natural environment becomes less and less in quantity. Note that this is quite different from talking about general price rises. But if we have reason to believe that any benefit or cost is likely to change its price significantly and relative to the general price level, then we should include that price rise in the analysis. This leads us to write the preservation benefit in year t as

$$P_t = P_0 e^{gt} \tag{20.10}$$

where P_0 is now the initial year's preservation benefit and g is the growth rate of the price of preservation benefits relative to the general price level. We could go further and argue that our development scheme will itself be subject to technological change which will render it less attractive through time. We discount the development benefits by a further factor, k, reflecting the rate of 'technological decay' of our project. Hence

$$D_t = D_0 e^{-kt} \tag{20.11}$$

If we now bring (20.10) and (20.11) together with the formula for NPV(D), we shall have

$$\text{NPV}(D) = -1 + \int_0^\infty De^{-(r+k)t}\text{d}t - \int_0^\infty Pe^{-(r-g)t}\text{d}t \tag{20.12}$$

which looks formidable, but reduces to

$$\text{NPV}(D) = -1 + \frac{D}{r+k} - \frac{P}{r-g} \tag{20.13}$$

Equation (20.13) takes on positive values if, and only if,

$$\sqrt{D} > (\sqrt{P} + \sqrt{k+g}) \tag{20.14}$$

If inequality (20.14) holds, then we find that the graph of present value against the discount rate, r, appears as in Figure 20.1. Here we see that net present value is positive only above a discount rate r_0 and below a discount rate r_1. In other words, the development project will succeed only if certain discount rates are adopted. High rates simply reduce the value of D in the normal way that discount rates affect benefits. Low rates tend to give the rate of growth g on the preservation benefits a chance to influence the choice against the development.

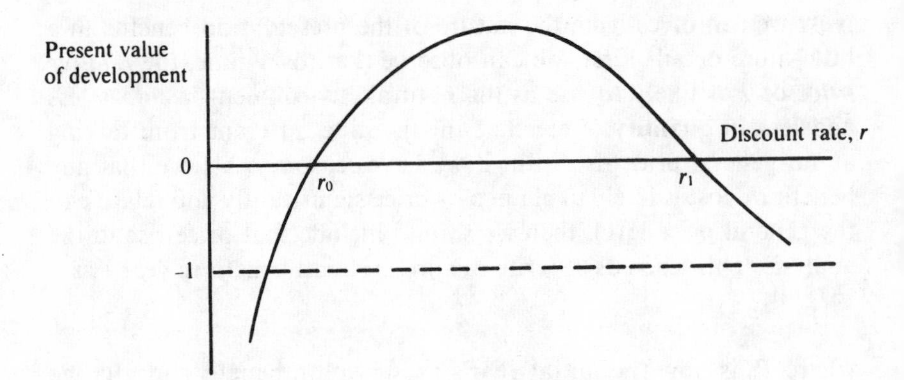

Figure 20.1 Development benefits and the discount rate.

Let P be (arbitrarily) 0.2 and let $k + g = 0.01$. Then inequality (20.10) tells us that, for the development to be worth while, D has to take a minimum value given by the inequality

$$\sqrt{D} = \sqrt{0.2} + \sqrt{0.01}$$

or

$$D = (0.547)^2 = 0.299$$

But this means that the ratio of D to P is $0.299/0.2 = 1.49$. That is, development benefits must be 43 per cent higher than preservation benefits for the development to be worth while. The result is therefore very sensitive to k and g. What the Krutilla–Fisher approach does, then, is to ensure that the benefits of the preservation option are correctly allowed for in a basic cost–benefit equation. The fundamental points are:

1. That the preservation benefits *foregone* are treated as part of the costs of the development.
2. That preservation benefits *increase* through time because of the relative price effect.
3. That development benefits have an offsetting discount factor, in addition to the 'basic' discount rate, because of 'technological decay'.
4. That the present value of development can be very sensitive to the preservation relative price effect and the technological decay factor.

There is a great deal more to the approach which brings in the concept of 'quasi-option' value (see Chapter 9) arising from improved information about preservation benefits – see the Further Reading to this chapter.

20.4 SAFE MINIMUM STANDARDS

An alternative approach, which has similarities with the Krutilla–Fisher algorithm, is based on the original ideas of the American economist, S.V. Ciriacy-Wantrup, and has been developed by Richard Bishop. It is known as the *safe minimum standards* (SMS) approach. The Krutilla–Fisher algorithm tends to put the benefit of doubt with the preservation option: it makes it harder to get the development option accepted compared to a more conventional analysis in which the g and k factors are absent. The SMS approach states quite explicitly that we should avoid irreversible environmental damage unless the social cost of doing so is unacceptably large. This rule sounds imprecise, but the SMS approach is deliberately 'fuzzy' because it does not rely on a single criterion for making discrete choices.

As with the wetlands example discussed previously, consider a development project which causes species loss in the wetland, but where we do not know the benefits of the species. They might be useful as genetic blueprints for the development of future crops, or they might not. They might have medicinal value, or they might not. There are thus two 'states of nature' corresponding to these two possibilities. Let us refer to then as Y ('yes' there are preservation benefits) and N ('no', there are not). We face a 'matrix' of possibilities as shown in Figure 20.2: we can develop (D) or preserve (P) and in each case we face the states of nature Y and N. We know the benefits of development B_D. *If* the species turn out to be useful we can say that they will have a benefit of B_P. If not, they have a benefit of zero. The 'cells' in the matrix record the *costs* of making a discrete choice if a certain state of nature occurs. (We ignore development costs and preservation costs, for convenience.) For example, if we choose to develop and the species turn out to have value after all, then the loss to society is the foregone preservation benefits B_P. If we develop and there is no species value, the loss is zero. If we preserve and the species turn out to have value then we have lost the development

	Y	N	Maximum losses
D	B_P	O	B_P
P	$B_D - B_P$	B_D	B_D

Figure 20.2 A matrix of development/preservation losses.

benefits but gained the preservation benefits, so the loss under combination (P, Y) is $B_D - B_P$. If we preserve and the species turn out to have no value, then the outcome is a loss of B_D.

Figure 20.2 describes a *game theoretic* situation, and which option we choose is going to depend on our attitude to the uncertainty involved. One rule which is already biased towards conservation is to avoid big losses – it aims to *min*imise the *max*imum losses, or *minimax*. The maximum losses are recorded in the final column of Figure 20.2. If we decide to develop, the maximum loss is the preservation benefit. If we decide to preserve, the maximum loss is the development benefit. This should occasion no surprise. The matrix does not seem to tell us much that we could not have deduced intuitively.

The SMS approach takes the game theoretic structure of Figure 20.2 as its starting point. Its first observation is that the lower is B_D the more is the minimax solution likely to be preferred. That is, we should ensure that B_D really is large before making irreversible decisions. This feature of the approach is similar to the requirement in the Krutilla–Fisher approach that we do not blindly accept the development benefit estimates – they must be scrutinised and shown to be large *relative to* the preservation benefits. Second, SMS points out that we often do not know the size of B_P. But if we do not know B_P we cannot compare it to the development benefits. In effect, we do not even know the matrix in Figure 20.2 which itself contains a further uncertainty in that we do not know the probabilities of the benefits occurring. Third, the minimax rule tells us nothing about *who* loses and gains. Chapters 14 and 15 have discussed the problem of intergenerational equity. We can note that irreversible costs are borne by the current generation and also by *all* future generations (unless their preferences are different to ours). If all generations were to count equally, it would seem odd to argue that it is fair for *one* generation to impose a cost on all others.

Because of these considerations – the tendency for optimism to get built into development proposals, the fact that we simply do not know what preservation benefits are likely to be, and the intergenerational issue – the SMS approach argues in favour of reversing the development bias, i.e. making preservation the preferred option *unless* it can be demonstrated that the social cost of preservation, i.e. the foregone development benefits, is unacceptably large. The resulting rules are thus imprecise and invoke criteria for choosing that are not present in the Krutilla–Fisher model.

20.5 IRREVERSIBILITY AND SUSTAINABILITY

The third approach to discrete choice has already been indicated in Chapter 14. There we suggested that sustainability considerations could be integrated into the development/preservation decision. This could be done by ensuring that the stock of natural capital was held constant over time by suitable compensating expenditures. The reasons for holding the capital stock constant are, in fact, very similar to those that the SMS approach would invoke for favouring the SMS solution: uncertainty about the future values attached to natural environments, and intergenerational justice. The sustainability approach expanded the uncertainty consideration to allow for the more generic functions of environments in protecting us from external shocks such as climatic change and loss of ecological function. SMS and the sustainability approach are thus similar in this respect.

But the essential difference between them is that sustainability appears inapplicable to the discrete choice problem. Discrete choice occurs because there is an irreversibility. The sustainability approach assumes that natural environments *can* be augmented and even created, so that, as some disappear, others compensate for the losses. The issue becomes one of just how far irreversibility is relevant. If it is an extensive feature of development then the sustainability approach does not have *general* application, and we require SMS or the Krutilla–Fisher approach (or some other approach yet to be developed). If irreversibility is not so important, then the sustainability approach has wider application.

21 · A CASE STUDY OF WETLANDS

21.1 INTRODUCTION

In this chapter we bring together some of the economic principles and methods we have examined in earlier chapters and apply them to the issue of *wetland ecosystem management*. Such ecosystems account for about 6 per cent of the global land area, and are considered by many authorities to be among the most threatened of all environmental resources. For much of the recent past, wetlands have been destroyed or altered, on a global basis, as human society sought to exploit the benefits provided by these natural systems. Extensive wetland resources, both in the industrialised and developing economies, have already been lost or are undergoing increasing change. These losses are occurring either as a direct result of conversion to intensive agriculture, aquaculture or industrial use (including waste disposal), or through more gradual degradation as the result of hydrological perturbation, pollution, recreation pressure or increasing grazing and fishing activities.

The indirect threat to wetlands often comes via external costs generated by economic activity in adjacent areas. The comparative failure (*vis-à-vis* 'point source' pollution) of 'non-point source' water pollution control policies in the industrialised economies has important implications for the continued health of wetlands. Agricultural run-off pollution is a major concern in this context. Other non-point source pollution threats to wetlands include the in-filling of wetlands by sediment and contamination by leachates from hazardous waste dumps and urban run-off.

21.2 SOCIAL INEFFICIENCY IN WETLAND RESOURCE USE

Today, there is a growing awareness that most wetlands are more valuable economic resources in their natural, or only slightly modified state, than if radically altered and intensively managed. Wetlands undoubtedly represent very valuable environmental assets with consequently high preservation/conservation value. Despite this, however, all the available evidence suggests strongly that wetlands are still not being managed in an economically optimal way. They are examples of what we could call economically inefficient habitat modification. Wetland resource users are lacking the guidance of appropriate economic signals which reflect the full social costs of their resource utilisation. The over-utilisation of wetlands (often leading to complete destruction of the existing ecosystem) has been the result of both *market* and *intervention failure* (see Chapter 4).

Social inefficiency in wetland use is connected to the fact that wetlands are *multifunctional resources* and that some of the multiple uses conflict with each other. In particular, because of the spatial location of the majority of wetlands, i.e. along rivers, on coasts and on level terrain with inherently fertile soils, multiple use pressure is inevitable. In this sense the '*natural*' use conflicts that arise can be considered to be in some sense almost inevitable. Nevertheless, use conflict and inefficient, socially costly, wetland management decisions also occur because of conflicting social objectives. Non-integrative government policy and even inefficient government interventions have resulted in *'created' use conflicts* and consequent sub-optimal wetland protection levels. Because wetlands cover the interface between aquatic and terrestrial ecosystems, they are somewhat problematic for traditional government resource management policies and institutional frameworks. Wetlands have tended to be neglected to some extent because of the sectora nature of established resource management and pollution control policies (see Chapter 11).

Conservation/preservation values often do not have any readily available market expression, unlike a number of the possible wetland development values (e.g. agricultural output value, residential housing and industrial facility value). Economically inefficient habitat modification has therefore been encouraged as natural and semi-natural wetlands have been completely or partially converted to other land uses. In the process it has almost certainly been the case that large, but unexpressed, social benefits have been sacrificed for smaller, but tangible and monetised, development benefits.

A further complication in a range of wetland use conflict situations, arises because of the potentially 'discrete' nature of the competing resource demands which precludes joint usage. Utilising a wetland as a 'natural' waste-water tertiary treatment process, for example, precludes recreation activities. In extreme situations it may be the case that such conflicts are also characterised by the presence of irreversibility. One use of the wetland, if sanctioned, may preclude another use not just in the present, but for all time. Thus the complete in-filling of a wetland to allow, say, housing development may represent an irreversible policy decision.

The true nature and extent of the 'discreteness' of wetland conflict situations and the irreversibility problem is as yet uncertain. Much depends on the structural and functional values (to be defined in the next section) of a given wetland, and on the scientific feasibility and practicability of wetland loss mitigation strategies (i.e. creation and restoration programmes).

In order to make some headway in this inherently uncertain resource management context, it is first necessary to look more closely at the definitional problems that surround wetland resources. Such problems have implications for any economic analysis of wetlands management and can become major legal and financial issues in the policy-making process.

21.3 WHAT ARE WETLANDS?

Although a universally acceptable definition of the term 'wetland' does not exist, the Ramsar Convention on Wetlands of International Importance (1971) offers the following definition:

> areas of marsh, fen, peatland or water, whether natural or artificial, permanent or temporary, with water that is static or flowing, fresh, brackish or salt, including areas of marine water the depth of which at low tide does not exceed six metres.

Because of the diversity of ecosystems within the wetland group a number of detailed classification systems are possible. Simplifying matters somewhat, Table 21.1 distinguishes four basic types and relates these to the output of goods/services and other functions provided by wetlands. Not all wetlands possess equally valuable structural aspects (i.e. tangible items such as plants, animals, fish, soil and water) or are likely to be equally productive in providing functional services (i.e. life-supporting services, pollution assimilation, cycling of nutrients and maintenance of the balance of gases in the air).

Table 21.1 Functions/services of the most threatened wetland types.

Wetland types	Functions/services
(1) Floodplains	Flood storage; flood protection; wildlife habitats; nutrient cycling/storage and related pollution control; landscape value; agriculture; recreation (incl. hunting); reduction of water erosion impact; storage of ground water and recharge
(2) Coastal wetlands	All those listed under (1) above, with the exception of aquifer recharge; shoreline protection/storm damage buffer zones; recreation; extended food web control; salinity balance mechanism; commercial goods output
(3) Wet meadows	High bio-diversity; hydrologic cycle control; landscape; water quality and aquifer storage; buffer zones for agricultural run-off; recreation
(4) Peatlands	Same as (3) above; global atmospheric cycle; resource extraction (energy and non-energy); specialised habitats

The benefits provided by wetlands often extend beyond the boundaries of the wetland itself and, for certain classes of wetlands, can be globally significant. Many wetlands support important, migratory fish and bird populations. It was this aspect of wetland value that led to the Ramsar Convention. Organic soil wetlands, under natural conditions are also net carbon sinks. They are therefore important links in the global cycling of carbon dioxide and other atmospheric gases. Drainage and despoliation of these wetlands over time has reduced or eliminated net carbon sinks, substituting instead net carbon sources. These wetland related carbon dioxide releases form one element in the complex 'greenhouse effect' and global warming trend (see Chapter 13).

21.4 TOTAL ECONOMIC VALUE OF WETLANDS

Some caution is, however, required in any rank-ordering of different wetlands. This exercise should, in principle, be based on a full appreciation of the total economic value (see Chapter 9) of wetlands. While some measurement and valuation of direct use value, expressed in terms of environmental commodities and amenities of direct benefit to humans, have been undertaken, indirect use value has remained difficult to quantify. Not enough is known about wetland functions such as, for example, life-support and pollution assimilation. Equally, the non-use value (existence and bequest) of wetlands has not been quantified, but as we pointed out in Chapter 10, studies of other environmental resources suggest that their existence values are positive and significant. It can therefore be argued that all the remaining wetlands stock ought to be protected, even though not all individual wetlands are considered to be uniformly valuable at present.

Table 21.2 Spatial location and threats to wetland resources.

Wetland types	Dominant causes of management failure
(1) Floodplains	Interference with hydrology; dams; river walls; rehabilitation of floodplains
(2) Coastal lagoons, river deltas and estuaries	Land reclamation; pollution; industrial development; abstraction of water for irrigation; reduction of aquifer supply and surface waterflow; recreation pressure; hunting activities
(3) Wet meadows	Drainage schemes (on- and off-site)
(4) Peatlands	Drainage schemes and agricultural reclamation; resource extraction (energy and non-energy); forestry

21.5 SOURCES OF INEFFICIENCY IN WETLAND RESOURCE USE: MARKET AND INTERVENTION FAILURE

Table 21.2 lists a number of threats to the main wetland types. In terms of market failure there is evidence in the wetlands context of information failure. Because of the lack of property rights and prices there is a lack of incentive to the full disclosure of information on wetlands. Clearly the full (use and non-use) value of wetlands is not widely known or appreciated. In some cases, the precise location and extent of wetlands are not known to public agencies, or wetland owners may be unaware of management programmes that are officially available.

Externality/public good failures are also relevant to wetlands.

They have suffered external cost damage due to pollution generated by agricultural and industrial activities located outside the wetland boundaries (in some cases many miles distant). Wetland conservation value has public good characteristics (non-exclusion and non-rivalry in consumption) across individuals now, and across individuals now and in the future. It is likely that the market will under-supply conservation benefits in the presence of such 'publicness'.

For wetlands that yield recreation opportunities, there is the danger that utilisation levels will result in congestion and subsequent decline in resource quality, as the area's carrying capacity is exceeded. For these '*congestible goods*', exclusion results in efficient use only if consumers pay for their use in accordance to their own willingness to pay for that use. There are, however, often obstacles to both collecting payment and eliciting full evaluations, or exclusion itself may not be feasible.

The development of wetlands can in a number of circumstances involve the irreversible loss of wetland preservation/conservation value. It may be that wetlands option value is a significant factor for many people and this would be lost in the presence of irreversibility. The desire to keep one's options on possible use of the wetland open will be partly individual, but may also involve bequest motivations towards one's children or future generations. Individuals may also place existence value on wetlands, i.e. the knowledge that the stock of wetlands will continue to be conserved, even though no individual current or future use is anticipated, may itself generate utility. The in-filling of wetlands may represent permanent removal and therefore a denial of opportunity for future generations to benefit from them (i.e. a failure to meet the intergenerational efficiency objective).

Because of the general absence of integrated resource management policies, intersectoral policy inconsistency frequently occurs and results in intervention failure. Table 21.3 presents a typology of the possible causes of intervention failures related to wetlands.

The interface between agricultural development policy and wetland conservation is an appropriate context in which to analyse such intervention failure. Examples can be drawn from Europe (e.g. saline soil marshes in the UK, France and Holland) and from North America (e.g. northern prairie pot-holes and southern bottom-land forested wetlands). In all these cases, high-value wetlands have suffered considerable losses due to conversion to intensive

Table 21.3 A typology of possible causes of intervention failures.

(1) *Intersectoral policy inconsistency*	
(*a*) Competing sector output prices:	Agricultural price fixing, and associated land requirements; non-market energy prices
(*b*) Competing sector input prices:	Tax breaks on agricultural land and forestry capital; low-interest loans; conversion subsidies (drainage, fill, flood protection, flood insurance); subsidies on other agricultural inputs (fertiliser, pesticide, energy, machinery) and on research and development
(*c*) Land-use policy	Zoning; regional development policy; direct conversion policy (roads, housing, tourism etc. plus secondary impacts).
(2) *Counterproductive wetlands policy*	
Institutional failure:	Lack of monitoring capacity and information dissemination; non-integrative agencies structure

agriculture. The agricultural conversion processes (up to the mid-1980s) have been artificially stimulated by a range of subsidies, price guarantees and tax incentives given to farmers.

21.6 METHODOLOGIES FOR THE MEASUREMENT OF WETLAND USE INEFFICIENCY

From the conservation viewpoint, because of the variation among individual wetlands, the significance of their structural/functional values can only be determined in detail on an individual or regional basis. Thus wetlands could be assessed in terms of the regional scarcity of affected habitat and landscape types. The commercial output value of wetlands could then be added to the habitat/landscape criterion. Finally, the presence of endangered or threatened species would augment the wetland's conservation value. In this way some broad wetland management classes could be distinguished once a comprehensive wetland resource inventory had been completed.

Economic analysis indicates that, because of the lack of substitution and the limited availability of mitigation measures, combined with general uncertainty, the development of high rank-order wetlands represents an irreversible policy decision. Such a decision should more often than not be postponed indefinitely.

For lower rank-order wetlands a number of mitigation possibilities may exist and development decisions are to some extent reversible. Mitigation may take the form of off-site measures (i.e. conserving and/or restoring similar wetlands elsewhere in the region; creation of artificial wetlands). On-site measures may also be available, such as new institutional rules covering the use of the wetlands or new, more intensive management strategies.

A basic requirement of any wetland resource management strategy is a better understanding of the value of wetlands both in use and non-use value terms. A range of economic techniques has been used to place monetary values on wetland goods, services and functions (see Table 21.4).

Table 21.4 Valuation of wetland structure and functions.

Type of benefit	Economic valuation techniques	Studies reported in the literature
Wetland conservation value *Direct output* (goods/services) *Benefits* (fish, commercial and sport; furs; recreation)	Public prices; marginal productivity value; market pricing; participation models with unit-day recreational values; hedonic pricing; travel cost models; contingent valuation	Lynne *et al.* (1981) Batie and Wilson (1979) Gupta and Foster (1975) Park and Batie (1979) Costanza *et al.* (1987) Mendelssohn *et al.* (1983) Thibodeau and Ostro (1981) Brown and Pollakowski (1977) Bishop and Heberlein (1980) Gosselink *et al.* (1974)
Indirect functional benefits (flood control; ground-water recharge; waste treatment; atmospheric and life-support functions)	Damage cost avoided analysis; alternative/substitute costs; energy analysis	Costanza *et al.* (1987) Gupta and Foster (1975) Tchobanoglous and Culp (1980) Fritz and Helle (1979) Williams (1980) Thibodeau and Ostro (1981) Kahn and Kemp (1985)
Option value and non-use value	Contingent valuation	Costanza & Faber (1989

Market prices have been used to value natural wetlands as a habitat for commercially harvested fish and animal species. Analytically, the problem is to determine the marginal productivity of an acre of wetland net of any human effort effect.

Studies estimating the value of coastal marshes in the USA have come up with quite a wide range of monetary marginal product and productivity figures; from $0.30 per acre (1981 prices) to $25.36 per acre per year (1983 prices) for shellfish and fish output.

Attempts have also been made to estimate wildlife and visual–cultural benefits of wetlands using market land prices as indications of the opportunity cost of wetland preservation. The economic benefits of these two service flows were implicitly derived from the prices paid by public agencies to purchase wetlands for conservation purposes, with due regard for the natural systems attributes of the land. Data on more than 8,000 acres of wetlands acquired by public agencies was analysed and a figure of $1,200 per acre was selected as representative of the capitalised value of wildlife benefits from the 'highest quality' land. The capital value of the 'highest quality' open-space land was estimated to be $5,000 per acre. Prices for lower grades of land were obtained by scaling to 'expert' quality-scoring indexes developed by ecologists and landscape architects. The main analytical weakness of the analysis was that consumer surplus was not estimated and therefore benefits were underestimated by the amount of consumer surplus that did exist.

The *hedonic pricing* approach requires data on, and makes a number of assumptions about, household mobility and informational requirements, as well as the operation of the residential property market. Analysts using this method attempt to estimate the first-stage marginal implicit price function for proximity to water and water-related open space (defined in the USA as 'setback'). In one such study, sample areas were all located close to one of three lakes within the Seattle city limits. The data used consisted of market sales information for dwelling units in these areas over the period 1969–74. The results suggested that variation in setback did have a substantial effect on property sales value.

Other US analysts have tried to construct an index of attribute factors to value coastal waterfront land. Data for land sales in Virginia Beach, Virginia, over the period 1953–76 have been used to empirically estimate a hedonic price equation. It was again found that an increased level of amenity gave increased values and that, over time, the annual value of the amenity increases (see Table 21.5).

The hedonic price approach can, at best, only capture part of the aggregate wetland value. Take, for example, the limiting case of a remote and unique wetland site threatened by mining development

Table 21.5 Annual values (in $) of waterfront amenity and present value, Virginia Beach, Virginia.

	Artificial channel		Natural bay	
Year	Land unit = 0.30 acre; frontage = 100 feet	Land unit = 0.75 acre; frontage = 150 feet	Land unit = 0.30 acre; frontage = 100 feet	Land unit = 0.75 acre; frontage = 150 feet
1955	182	461	466	515
1965	192	473	568	763
1975	203	557	699	1,064
1985	216	652	848	1,402
1995	231	760	1,016	1,783

Source: Adapted from Shabman and Bertelsen (1979).

and pollution. The scarcity of local residents – as opposed to recreationists – would preclude the use of either wage or property data, and further, the scenic vistas and other elements of the site are unique. Non-use values are also, of course, not being captured by this method.

Wetlands have significant values as recreational areas, and three valuation techniques have been deployed in order to capture this aspect of the ecosystem service. A *participation model* based on national unit-day values of recreational activities undertaken in the Charles River wetlands near Boston (USA) has been used to value five types of recreation activity; Table 21.6 summarises the results. The limitations of this type of approach include the lack of meaningful coefficients linking water quality and recreational activity, as well as the artificiality of the 'recreation day value'. It is doubtful whether the latter represents an adequate estimation of the value of recreation sites to the average user.

Travel cost methods (in which travel costs are taken as surrogates for visit market prices) have been applied in cases where natural wetlands provide recreation services. A recent (1987) example contained a survey of recreationists using wetlands in Terrebonne Parish, Louisiana, over a period of a year. The willingness-to-pay results are, however, questionable because the number of visitors who were non-local was quite small. The value of travel time was expressed in terms of a cost of travel time quantified via foregone wages. The typical user group (2.72 persons) was estimated to forego average total hourly wages of $26.90. Annual total willingness to pay ranged between $2 million and $5 million.

Table 21.6 Value of recreational activities in Charles River wetlands. (The total is $187.74 (PV = $3,130).)

(A)	Days of use/acre/year	1.27	0.46	0.48	0.69	9.05
(B)	Expenditure/user/year	$128.59	$262.45	$138.34	$132.48	$128.59
(C)	Pay lost/user/year	$13.09	$35.71	$12.61	$2.97	$13.09
(D)	Extra willingness to pay/user/year	$173.98	$277.04	$202.29	$121.26	$173.98
(E)	Days of activity/user/year	12.5	8.3	12.7	21.0	28.0
(F)	Total recreational value/acre/year	$32.07	$31.87	$13.35	$8.43	$102.02

Notes: Rows B, C and D yield the cost and consumers' surplus per person per year; row F gives total recreational value as A (B + C + D) divided by E.
Source: Thibodeau and Ostro (1981).

Another travel cost study examined the value of Canadian goose hunting in the Horicon Marsh area of central Wisconsin during 1978. Using the traditional zonal variant of the travel cost method, they found individual willingness-to-pay estimates ranging from $8 to $32 depending upon whether travel time and time at the site were included.

The travel cost method works best when recreation visitors travel from a wide range of distances to a wetland site and only visit that one site. This being the case, it is fair to conclude that the method is not adequate as an estimator of total wetland conservation value. Wetlands are not homogeneous resources. Structural and functional characteristics vary, to some extent, from site to site and certainly from type to type.

In Broadland, Norfolk, Halvergate marshes have become an environmental *cause célèbre*, as they represent a strategic part of the last remaining extensive stretch of open grazing marsh in eastern England. The conservation-versus-development conflict that raged between 1980 and 1985 over these marshes would not, however, have been reduced by the introduction of travel cost type analysis and valuation into the decision process. For nine months of the year Halvergate Marshes are practically inaccessible. Non-use values are presumably very important in this case, as well as political factors and the test-case nature of the dispute. In principle, non-use values can be encompassed within so-called contingent valuation methods.

Contingent valuation methods have proved popular in recent years. The results of a simulated market method and a contingent valuation method (administered via a mail questionnaire) have recently been compared, in order to estimate the recreational value (duck hunting) of Horicon Marsh, Wisconsin. It was found that the contingent valuation method yielded a willingness to pay of $21, as against a simulated market value of $63.

Overall, too little attention has so far been paid to the question of the comparative validity of estimates of recreational value, derived from alternative valuation methods under similar conditions or problem settings. Encouragingly, when two contingent valuation models (close-ended and open-ended question format) and a travel cost valuation model (regional model) were compared in 1985, in the context of recreational boating on freshwater in the Four Lakes region of Texas, it was found that the close-ended form of the contingent valuation method and the travel cost method provided comparable estimates of consumers' surplus for some (but not all) of the lake sites studied (see Table 21.7).

Table 21.7 Average consumer's surplus estimates (in $): travel cost and (close-ended) contingent valuation methods, Four Lakes region, Texas.

Lake	Travel cost	Contingent valuation
Control	32.06	39.38
Livingstone	102.09	35.21
Houston	13.01	13.81

Source: Adapted from Seller *et al.* (1985).

Given the inherent difficulties in this validation context it seems reasonable to expect that at least a 'second best' validation procedure be adopted. That is, that alternative method results should be compared for approximate consistency under similar conditions or problem settings. Only then can the analyst be reasonably sure that the monetary values derived are in the right 'ball-park' range.

Alternative/substitute costs approaches have been used to value the indirect benefit of wetlands as municipal water supply sources. The cost of delivering water from a wetland wellhead was estimated to be 0.773 cents per 1,000 gallons per day cheaper than the alternative supply source. The estimated capitalised value of a 10-acre wetland supplying 1 million gallons of water a day was $52,000 per acre, on the basis of annual benefits of $2,800 per acre. This same approach has also been used to value the water supply benefits supplied by the Charles River wetlands in Massachusetts. The results worked out at a daily saving of $16.56 per acre or $6.044 per acre per year ($100,730 present value, at 6 per cent discount rate, per acre).

Houghton Lake Marsh in Michigan has been tested as a wastewater treatment facility over two irrigation seasons (4–6 months long). In 1978–79 some 85 million gallons of effluent were applied to the 500-acre peat marsh. It is claimed that this natural facility saved more than $1 million as compared with the construction of an upland spray irrigation facility. Substantial operation and maintenance cost savings were also said to be achievable. Elsewhere it has been estimated that an acre of marsh substitutes for treatment plant costs of $85 and annual operation and maintenance costs of $1,475. The total cost savings had an annual present value of $16,960 (at 6 per cent discount rate).

The validity of the alternative cost approach depends critically on the satisfaction of three basic conditions.

1. That substitutes can provide similar functions or services as the natural wetland.
2. That the alternative chosen and costed is the least-cost alternative.
3. Willingness-to-pay evidence indicates that per capita demand for services would be the same at the two different levels of cost.

In Massachussetts, the US Army Corps of Engineers has compared the Charles River, a stream of relatively slow run-off and extensive wetlands, with the Blackstone River, which is characterised by rapid run-off. In a 1955 flood event, nearly 60 per cent of the flood volume passed a point on the Blackstone, while at a comparable point on the Charles, only 10 per cent passed in two days. The wetlands of the Charles River reduced the peak river flows by 65 per cent and desynchronised the peak hours from the storm itself by three days.

The value of this wetland function could be interpreted in terms of the cost of the property damage which would occur if the wetlands were lost, i.e. on the *damage costs avoided* principle. The Corps predicted the annual monetary loss at various amounts of reduction in wetland storage. Thus a 10 per cent wetland loss produced annual property losses of $707,000, while a 40 per cent loss produced damages of $3,193,000. Other analysts have predicted that the total loss of the Charles River wetlands would increase flood damage costs by $17 million, some $2,000 per acre of wetland (PV = $33,000, at 6 per cent discount rate).

In some regions marshes also offer protection from tidal surges and wind velocities in storms. It has been calculated that if the Terrebonne wetlands in Louisiana receded by one mile, expected damages in a four-parish area would increase by $5,752,816 annually. The loss of a one-acre strip running the length of Terrebonne parish coastline would increase expected damages by $128.30 per acre annually (PVs, at 8 per cent discount rate = $71,910,200 for the one mile strip; and $1,604 for one acre). If it is assumed that population grows at 1.3 per cent annually, the PV of damage loss is $1,915 per acre or $85,900,000 in total. If the wetlands were allowed to recede at their present rate, increased property damage would have a PV of between $2,136,092 and $3,133,440.

A number of estuarine and delta wetlands have been heavily polluted by agriculture and industry located upstream. One recent study has tried to establish a lower bound of the marginal damage function for reduction in the level of submerged aquatic vegetation in Chesapeake Bay, USA. The loss of the vegetation, due to nutrient enrichment, has an adverse impact on the wetlands fisheries. A 20 per cent reduction in the vegetation results in total annual losses for finfish and shellfish of approximately $1 million (1978) prices. Total loss of vegetation would produce fisheries damage costs of $13 to $14 million. These estimates are only lower bound figures because no account is taken of damage to waterfowl, hunting, bird watching and other recreational activities. Since the Bay represents a unique and important ecosystem, existence values are also likely to be quite large.

The *opportunity cost approach* represents a pragmatic perspective on the valuation dilemma, without itself being a valuation technique. Thus the opportunity cost of unpriced wetland functions and services can be estimated from the foregone income of potential development uses.

In the case of Broadland Marshes, Norfolk, for example, more intensive agricultural uses are the only feasible development options. Studies undertaken in the UK suggest that such agricultural drainage and conversion schemes have often produced negative net benefits and, therefore, that the social opportunity cost of conservation, i.e. the costs of retaining the area as wetland, is unlikely to be very high, and may be zero.

In the USA the opportunity costs of preserving coastal wetland sites have also been quantified. Two of the major pressures for development of Virginia's wetlands, for example, have been for residential home development in the urbanising coastal areas and for water access (marinas) and recreation home development in the more rural areas of the state. Two studies undertaken in the 1970s, used a hedonic price equation, in order to regress land sale prices on a set of explanatory variables representing individual land unit characteristics. The characteristics included measures of water access and waterfront location created from filled wetlands. The results are summarised in Table 21.8.

Table 21.8 Wetland development values: Virginia Beach[a] and Captain's Cove[b], Accomack County, Virginia.

	Virginia Beach land units with frontage to open water (¾-acre unit with 150ft frontage)	Virginia Beach land units with frontage to man-made channel (¾-acre unit with 150ft frontage)
Value per unit[c] (gross benefits)	$14,000	$6,500
Value per acre (gross benefits)	$19,000	$8,600
	Captain's Cove (no fastland alternative)	Captain's Cove (fastland alternative)
Marina (5 acres)		
PV total net benefits	$29,000	$-600
PV marginal net benefits	$5,800/acre	$-120/acre
Housing (14 acres)		
PV total net benefits	$30,000	$400
PV marginal net benefits	$111/acre	$111/acre
Housing (136 acres)		
PV total net benefits	$35,000	$5,400
PV marginal net benefits	$37/acre	$37/acre

[a]Area where heavy pressure and in-filling has already taken place, i.e. high marginal preservation value.
[b]Located in rural area.
[c]Net development benefits = (gross benefits) – (dredging + filling costs).
Source: Shabman *et al.* (1979).

In Virginia Beach, the opportunity costs of preservation, if development is designed for residential units, are not particularly high. These gross benefit estimates have to be further reduced by the development costs incurred to fill and develop the wetlands. In Captain's Cove, five acres of marsh were initially destroyed to construct a marina and common recreation area. This marina provided water access to 3,700 interior land units. The development benefits depend crucially on whether so-called fastland sites are available as alternative development options. The marina had a marginal value of £5,800,000 per acre, but subsequent wetland loss to provide housing produced only minor benefits (see Table 21.8).

These results all assume that no fastland alternative site is available. When a fastland alternative location that provides water access is available without necessitating wetlands loss, and is used as the next best alternative for comparison, net wetland development values are only positive when large areas of marsh are converted.

The opportunity cost approach does show that estimates of wetland development benefits can provide a focus for an application of a *safe minimum standard* rule for wetland conservation (see Chapter 20). It is also in line with an argument made earlier in this chapter that, given the general, on-going loss of wetlands and the uncertainties surrounding the precise magnitude and significance of wetlands services/functions, a positive case for wetlands development is required, rather than vice versa. The use of the development value has great utility in this uncertain policy context.

21.7 MECHANISMS FOR SOCIAL COST INTERNALISATION

The previous sections of this chapter have shown that although not all wetlands are equally valuable, such ecosystems represent sources of significant service and function value. It has also been demonstrated that wetlands are being lost at an unacceptable and economically inefficient rate, due to a combination of market and intervention failures. Following the schema set out in Figure 21.1 the related questions now to be addressed are:

1. What policy instruments are available in order to form the basis of a sustainable wetland management strategy?
2. What can be done, and is being done, to slow down or reverse the wetlands loss trend?

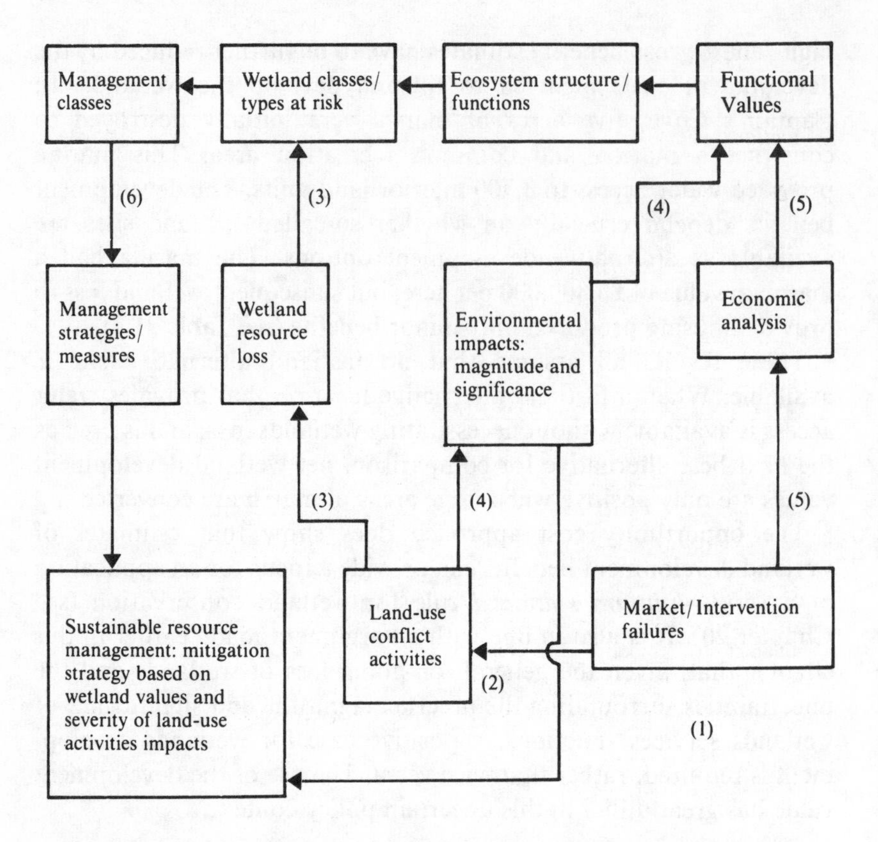

Figure 21.1 Factors leading to ecosystem change and the need for a more rational management strategy.

A successful management strategy will need to be based on data indicating wetland value, the sources of land use conflict and the severity of land use activities impacts (see Table 21.9). The policy instruments package will need to be tailored, to some extent, to fit the wetlands loss contexts being examined.

Any loss mitigation strategy would have to be based on a comprehensive wetland inventory and practicable restoration or artificial creation techniques. The development of some wetlands may then be 'acceptable', as long as development net benefits remained positive despite the inclusion of all mitigation costs. Table 21.10 presents a wetlands loss mitigation matrix which could play an important role in the initial gross screening stages of any management strategy.

Table 21.9 Development activity environmental impacts.

Activity type and location	Magnitude and significance of impacts
On wetland site	
1. Agricultural conversion (to higher-intensity regimes); mining excavation (phosphates, peat, etc); industrial/residential developments (housing, ports and marinas, etc.)	Immediate, total and essentially irreversible (either on technical and/or cost grounds) impact; total wetland habitat loss
2. Causeway/road construction; tidal power barrage; canalisation; oil/gas well-drilling	Permanent, often extreme and potentially irreversible; major change to hydrologic regime in wetland, resulting in loss of wetland vegetation
3. Disposal of dredging spoil; disposal of municipal waste and some industrial wastes	Persistent but gradual change to soils and substrate, often irreversible on cost grounds
4. Sewage effluent disposal; agricultural run-off; drainage from mining and afforestation on peat and other upland bogs, recreation pressure (people, housing and other facilities)	Chronic impact, sometimes persistent and widespread; significant water quality deterioration and loss of ecological diversity due to nutrient over-enrichment
5. Pipe-laying; geophysical oil survey	Temporary and usually localised damage
Off-wetland site	
2. Hydroelectric dam upstream of wetland; upstream water diversions or withdrawals	Significant impact on water levels in downstream wetland
4. Agricultural run-off; industrial waste discharges to water courses (especially chemicals)	Water quality deterioration and species loss

Source: Adapted from Nelson and Logan (1984).

Table 21.11 summarises the main potential management measures that have so far been formulated. For unique irreplaceable and complex wetlands, outright development should be precluded and any on-going usages strictly regulated via institutional rules. It is these wetlands that should be subject to the strongest regulatory and fiscal management measures. Any activity likely to cause wetland displacement or permanent hydrologic and/or soil or substrate change should be prohibited. Other activities could be controlled by site-specific permits or multiple-use management regimes operated, for example, via management agreements such as those worked out

Table 21.10 Wetland loss mitigation matrix.*

Wetland management class (by relative value and scarcity)	1 Wetland displacement	2 Permanent hydrological change	3 On-going soil or substrate change	4 Persistent water quality decline	5 Temporary damage
1. Very high value wetlands; unique and irreplaceable	NMF/NOC	NMF/NOC	NMF/NOC	MF/NOC	MF/NOC
2. High value wetlands; regionally scarce or becoming scarce	NMF/NOC	MF/OC	MF/OC	MF	MF
3. High-to-medium value wetlands; regionally/nationally abundant	OC	MF	MF	MF	MF
4. Medium-to-low value wetlands	OC	MF	MF	MF	MF

*Notes: NMF=no mitigation of impacts is feasible; NOC=no off-site compensation is acceptable; MF = impact mitigation feasible on-site and/or off-site; OC = off-site compensation possible.

Source: Adapted from Nelson and Logan (1984).

under the UK's Environmentally Sensitive Areas Policy (1986). This class of wetlands is top priority for public acquisition and management, especially in regions where rates of wetland conversion have been high and a relatively small wetlands reserve remains intact. Pollution control measures off-site may well be required to maintain and improve water quality levels. Public and private land banking is required.

Table 21.11 Wetland management policy instrument options.

	Instruments		Comments
1. *Regulation*			
Planning designations	Prohibitions zoning and designation Permissions	subject to licence/permit	To regulate wetland uses and activity impact mitigation, with or without compensation. National, regional or local permits with uniform conditions. Zoning and designation of wetlands by permitted use or activity; UK SSSIs, nature reserves, national/regional parks, global biospherical reserves (Ramsar Convention); varying degrees of site protection in practice
Pollution abatement	Specific controls over land use Ambient quality standards	subject to licence/permit	Increased stringency in pollution control policy; ambient environmental quality standards/objectives and emission limits
2. *Acquisition and management*			

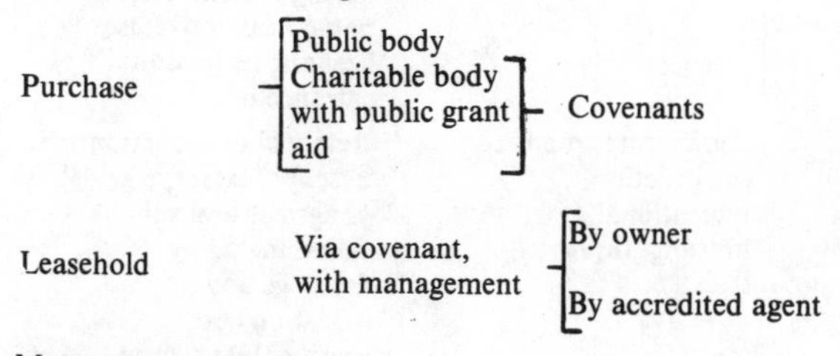

	Instruments		Comments
Purchase	Public body Charitable body with public grant aid	Covenants	
Leasehold	Via covenant, with management	By owner By accredited agent	
Management agreement with landowner subject to agreement			

Table 21.11 *cont*

3. *Incentives and charges*

Subsidies for conservation management	ESAs; management agreements Compensation for wetland, wildlife, crop damage Conservation practices (headlands hedgerows, etc.)	
Tax incentives for conservation management	On land On inputs and other costs	Income, capital gains and estate tax exemptions for protected wetlands; deductions or credits on wetland donations or sales for conservation; property tax relief for protected wetlands
Wetland loss mitigation charges		Wetland development fees and related public trust fund for conservation; mitigation land banks (unadulterated or restored wetlands)
User charges	Entrance fees Licences	Wetland hunting, fishing licence fees; non-consumptive use licences; recreation entrance fees
Development activity subsidies	Agriculture, road construction, recreational housing, forestry, etc.	Removal or reduction in scope/extent, e.g. of agricultural subsidies, including drainage and irrigation cost sharing, loans, crop flood insurance, commodity price supports; tax deductions for development costs

For other wetlands, general regional permits and local zoning, subject to uniform conditions, are appropriate. Some areas could be covered by temporary public acquisition by leasehold and multiple-use management systems. User charges could finance on-site mitigation measures. Some off-site restoration of degraded wetlands via mitigation banking is feasible, if the development activities proposed fall short of a wetland displacement impact. Public and private land banking could be encouraged.

In certain cases, incentives and charges may offer the most cost-effective management control. Conservation tax benefits and subsidies could be encouraged in this context, and agricultural tax benefits or subsidies removed. Wetland developer mitigation charges and other user charges could be used to generate revenue for on-site and off-site mitigation, as well as for off-site amelioration in the form of unaltered and/or restored wetlands elsewhere in the region.

22 · ENVIRONMENT AND THE DEVELOPING COUNTRIES

22.1 THE PROBLEM

Natural and man-modified resources in developing countries exhibit extensive interdependence. The extent of tree cover has a direct influence on soil fertility and water run-off rates: the trees serve both a protective and moisture-retaining function. As tree cover is reduced through clearance for agriculture, timber production and fuelwood demand, so the carrying capacity of the recovered and surrounding soils may be reduced, depending on the land management regime that succeeds deforestation. Agricultural productivity declines and livestock may be moved to more marginal grazing land. As soils dry out, the risks of flood and wind erosion increase. In the limit, desertification occurs. The water run-off takes soil from the land to produce sediment in rivers, irrigation channels, reservoirs and natural lakes, reducing the water resource input to agriculture, polluting drinking-water supplies, reducing the capacity of hydropower reservoirs, raising water levels and increasing the size of flood plains, and damaging fishery resources. As the demand for woodfuels grows and tree cover is reduced, so the stock of trees diminishes as the resource is 'mined' instead of being managed on a sustainable basis. With fuelwood scarcity comes the diversion of animal dung from its traditional role as soil nutrient to direct burning for fuel. Reduced dunging of the land further impairs soil fertility.

Soil fertility, water as irrigation and trees as a protective input to fertility and the source of fodder inputs to livestock are thus closely interrelated resources. In the absence of imported food or large supplies of fertiliser, the impact of any mismanagement of this resource base shows up in reduced output of food and reduced quality, and perhaps quantity, of water for human consumption. Of equal importance, resource mismanagement is recursive: damage to the resource base now makes the ability to recover from that damage, or from exogenous events such as reduced rainfall, less in the future. Thus as soils reduce in humus so hitherto manageable variations in rainfall become catastrophic events, and previous catastrophes increase in size. The adaptability and renewability of the resource system is impaired, with consequent costs to be borne in the near- and long-term future by existing and future generations.

Developing countries, especially poor ones, have a more immediate dependence on their renewable resources than have developed countries. In the latter case, dependency is still present, but the role of technology and the application of capital makes the linkage more 'roundabout'. The sophistication of water treatment plants, for example, makes the issue of water pollution less immediately damaging, important though it is. Since rural communities in developing countries drink water directly from its source, the reliance on the self-cleansing properties of natural rivers and lakes is more direct. The pollution of those supplies thus has immediate health effects. The same contrast between developing and developed countries holds for forests and woodland. The transition to fossil fuels in developed countries is largely complete, but woodfuels and crop residues are a basic and extensive source of fuel in developing countries. As trees are depleted so the rural household (and many urban ones too) has less capability to cook food and provide a source of heat. That reduced capability is not offset, other than at the margin, by substitution of fossil fuels, due to income constraints.

While there is a complex of reasons at work, it is primarily the pressure of increased population that has led to a situation in many developing countries where the rate at which land and water renewable resources are being used is in excess of the regenerative capacity of those resources. In short, renewable resources are being 'mined' in much the same way as exhaustible resources: their stock is being reduced. But the rate of depletion of renewable resources causes greater concern for the reasons given above: the depletion of any one of them impacts on the availability of the others; the rates of depletion feed back on the ability of the resources to renew themselves; and the resources are directly consumed as water and fuel, and both directly and indirectly as food. It is these three features:

- directness of impact
- resource interconnectedness
- damage to regenerative capacity

that provide the fundamental rationale for concern about natural resource mismanagement in developing countries. The rest of this chapter is devoted to an expansion of this theme.

22.2 THE DEPENDENCE ON NATURAL RESOURCES

Fuel

Many developing countries have a substantial reliance upon woodfuels (fuelwood and charcoal) and, to a lesser extent, crop wastes as their source of energy. Table 22.1 shows the percentage degree of reliance on these 'traditional' fuels. Countries with the highest reliance on traditional fuels also tend to be the poorest. Indeed, in many ways, the extent to which traditional fuels are used

Table 22.1 Developing country dependence on traditional fuels (traditional fuels as a percentage of total primary energy)[a].

Nepal	93	Senegal	60
Malawi	92	Yemen AR	58
Tanzania	91	Fiji	55
Guinea-Bissau	89	Indonesia	49
Ethiopia	89	Sri Lanka	45
Sudan	83	Botswana	45
Paraguay	83	St Vincent	44
Niger	80	St Lucia	39
Uganda	71	Costa Rica	38
Kenya	71[b]	Bolivia	35
Gambia	70	Morocco	35
Haiti	70	Zambia	35
Bangladesh	70[c]	Zimbabwe	30
Solomon Islands	66	Turkey	24
Liberia	64		

[a]Defined as woodfuels = fuelwood and charcoal. In some cases bagasse and other crop residues are included.
[b]Authors' estimate.
[c]A range of 69–76 per cent is implied in the Bangladesh Energy Assessment.
Source: Computed from World Bank / UNDP *Energy Assessments* for each country, except Kenya. Estimates are for 1980–83 except for Indonesia (1978), Haiti (1979) and Kenya (1984).

can be regarded as an (inverse) indicator of the stage of economic development. Table 22.1 reveals two fundamental facts. First, if the sources of woodfuels are not sustained, there will be a major direct impact on the population. Second, that problem will be greatest in the poorest countries who can least afford to import petroleum product substitutes for direct household use of woodfuels, such as kerosene and liquid petroleum gas (LPG). (Note that electricity is also a very limited substitute since it is not used for cooking.)

There is in these statistics alone a cause for concern. The extent to which existing wood stocks are being depleted is more difficult to gauge, but Table 22.2 assembles some of the data and indicates that depletion of potentially renewable stocks is already a serious problem in a number of developing economies. For example, Nepal's annual consumption of wood appears to be more than 130 per cent greater than the mean annual increment in wood stocks (the 'sustainable yield'). Self-evidently, such situations must, unless

Table 22.2 Illustrative examples of excess harvesting of wood over sustained yields[a].

Uganda	+21	Ethiopia	+150
Malawi	+31	Tanzania	+151
Sahel[b]	+30	Niger	+193
Sudan	+71	Haiti	+220
N. Nigeria[b]	+73	Yemen	+300
Nepal	+132	Mauritania	+893

[a]That is, harvest minus mean annual yield from existing stock. All estimates relate to a year in the period 1979–81.
[b]From D. Anderson and R. Fishwick, *Fuelwood Consumption and Deforestation in African Countries*, World Bank Staff Paper, No. 704, World Bank Washington DC, 1984.
Source: Apart from Note b, taken from data in World Bank/UNDP *Energy Assessments* for these countries.

corrected, lead to the eventual loss of the resource. If initial stocks are very high then excess consumption at levels shown in Table 22.2 could be argued to be tolerable for the near future. But the gravity of the situation can be revealed with two examples.

In Haiti, the United Nations Development Programme has estimated a reduction in the forest area of 59 per cent between 1956 and 1977. The World Bank conservatively estimates the annual current rate of decline at 5 per cent. Yet over 70 per cent of Haiti's energy demands are met from domestic wood supplies and 98 per cent of household energy demand is met from fuelwood and charcoal. With only some 200,000 hectares of forest left, Haiti illustrates the seriousness of the problem of direct consumption of a renewable fuel resource. In Nepal, forest area has shrunk from 6.4 million hectares in 1963–1964 to about 4.0 million in 1980 and the forests are disappearing at the rate of 100,000 hectares per annum. Fuelwood demand in Nepal could be met by an additional 1.2 million hectares of high-yield plantations, a planting rate of 50,000 hectares per annum by 1990 and 100,000 hectares per annum in the 1990s to get into reasonable balance by 2000.

Water

Water serves several functions. In terms of direct use it is required for human consumption and livestock, comprising some 5 per cent and 3 per cent of world water use, respectively. Irrigation accounts for 70 per cent of world use of water, and industry and mining for 22 per cent. Table 22.3 shows water use for selected developing countries, and reveals that irrigation use generally accounts for even higher proportions of total water use in developing countries. Estimates of per capita drinking water suggest ranges of 65–290 litres per day per capita for Africa for urban households with connections, 20–45 litres per day for urban households with public standposts, and 15–35 litres per day for rural households. The corresponding ranges for Southeast Asia are 75–165, 25–50 and 30–70. Table 22.4 shows estimates of the percentage of the population with access to safe drinking water in countries where access is less than 20 per cent.

Table 22.3 Percentage water use in selected developing countries.

Country	Energy production	Irrigation	Industrial	Domestic consumption	Total water use (10^9 m^3)
Egypt	1	91	3	5	30
India	negative	92	1	7	386
Indonesia	negative	86	2	12	42
Pakistan	0.4	97	0.3	2	118
Phillipines	10	72	4	14	14
Thailand	4	90	1	5	20
Brazil	5	20	30	50	10

Source: Computed from *The Global 2000 Report to the President*, Penguin Books, London, 1982, p. 147. Note that 'use' is not the same as 'consumption' since industrial water use is recycled to a considerable extent. Irrigation and drinking water are examples of consumptive uses.

The overall picture is clear. Developing countries have a marked direct dependence on natural resources – on wood for fuel and on untreated water for human consumption and cooking, for livestock,

Table 22.4 Ranking of developing countries according to access to drinking water 1975–1980 (expressed as a percentage of population).

Ethiopia	4	Lesotho	14
Mali	6	Central African Republic	16
Bhutan	7	Mauritania	16
Afghanistan	10	Papua New Guinea	16
Guinea-Bissau	10	Uganda	16
Nepal	11	Paraguay	18
Mozambique	13	Sri Lanka	19
Sierra Leone	14		

Source: UNICEF, *The State of the World's Children 1985*, UNICEF, Paris, 1985.

and for irrigation. Other resources play a similar role: fodder for livestock comes from trees, shrubs and pastures; crops depend on organic manures from animals and from crop residues; a considerable amount of food comes from the wild; medicines are frequently based on plants; the forests supply building materials, and so on. On this count alone, managing natural resources in a sustainable fashion is vital to the well-being and development prospects of the people of the developing world.

22.3 RESOURCE INTERCONNECTIONS

Land and water resources are interconnected in developing economies in ways which are more fragile than in the developed countries. While the same ecological cycles are at work, the capacity to distance people from the effects of interrupting those cycles is greater the richer the country is. The main linkages in resource systems are shown in Figure 22.1. The interconnectedness is shown by considering what happens if a single 'shock' to the system occurs, in this case removal of tree cover in excess of the sustainable yield. It is important to note that the links are not *necessary* ones, i.e. the consequences shown are not inevitable. In the first place, the negative impacts may be ameliorated by deliberate policy designed to reduce damage, e.g. by pollution clean-up, dredging of water channels, relocation of populations at risk from flood, and so on. Second, much will depend on what activity takes place when the removal of

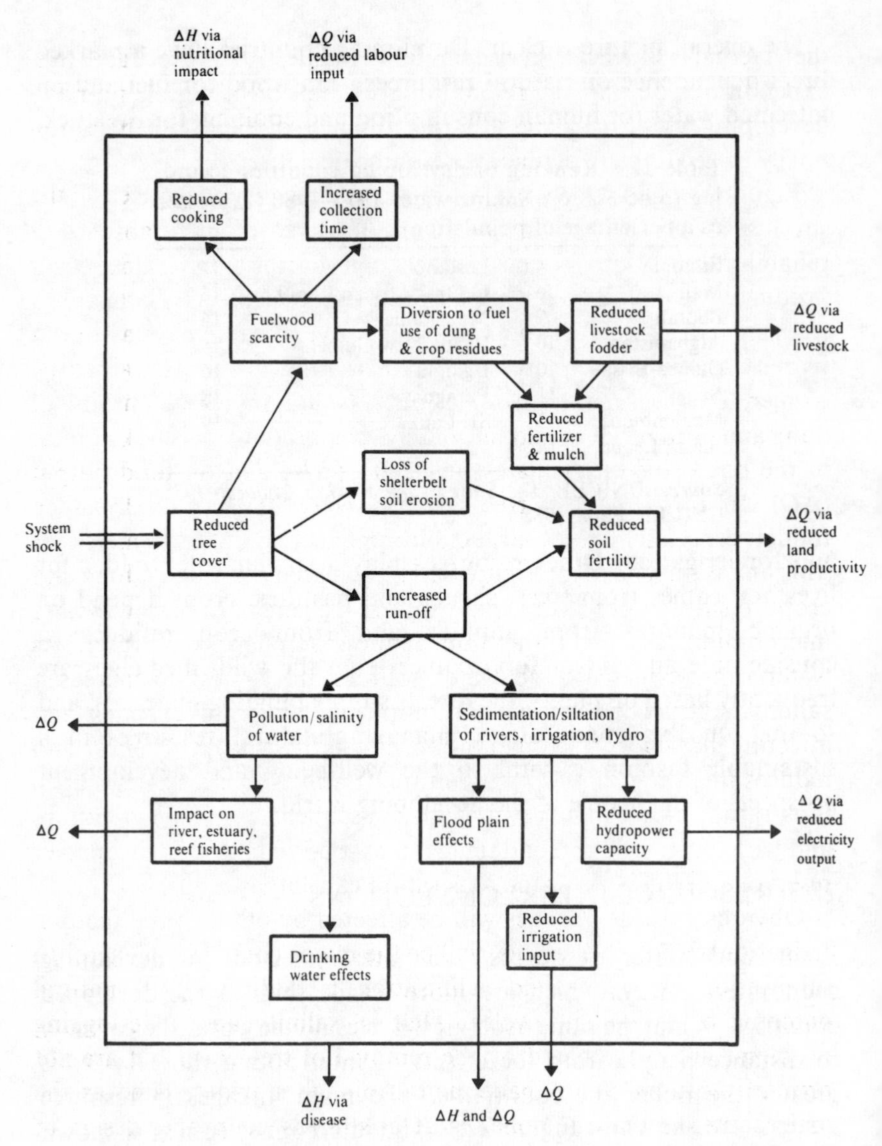

Figure 22.1 Resource interconnections in developing countries.

tree cover takes place. Left alone, soil erosion may well occur, as it will if indiscriminate grazing is allowed. But if the land is terraced and farmed on a sustainable basis, soil run-off need not increase and many of the negative consequences of tree cover removal will not occur. The reasons for thinking that the 'negative consequence' model of resource interconnection is still appropriate arise from (*a*) the factual evidence on depletion episodes, (*b*) the fact that expenditures on damage reduction are small or negligible in many developing countries, and (*c*) the absence of land rehabilitation policy to accompany deliberate resource depletion.

Figure 22.1 shows various linkages. As tree cover is reduced so the soil loses wind shelter and water run-off increases in the absence of rehabilitation. Both effects reduce soil fertility and hence land productivity. These direct impacts are supplemented by others. As trees are reduced so fuelwood becomes scarce. Adaptive behaviour to scarcity tends to take three forms: increasing distances walked by women to collect remoter sources of fuelwood, increased burning of dung and crop residues and, utimately, abandoning one cooked meal in the day. This has impacts on health (ΔH) and on food output (ΔQ). In turn, reduced tree cover directly impacts on livestock fodder, especially in semi-arid environments and upland watersheds (this link is not shown in Figure 22.1). Increased burning of dung and crop residues deprives the soil of chemical, physical and microbiological inputs and thus reduces soil fertility further. As water run-off increases so water sources are subject to increased salinity, so affecting agricultural output via irrigation channels and affecting potability. Sedimentation affects irrigation channels and reduces the water input to agriculture. It affects rivers and estuaries, thus raising flood plain levels with consequent costs to agriculture and with increased risks of damage to property and life. Hydropower reservoirs become silted and generation capacity is reduced.

Obviously, these linkages will be affected by other causal factors. Irrigation alone, for example, will be the major contributing factor to salinisation of water, and salinisation is additionally a natural outcome of hydrologic cycles. That is, salinity and waterlogging problems are exacerbated by the removal of tree cover but are not primarily caused by it. Inadequate drainage is widely cited as the main cause, so that the policy issue at stake is the proper design of new systems and rehabilitation of defective sysstems.

22.4 THE ECONOMIC COST OF RESOURCE DAMAGE

The combination of the close dependence on natural resources and the fragility of the interconnections between them does much to explain why natural resource management is important in the developing world. The fragility issue should not be exaggerated. Anyone who has worked in arid zones, for example, will know that even these areas, which look like traditional picture images of the desert in the dry season, can be transformed into lush greenness in the wet season. But the opposite is also possible. Lush forests can be quickly converted into infertile land after clearance and used for intensive livestock farming; whole mountain-sides can be stripped of vegetation as erosion removes top soil. This environmental damage has an economic cost. Some examples follow.

Fuelwood

As fuelwood becomes scarce due to deforestation, so the time taken to collect the wood increases due to the longer distances that have to be travelled. Increased collection times have a clear opportunity cost in terms of foregone farm output or other activity. There were an estimated 2.25 million rural households in Kenya in 1980. In that year, it is estimated that 1.1 million of these were in regions where tree stocks were beginning to be 'mined'. Taking the daily unskilled labour rate at that time to be US $2.6 per day, then even an extra one hour per day per household in collection time could be valued at 26 cents per household, or $290,000 per annum. Such procedures raise issues about the economic valuation of time but offer some guideline to the orders of magnitude involved. If anything, the procedure will exaggerate costs since foregone farm output will not be valued at the unskilled labour rate and because there are other adjustments to fuelwood scarcity such as choosing more 'efficient' woods, brewing less beer, and so on.

The primary use for woodfuels is cooking. As wood becomes scarce, so cooking habits may change, sometimes to the detriment of nutrition. Fuel scarcity also inhibits the introduction of new nutritious foods such as soya beans, due to the extra cooking time required. The cost of such dietary changes, and the foregone benefits from the inability to introduce new foods, is obviously complex to estimate and no empirical work appears to be available. But it is a positive cost which must be debited to resource mismanagement.

Apart from collection time and nutrition effects, fuelwood scarcity has a third impact. As the wood becomes scarce, so animal dung and crop residues are burned rather than being applied to the soil as fertiliser and soil conditioners. A study by Newcombe (1984) estimates that up to 90 per cent of cattle dung produced in Eritrea, and 60 per cent in Tigrai and Gondar, Ethiopia, is used as fuel. By estimating the crop-response functions, it is possible to calculate the incremental grain production that would result if the dung was applied to the land. Multiplying this incremental yield by the market price of the grain gives an implicit price for the dung. A second approach to valuation is to estimate the nutrient value of the dung and value this at the price of imported fertiliser (this is a minimum value since there are non-chemical benefits to soil from using dung). A third approach is to observe what prices are actually paid for dung – in Ethiopia dung is sold to urban markets. The resulting estimates produced by Newcombe were, per tonne of dung:

	Ethiopian Birr	US$ (1983)
Market price	99–238	47–114
Fertiliser substitute	45	22
Grain response	128–189	61– 91

As an illustration we can take Newcombe's estimate of dung burned in Ethiopian households at 7.9 million tonnes per annum. It follows that the foregone agricultural output by diverting dung to fuel uses is some US$600 million (an average grain-response value of $76 multiplied by 7.9 million tonnes) in a year.

Sedimentation of hydro-reservoir capacity

A major cause of sedimentation of rivers, irrigation channels and dams is deforestation. Sediment loads carried in major rivers can be very large indeed. The Lo and Ching rivers in the People's Republic of China carry sediment that corresponds to an erosion rate for the drainage area of some 1,500 tonnes per hectare per year. Rates in the five major river basins of the world vary from 2.8 to 302 tonnes/hectare/year, compared with so-called 'tolerable rates' of 10 tonnes. One recipient of sedimentation is dams. Sfeir-Younis (1985) has simulated the effects of sedimentation on 200 major dams built from 1940 onwards. At a 1 per cent constant sedimentation rate per annum, the 'live' storage capacity of the dams will be reduced by *one-third* by the year 2000. At increasing sedimentation rates (1 per cent to 4 per cent per annum), some two-thirds of capacity would be lost.

Again, as a purely illustrative calculation, we can estimate the direct cost of foregone electricity production through sedimentation. Hydropower capacity in 100 developing countries contributed 445,000 gigawatt hours (GWh) out of a total electricity demand of 1,090,000 GWh in 1980, i.e. 41 per cent. Taking the *lower* of Sfeir-Younis' scenarios the losses to these dams in the year 2000 would be 148,000 GWh. The amount of oil needed to produce this amount of electricity is approximately 37 million tonnes. At a 1988 world price of US$15 per barrel, this would amount to a direct cost due to sedimentation of over US $4,000 million. This is for a *single* year (2000).

22.5 ECONOMIC INCENTIVES AND NATURAL RESOURCE MANAGEMENT POLICY

Given that natural resource management is important in the developing world, how might economic analysis help in improving it? The first way in which it can help is to demonstrate the economic value of good resource management. In large part, decision-making in developing economies is very much focused in ministries of finance and economic planning. Indeed, this bias is fairly familiar in the developed world where treasury departments often seem to wield more power than the government departments that have spending programmes. Environmental policy is not likely to be seen to be important unless its economic dimensions can be indicated. As we have seen, this is largely an exercise in the valuation of environmental functions.

The second way in which economics can help is to indicate the effects of devising incentives to manage resource sustainably. Some of the relevant considerations are outlined below.

Because environmental economics is rooted in the foundations of neoclassical economics, it tends to emphasise the role that incentives can play in securing a more rational management of natural resources. Its presupposition is that 'rational economic man' is as helpful an abstraction for rural Africa as it is for a Western consumer. Rational economic man is assumed to respond to economic signals in a continuous fashion so as to maximise his (or her) own welfare. One way of testing the reasonableness of this construct is to see in practice how resource management responds to changes in incentives. In reality this is difficult because of the evidence which suggests that changes in *individual* incentives may do little to encourage wise resource management. It seems more likely that what has to be manipulated is a *package* of incentives. This is very likely to consist first of enabling incentives aimed at the resource user, such as more secure land tenure, clarification of resource rights, access to credit (which frequently follows on from security of tenure), and information (extension) projects. Once the user is 'enabled' to respond, the issue becomes one of altering the economic signals to the resource user, e.g. producer prices. The implication is that 'wrong' signals induce environmental degradation and that the resource user may well perceive and understand that the degradation process is coming about because of his own actions. But those actions have to be seen as responses to external incentives. Given that set of signals, resource degradation may, perversely, be a rational economic response.

The producer price argument

It is widely argued that the most efficient and administratively simple alteration to incentives is the raising of producer prices. Most developing countries have some form of price control in official markets, although the existence of widespread black markets reduces the impact of the controls. The result is that farmgate prices are often markedly below the 'border' price for internationally traded goods, the price received (f.o.b.) for exports or the price paid (c.i.f.) for imports. Table 22.5 shows illustrative ratios for selected widely traded commodities in Kenya, Malawi and Tanzania. It will be observed that Kenya maintains producer prices very close to world prices for coffee and tea, but Malawi tobacco growers receive less than one-third of the world price and Tanzanian tobacco growers receive 50–70 per cent of the world price. A similar picture arises for Tanzanian cotton and coffee, with procurer prices being around 50 per cent of world prices (the exception being cotton in 1985 due to world price changes).

The argument might be that raising producer prices close to world prices will increase the incomes of farmers, enabling them to secure surpluses which can be re-invested in resource conservation, and to expand supply. But reality is likely to be far more complex and the current state of research is simply not adequate to pronounce on the

Table 22.5 Ratios of producer prices to international prices.

	Kenya		Malawi	Tanzania		
	Coffee	Tea	Tobacco	Tobacco	Cotton	Coffee
1970	0.91		0.30	0.78	0.73	
1971	0.90	0.79	0.33	0.84	0.61	
1972	0.98	0.77	0.29	0.84	0.57	0.57
1973	0.96	0.77	0.27	0.84	0.35	0.43
1974	0.97	0.67	0.20	0.68	0.33	0.43
1975	1.01	0.75	0.20	0.70	0.52	0.36
1976	0.96	0.74	0.23	0.65	0.42	0.30
1977	0.93	0.89	0.22	0.63	0.46	0.35
1978	1.02	0.85	0.30	0.70	0.56	0.39
1979	0.99	0.75	0.24	0.51	0.51	0.29
1980	1.04	0.83	0.23	0.47	0.53	0.41
1981	0.89	0.89	0.20	0.50	0.62	0.53
1982	0.82	0.86	0.18	0.50	0.73	0.52
1983	0.90	0.68	0.31	0.70	0.67	0.47
1984	0.83	0.98	0.28	0.55	0.65	0.47
1985			0.29	0.72	1.03	0.50

Source: U. Lele, 'Agricultural growth, domestic policy and extension assistance to Africa: Lessons of a quarter century', Paper presented at the 8th Agricultural Sector Symposium, Washington DC, January 1988.

nature of the linkage from producer price to agricultural supply response to natural resource effects. Some authorities argue that price increases encourage switches between crops, but have no effect on aggregate output. Others suggest that the aggregate response is also positive.

The natural resource dimension adds further complexity. Positive supply response is likely to be through an extension of agricultural margins rather than an intensification through application of technology, fertilisers, etc. This will be more obviously the case where there is limited access to technology or limited capability to adopt it. But extended margins may involve clearance of hitherto forested land or movements into previously marginal arid areas. In both cases there are environmental costs in terms of increased desertification risks which affect medium- to long-run agricultural yields. Even with intensive responses there are issues of the type of crop which is encouraged. The presumption is that bush and tree crops provide extensive root structures and canopy cover and are therefore less erosive than many root crops such as cassava, maize, millet and sorghum. If so, policy has to consider the social costs of encouraging expansion of erosive crops.

The portfolio argument

Economic theory suggests a link between undeveloped capital markets and natural resource degradation. Given the essentially unequal distribution of incomes in developing countries, there are always institutions and individuals seeking outlets for money surpluses. With the widespread presence of exchange control and the absence of well-developed property and capital markets, such surpluses tend to gravitate to a few legitimate outlets and, of course, to some illegitimate ones. Livestock ownership is an example of legitimate investment, as is land purchase by absentee landlords. As long as this encouragement exists there is a tendency to over-stock on common land, and hence an incentive to degrade pasture. If land titlement also depends on clearance of vegetation, as it does in many cases, the same incentive exists to deforest and devegetate.

It is difficult to assess how important this portfolio argument is in the current state of knowledge. It would appear to be significant in some Latin American countries.

Exchange rates

Exchange rates are simply prices of one currency in terms of another, but the rate set for this price can have effects on natural resources. A typical situation in developing countries is the over-valuation of exchange rates, i.e. holding them above what the market rate would be if market forces were allowed to operate. Over-valuation tends to come about because of expansionary monetary and fiscal policies which push up domestic prices relative to those of principal trading partners, and through policies of protecting indigenous industrial sectors with import quotas and tariffs which keep domestic industrial prices high.

If the exchange rate is over-valued, agricultural exports are discouraged because the exporter receives less in local currency for any exports than would be the case with a lower exchange rate. Effectively, then, an over-valued exchange rate will have the same effect as maintaining low producer prices. Over-valuation thus produces an implicit tax on domestic production. Moreover, over-valuation makes imported commodities artificially cheap, so that domestic farmers find themselves competing with effectively subsidised imported foodstuffs. What evidence there is suggests that over-valuation – measured by real rates of currency appreciation – does result in lower rates of growth of agricultural output. Although the 'automatic' response to over-valued exchange rates is to suggest a devaluation, this may have little impact on farmers' incentives due to the presence of middle men, district traders, processors and trading monopolies. This network of intermediaries may soak up the benefits to exports of any devaluation, leaving producer margins little changed. In effect, the devaluation creates a 'rent' which is dissipated by those with economic power, leaving the farmer unaffected.

What does this mean for natural resource management? As with domestic prices, the issue is one of how to increase agricultural output without generating environmental degradation which then feeds back as lower yields in the medium and long term. The exchange rate evidence suggests that over-valued rates do harm agricultural output, but the impact on natural resources will depend on the types of crops encouraged (so that relative prices matter as well as absolute prices) and on the type of farmer-response. These have not yet been systematically studied.

Non-price incentives

Many other incentives are likely to be important in a package of measures designed to stimulate sustainable increases in agricultural production. Measures are also likely to be location-specific. But one type of incentive is very widely recommended in many studies – a change in the form of land tenure.

The basic argument is that tenure insecurity reduces the farmer's incentive to invest in land improvement. The sequence of relationships is shown in Figure 22.2. But there is a debate over the effect of tenurial change on natural resources and there is a great deal more to be learned before any significant conclusions can be reached. The basic contrast is made between common property systems and sets of individual land rights. In the former, land is managed by the community as a whole. In the latter, land is effectively 'privatised' by individuals. It is widely mooted that 'the tragedy of the commons' ensues in common property regimes since each individual has an incentive to increase personal gains, e.g. by adding more personal

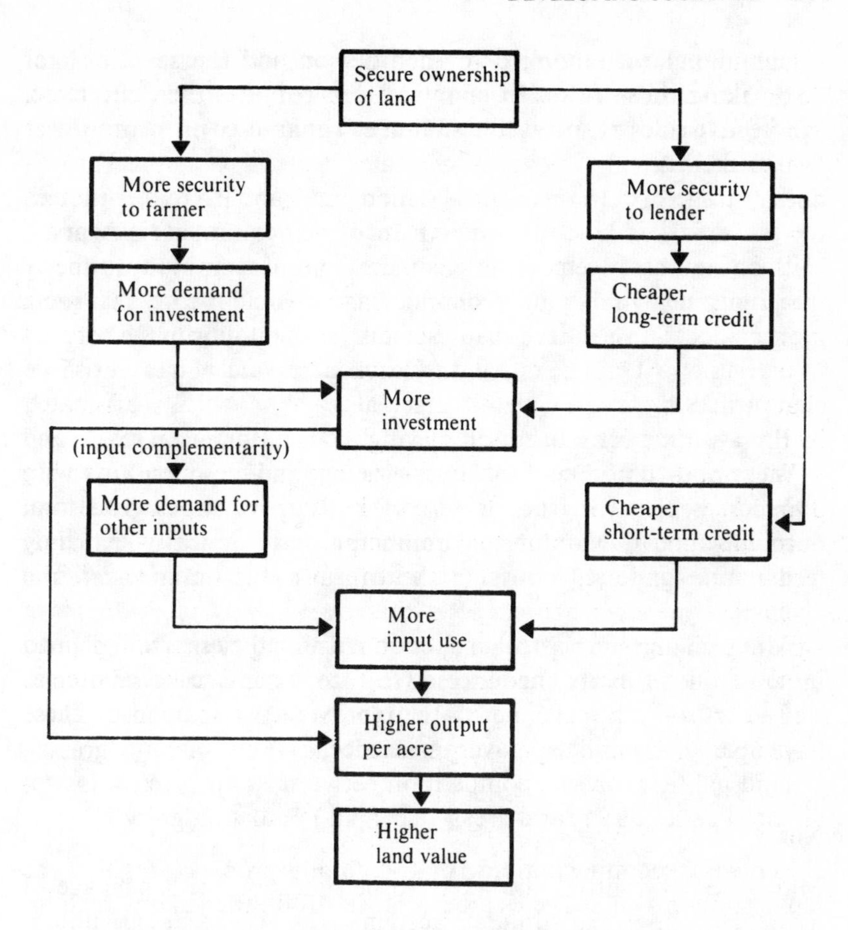

Figure 22.2 Land tenure and agricultural output.

livestock to common grazing land, knowing that the costs of environmental degradation will be shared with others. That is, a personal cost–benefit calculation is inconsistent with the common good (see Chapter 16).

Much of the literature based on this idea is overly simplistic, but the main germ of the idea has validity. Several observations are in order:

1. Common property regimes are not static or homogeneous. There are many examples of them working well and, in particular, working well in terms of sustaining the stock of natural environmental capital on common land. What matters is the set of

communal rules applied to each person and the sanctions for breaking those rules. In many African contexts there are close, effective rules of social cohesion which guarantee proper resource management.

2. The 'tragedy of the commons' is more aptly applied to *open-access* resources, i.e. those for which there are no communal rights at all. Such an arrangement is distinct from a common property arrangement, and while economic theory would predict that some open-access resources can achieve an equilibrium short of extinction, it is clear that these resources are at grave risk of destruction.
3. Privatisation often occurs 'naturally' as land starvation sets in and the opportunity cost of over-grazing and resource over-use becomes clear. Privatised ownership is more widespread than might appear. Moreover, common-property systems frequently contain significant household and farm rights on an individual basis.
4. Many changes in pastoral societies are nonetheless altering some common property resources *back* to open-access resources. Factors of relevance here are improvements in transport for people and animals, government declarations making grazing land public property, competition between ethnic groups, loss of social cohesion in rural areas and rapid population growth.

Blanket recommendations that common-property regimes be replaced by privatised ones are simplistic. Too little is known about the costs and benefits of changing tenurial regimes and there is little option but to approach the issue on a case-by-case basis. There must, however, be a presumption, on the evidence available, that a more efficient agriculture would ensue from clearer titling, leasing and registration of land.

22.6 CONCLUDING NOTE

That part of environmental economics which deals with developing country problems is in its infancy. This chapter has attempted to give no more than the flavour of the kinds of issues that need to be tackled. There are a great many others. The significant point, however, is that the interconnections between environment and economy are far more direct and, arguably, more important in the developing world than they are in the developed world. This may show up in the importance of the concept of sustainable development (see Chapter 3) for the developing world, although we regard it as important also in the developed countries where economies and environments may only superficially appear to be 'decoupled'.

REFERENCES AND FURTHER READING

The literature on **environmental economics** is massive, something that often comes as a surprise to those who think economists have made little effort to tackle environmental issues. Along with references cited in the chapters, we highlight below some of the seminal contributions and point to a few modern sources which the reader may find useful. Academics dispute the value of reading the original contributions – styles change and, more importantly, the analysis improves with time and authors find briefer ways of saying things. To this end we believe it is always best to go back to the original only after a modern synopsis or contribution has been read.

CHAPTER 1

The standard work in the **history of economic thought** is Marc Blaug's *Economic Theory in Retrospect*, Cambridge University Press, Cambridge, 1978. A very readable and different survey of the same history can be found in Ken Cole, John Cameron and Chris Edwards, *Why Economists Disagree: The Political Economy of Economics*, Longman, London, 1983. M. Lutz and K. Lux's *The Challenge of Humanistic Economics*, Benjamin Cummings, New York, 1979, surveys the humanistic economics paradigm. See also R. D. Hamrin: *Managing Growth in the 1980s: Toward a New Economics*, Praeger, New York, 1980. Some of the wider aspects of **environmentalism** are explored in Tim O'Riordan and R. Kerry Turner (eds). *An Annotated Reader in Environmental Planning and Management*, Pergamon Press, Oxford, 1983. The **'limits to growth' debate** has spawned an immense literature; the volume that acted as a catalyst for much of the modern contribution was D. H. Meadows *et al.*, *The Limits to Growth*, Universe Books, New York, 1972. The 'social limits to growth' thesis is explored in the following texts: Michael J. Boskin, *Economics and Human Welfare*, Academic Press, New York, 1979; Fred Hirsch, *Social Limits To Growth*, Routledge and Kegan Paul, London 1977; Tibor Scitovsky, *The Joyless Economy*, Oxford University Press, London, 1976. The **sustain-**

ability debate of the 1980s has been characterised by definitional confusion and a lack of precision – see R. Kerry Turner (ed.), *Sustainable Environmental Management: Principles and Practice*, Belhaven Press, London and Westview, Boulder, 1988; G. O. Barney *The Global 2000 Report to the President of the US* (2 vols), Pergamon Press, Oxford, 1980; Julian Simon and Herman Kahn, *The Resourceful Earth: A Response to Global 2000*, Basil Blackwell, Oxford, 1984; World Commission on Environment and Development, *Our Common Future*, Oxford University Press, London, 1987; Robert Repetto (ed.), *The Global Possible*, Yale University Press, New Haven, 1985; and International Union for the Conservation of Nature, *World Conservation Strategy*, IUCN–UNEP–WWF, Gland, Switzerland, 1980. Possible **ecological models** for economic development have been explored by: Michael Common, 'Poverty and progress revisited', in David Collard, David Pearce, and David Ulph (eds), *Economic Growth and Sustainable Environments*, Macmillan, London, 1988; and by Richard Norgaard, 'Coevolutionary development potential', *Land Economics*, 60 (2), 1984. The view that markets in environmental damage will emerge and secure the optimal level of damage was first elucidated in R. Coase, 'The problem of social cost', *Journal of Law and Economics*, **3**, 1960.

CHAPTER 2

The relevance of the **laws of thermodynamics** to economics was first demonstrated by Kenneth Boulding, 'The economics of the coming spaceship Earth', in H. Jarrett (ed.), *Environmental Quality in a Growing Economy*, Johns Hopkins University Press, Baltimore, 1966. This volume is quite difficult to find now. Fortunately, Boulding's essay has been reprinted in a number of places. See, for example, the interesting collection of essays assembled by Herman Daly, *Economics, Ecology, Ethics: Essays Toward a Steady-State Economy*, W. H. Freeman, San Francisco, 1980. This volume contains many other interesting readings, including two extracts from the work of Nicholas Georgescu-Roegen who has stressed the relevance of the Second Law of Thermodynamics, the 'entropy' law. Georgescu-Roegen's *magnum opus* is *The Entropy Law and the Economic Process*, Harvard University Press, Cambridge, Massachusetts, 1971. Because it has been reprinted a number of times it is easy to get hold of but, be warned, it is difficult reading. The first-time reader is better advised to consult the extracts in Daly's collection of essays. Unfortunately, Georgescu-Roegen's work has given rise to many misunderstandings, including the idea that energy is somehow a measure of 'value', a misconception widely disseminated in the literature on so-called 'energy analysis'. Some idea of the debate can be found by looking at the paper by Stuart Burness, Ronald Cummings, Glen Morris and Inga Paik, 'Thermodynamic and economic concepts as related to resource-use policies', *Land Economics*, **56** (February 1980). The misconceptions in this paper are corrected by Herman Daly in his

'Comment' in the same journal in August 1986. The formalisation of the **materials-balance approach** to the economy–environment interaction is to be found in R. U. Ayres and A. Kneese, 'Production, consumption and externality', *American Economic Review*, **LIX**, (June 1969), and in A. Kneese, R. U. Ayres and R. d'Arge, *Economics and the Environment*, Johns Hopkins University Press, Baltimore, 1970. The **input–output approach** to integrating environment into economic analysis was pioneered by the 'father' of input–output analysis itself, Wassily Leontief, 'Environmental repercussions and the economic structure: an input–output approach', *Review of Economics and Statistics*, **II**, (August 1970). But the best and most detailed exposition, with a literature review and an application to Canada, is undoubtedly Peter Victor's *Pollution: Economy and Environment*, Allen and Unwin, London, 1972. A recent, excellent, if occasionally difficult volume which brings materials-balance together with **externality theory** (see Chapter 4–8 of this book) is Charles Perrings, *Economy and Environment*, Cambridge University Press, Cambridge, 1987.

CHAPTER 3

The literature on the concept of **sustainability** is extensive but both variable in quality and not always very helpful. The interpretation in this chapter is our own. Sustainable development was popularised by the 'Brundtland Commission' – a commission of experts assembled under the leadership of Gro Harlem Brundtland, Prime Minister of Norway. Its report, which is easily readable, is World Commission on Environment and Development, *Our Common Future*, Oxford University Press, London, 1987. The reader will not, however, find it easy to identify our discussion of 'ecological rules' and constant capital stock requirement in the Brundtland Commission report. An elaboration of our chapter can be found in D. W. Pearce, E. Barbier and A. Markandya, *Sustainable Development: Economics and Environment in the Third World*, Edward Elgar, London, 1989. The view that substitution between resources and between types of capital can achieve a sustainable outcome is given in a very readable essay by Robert Solow, 'The economics of resources and the resources of economics', *American Economic Review* (May 1974). This is reprinted in R. Dorfman and N. Dorfman, *Economics of the Environment: Selected Readings*, Norton, New York, 1977. The idea that valuation functions may be 'kinked' is due to Jack Knetsch and John Sinden, 'Willingness to pay and compensation demanded: experimental evidence of an unexpected disparity in measures of value', *Quarterly Journal of Economics* (August 1984). The relevance of this idea to the constant capital stock intepretation of sustainability is ours, however.

CHAPTER 4

The origin of the idea of **optimal externality** comes from the work of Arthur Pigou, *The Economics of Welfare*, Macmillan, London, first published in 1920 and reprinted many times. But the modern reader will find it very dull, however important it has been in the development of externality theory. The diagram we use in this and subsequent chapters was first introduced by Ralph Turvey, 'On divergences between social cost and private cost', *Economica* (August 1963). Many textbooks have used it since. Turvey's paper is reprinted in a useful set of readings brought together by Robert and Nancy Dorfman, *Economics of the Environment*, 2nd edn, W. W. Norton, New York, 1977.

CHAPTER 5

The seminal paper on **market solutions to externality** is Ronald Coase, 'The problem of social cost', *Journal of Law and Economics* (October 1960) which spawned a large literature all on its own. The original is still worth reading although it is dry in style and may appear overly legalistic to many readers. A conveniently shortened version of it is published in the readings by Dorfman and Dorfman (above). The attempt to make the **Coase theorem** relevant to conditions of imperfect competition was made by James Buchanan, 'External diseconomies, corrective taxes and market structure', *American Economic Review*, (March 1969). The **non-convexity** issue is most clearly dealt with in William Baumol and David Bradford, 'Detrimental externalities and non-convexity of the production set', *Economica* (May 1972). This paper is modified as one of the chapters in the best theoretical (but nonetheless quite difficult) text on pollution economics – William Baumol and William Oates, *The Theory of Environmental Policy*, 2nd edn, Cambridge University Press, Cambridge, 1988.

CHAPTER 6

The best treatment of the **Pigovian tax** is in Baumol and Oates (see above). Baumol and Oates also formulated the least-cost theorem for pollution charges in their paper 'The use of standards and prices for the protection of the environment', *Swedish Journal of Economics*, **73**, 1971. The problems with the tax solution need to be viewed from the point of view of the administrator as well as the academic. A rare combination of the two is John Pezzey's essay, 'Market mechanisms of pollution control: "polluter pays", economic and practical aspects', in R. Kerry Turner (ed.), *Sustainable Environment Management: Principles and Practices*, Belhaven Press, London and Westview Press, Boulder, 1988.

CHAPTER 7

Again, the best reference here is Baumol and Oates (see above). The debate about the **relative efficacy of taxes and standards** is part of a wider debate about 'prices versus quantities', i.e. whether it is better to regulate by using the market mechanism and adjusting prices as incentives to achieve the purpose of the regulation, or whether it is better to regulate by setting quantity targets. The earliest formation of the problem in terms of the relative slopes of what we call **MEC** and **MNPB** is Martin Weitzman's essay, 'Prices vs. quantities', *Review of Economic Studies*, **41** (October 1974). A generalisation is Marc Roberts and Michael Spence, 'Effluent charges and licences under uncertainty', *Journal of Public Economics*, **5** (April–May 1976). Both contributions can only really be appreciated by someone numerate and with a knowledge of economics, however. A useful discussion, but all in the American context, is Thomas Schelling's edited volume *Incentives for Environmental Protection*, MIT Press, Cambridge, Massachusetts, 1983.

CHAPTER 8

The idea of marketable permits was first formulated in J. H. Dales, *Pollution, Property and Prices*, University of Toronto Press, Toronto, 1968. Baumol and Oates (see above) have a helpful chapter on **marketable permits**, but the major work on the subject is Tom Tietenberg's *Emissions Trading*, Resources for the Future, Washington DC, 1985. This volume is clearly written and has a wealth of institutional detail on the USA experience. Many of the basic arguments for marketable permits are elegantly set out in Susan Rose-Ackerman, 'Market models for pollution control: their strengths and weaknesses', *Public Policy*, **25** (1977).

CHAPTER 9

The underlying theory to the idea of using **WTP** to measure utility or welfare is contained in textbooks on welfare economics. A comprehensive volume is Richard Just, Darrell Hueth and Andrew Schmitz, *Applied Welfare Economics and Public Policy*, Prentice Hall, New Jersey, 1982. A popular text is Robin Boadway and Neil Bruce, *Welfare Economics*, Blackwell, Oxford, 1984. The idea of **total economic value** as a summation of various types of value seems to have emerged in the journal literature as the component parts of TEV were being developed. The original paper which stimulated the literature on **option value** is Burton Weisbrod's 'Collective-consumption services of individual consumption goods', *Quarterly Journal of Economics* (August 1964). The 'state of the art' on option value is well summarised in Richard Bishop's excellent essay,'Option value: an exposition and extension', *Land Economics* (February 1982). On quasi-option value, see Anthony Fisher and W. Michael Hanemann, 'Quasi-option value: some misconceptions dispelled', *Journal of Environmental Economics and Management* (June 1987). At the time of writing much of the literature on **existence value** is unpublished, but a very clear and fascinating paper is David Brookshire, Larry Eubanks and Alan Randall, 'Estimating option prices and existence values for wildlife resources', *Land Economics* (February 1983). The paper that started it all is John Krutilla's 'Conservation reconsidered', *American Economic Review* (September 1967). The study of the Grand Canyon is D. Brookshire, W. Schulze and M. Thayer, 'Some unusual aspects of valuing a unique natural resource', University of Wyoming, *mimeo*, 1985. Strand's study is reported in J. Strand, 'Valuation of freshwater fish as a public good in Norway', University of Oslo, Department of Economics, *mimeo*, 1981.

CHAPTER 10

There are several comprehensive surveys of **valuation methodologies**. See, for example, David Pearce and Anil Markandya, *The Benefits of Environmental Policy*, Organisation for Economic Cooperation and Development (OECD), Paris, 1989, and Per-Olov Johansson, *The Economic Theory and Measurement of Environmental Benefits*, Cambridge University Press, Cambridge, 1987. A brief, fairly non-technical analysis is Allen Kneese's *Measuring the Benefits of Clean Air and Water*, Resources for the Future, Washington DC, 1984. A very detailed analysis relating to **water quality benefits** but with excellent expositions of the CVM and travel cost approaches is V. Kerry Smith and William Desvousges, *Measuring Water Quality Benefits*, Kluwer Nijhoff, Dordrecht, 1986. The **CVM** is exhaustively treated in Ronald Cummings, David Brookshire and William Schulze (eds), *Valuing Environmental Goods: An Assessment of the Contingent Valuation Method*, Rowman and Allenheld, Totowa, NJ, 1986. The US experience of **cost–benefit approaches** to the environment is discussed in V. Kerry Smith (ed.), *Environmental Policy Under Reagan's Executive Order*, University of North Carolina Press, Chapel Hill, 1984. The **willingness-to-pay versus willingness-to-accept** debate is unresolved. A useful paper is David Brookshire and Don Coursey, 'Measuring the value of a public good: an empirical comparison of elicitation procedures', *American Economic Review* (September 1987). The Grand Canyon paper referred to is W. Schulze *et al.*, 'The economic benefits of preserving visibility in the National Parklands of the South West', *Natural Resources Journal*, **23**, 1983.
Problems of validity and reliability of CVM studies are exhaustively examined in R. Mitchell and R. Carson, Using Surveys to Value Public Goods: The Contingent Valuation Method, Resources For The Future, Washington DC, 1989.

CHAPTER 11

Given the nature of the topic, literature analysing **pollution control policy** tends to date rather rapidly; two publications – the *ENDS Data Services Report*, London (published monthly) and *Resources* (published quarterly by Resources for the Future, Washington) – provide excellent up-to-date coverage of pollution policy. Many of the issues in the direct regulations versus economic incentive debate are covered in two international surveys: Peter Baker and Clifford Russell, 'Alternative policy instruments', in Allen Kneese and John Sweeney, '*Handbook of Natural Resource and Energy Economics*', **1**, North-Holland, Amsterdam, 1985, and Peter Downing and Kurt Hanf (eds), *International Comparisons in Implementing Pollution Laws*, Kluwer Nijhoff, Boston, 1983. The concept of 'Best Practicable Environmental Option' is outlined in UK Royal Commission on Environmental Pollution, 12th Report *Best Practicable Environmental Option*, HMSO, London, 1988. The references cited for Chapters 6, 7 and 8 also contain useful information relating to the overall pollution control policy context. For an excellent review of the current role of economic instruments, their effectiveness and future prospects see J.B. Opschoor and H.B. Vos, *Economic Instruments For Environmental Protection,* OECD, Paris, 1989.

CHAPTER 12

Accurate and unbiased information on the **state of the environment in the USSR** is not very plentiful. The following journals carry articles on Soviet environmental concerns on a fairly regular basis: *New Scientist*, *Soviet Studies* and *Soviet Geography* (try to read Western and Soviet analysts on the same topic to get a 'flavour' of the issues and the politics involved). A debate on the environmental efficiency of **centrally planned and market-based economic systems** appeared in the journal *Soviet Studies* over the period 1978–1982. The contributions cover a number of important conceptual and empirical difficulties and also give the reader a sense of the 'political' nature of the dispute. See Robert McIntyre and James Thornton, 'On the environmental efficiency of economic systems', *Soviet Studies* (April 1978), and a critical reply by Charles Ziegler, 'Soviet environmental policy and Soviet central planning: a reply to McIntyre and Thornton', *Soviet Studies* (January 1980); the debate continued with a reply by McIntyre and Thornton, *Soviet Studies* (January 1981) and a rejoinder by Ziegler, *Soviet Studies* (April 1982). A book reputedly written by a 'mole' in the Soviet bureaucracy and very critical of Soviet environmental policy is Boris Komarov, *The Destruction of Nature in the Soviet Union*, Pluto Press, London. A more balanced account can be found in Charles Ziegler, *Environmental Policy in the USSR*, Pinter, London, 1987. The Soviet river-diversion controversy is analysed in Philip Micklin, 'The diversion of Soviet rivers', *Environment* (March 1985) and Michael Kelly *et al.* 'Large-scale water transfers in the USSR', *Geo Journal* (March 1983).

CHAPTER 13

There are many books about **acid rain**. A readable account is John McCormick's *Acid Earth*, Earthscan, London, 1985. The Worldwatch Institute in Washington DC publishes a series of 'Worldwatch Papers' which are generally well-researched and presented in a non-technical fashion. Sandra Postel's *Air Pollution, Acid Rain and the Future of Forests*, Worldwatch Paper 58, 1984, is useful on acid rain. The two volumes produced by the UK House of Commons Environment Committee, (Session 1987–1988), *Air Pollution*, 1988, are very valuable and cover CFCs, CO_2 and acid rain. A popular account of the **ozone layer problem** is John Gribbin, *The Hole in the Sky: Man's Threat to the Ozone Layer*, Corgi, London, 1988. More technical and written before the Antarctic 'hole' discovery is John Cumberland, James Hibbs and Irving Hoch (eds), *The Economics of Managing Chlorofluorocarbons*, Resources for the Future, Washington DC, 1982. An excellent non-technical explanation of the **greenhouse effect** and its potential impact can be found in Jackie Karas and Michael Kelly's *The Heat Trap: The Threat Posed by Rising Levels of Greenhouse Gases*, Friends of the Earth, London, 1988 (Sections 7 and 8 are not part of the Karas and Kelly analysis and should be read as campaigning statements on behalf of the FOE pressure group itself). An authorative review of the greenhouse effect and sea-level change impacts can be found in Richard Warrick and Philip Jones, 'The greenhouse effect: impacts and policies', *Forum for Applied Research and Public Policy*, Fall 1988. At a more technical level, see B. Bolin *et al.*, '*The Greenhouse Effect, Climatic Change and Ecosystems*, SCOPE 29, Wiley, Chichester, 1986.

CHAPTER 14

Simple expositions of the derivation of the **SOC** and **STPR discount rates**, and the weighted average (or 'synthetic') rates which are not discussed in this chapter, can be found in David Pearce, *Cost–Benefit Analysis*, 2nd edn (revised), Macmillan, London, 1986. The environmentalist critique is best dealt with in three papers by philosophers: Brian Barry, 'Justice between generations', in P. Hacker and J. Raz (eds), *Law, Morality and Society*, Clarendon Press, Oxford, 1977; and Robert Goodin, 'Discounting discounting', *Journal of Public Policy*, **2** (1982); and Derek Parfit, 'Energy policy and the further future: the social discount rate', in Donald Maclean and Peter Brown (eds), *Energy and the Future*, Rowman and Littlefield, Totowa, NJ, 1983. Robert Goodin's *Protecting the Vulnerable*, University of Chicago Press, Chicago, 1987, is also useful in this respect. A survey of the environmentalist critiques, and on which we have partly relied for this chapter, is David Pearce and Anil Markandya, *Environmental Considerations and the Choice of Discount Rate in Developing Countries*, Environment Department, World Bank, Washington DC, May 1988. The *magnum opus* on **discounting** remains Robert Lind (ed.), *Discounting for Time and Risk in Energy Policy*, Johns Hopkins University Press, Baltimore, 1986. While some of the essays in this volume are very mathematical, Lind's editorial introduction is a masterly overview of the contributions. A rich and challenging book which covers ways in which economic evaluation might be modified in light of intergenerational justice is Talbot Page's *Conservation and Economic Efficiency*, Johns Hopkins University Press, Baltimore, 1977. This is an important book. Some sections are a bit hard-going and some now seem somewhat dated, but the central idea of an economy 'bounded' by considerations of intergenerational justice and within which the present value criterion operates has many parallels with some of the themes of this text. On pure time preference see Mancur Olson and Martin Bailey, 'Positive time preference', *Journal of Political Economy*, **89** (1981).

CHAPTER 15

For a good account of why interpretations of the notion of **environmental value** differ, see Thomas Brown, 'The concept of value in resource allocation', *Land Economics* (Fall 1984). The main philosophical positions in the field of **environmental ethics** have been surveyed by R. Kerry Turner, 'Wetland conservation: economics and ethics', in David Collard, David Pearce and David Ulph (eds), *Economics, Growth and Sustainable Environments*, Macmillan, London, 1988. A good philosophical survey of the subject is Robert Elliot and Aron Gare (eds), *Environmental Philosophy*, University of Queensland Press, St. Lucia, Australia, 1983; see also P. G. Brown and D. Maclean (eds), *Energy and the Future*, Rowan Littlefield, Totowa, NJ, 1983. The **Gaia hypothesis** is explained by its creator James Lovelock in *Gaia: A New Look at Life on Earth*, Oxford University Press, London, 1979. Finally, the journal *Environmental Ethics* is entirely devoted to publishing articles (mostly supportive) on aspects of a new environmental ethic.

CHAPTER 16

The economic analysis of **renewable resource use** quickly becomes complicated, far more so than we have indicated in this chapter. Different authors have tried different ways of making the subject more easy to handle, but it seems fair to say that the reader who wishes to get a grasp of the subject will need to tackle some of the mathematical literature. In this respect the undoubted classic work is Colin Clark's *Mathematical Bioeconomics*, Wiley, New York, 1976. This can now be supplemented with Jon Conrad and Colin Clark, *Natural Resource Economics: Notes and Problems*, Cambridge University Press, Cambridge, 1987. Another classic treatment which mainly covers **exhaustible resources** is Partha Dasgupta and Geoffrey Heal's *Economic Theory and Exhaustible Resources*, Cambridge University Press, Cambridge, 1979. It is also worth persisting with Partha Dasgupta's *Control of Resources*, Blackwell, Oxford, 1982, which uses less maths, but which is still slightly hard going in places. The first paper to make the **analytics of open-access solutions** clear was H. Scott Gordon, 'The economic theory of a common property resource: the fishery', *Journal of Political Economy* (1954).

CHAPTER 17

Colin Clark's *Mathematical Bioeconomics*, Wiley, New York, 1976, provides the theoretical basis for the idea that **extinction** might be 'optimal', although Clark himself is at pains to point out that other values are relevant to the decision to make a resource extinct. A useful compendium of potential uses for species threatened with extinction is Norman Myers' *A Wealth of Wild Species*, Westview Press, Boulder, 1983.

CHAPTER 18

The general texts listed in the further reading for Chapter 16 are relevant. The original paper is Harold Hotelling, 'The economics of exhaustible resources', *Journal of Political Economy* (1931). An elegant essay is Shantayan Devarajan and Anthony Fisher, 'Hotelling's "Economics of Exhaustible Resources": 50 years later', *Journal of Economic Literature*, **XIX** (March 1981). A useful diagrammatic presentation of the implications for the **Hotelling rule** of relaxing its many assumptions is Michael Toman, ' ''Depletion effects'' and non-renewable resource supply: a diagrammatic exposition', *Land Economics* (November 1986).

CHAPTER 19

The Meadows *et al. Limits to Growth* publication contains a pessimistic static reserve index calculation of resource lifetimes; this can be contrasted with the H. E. Goeller and A. Zucker, 'Infinite resources: the ultimate strategy', *Science* (February 1984), optimistic calculations, (note the energy backstop assumption). The strategic materials question is analysed by Derek Deadman and R. Kerry Turner, 'Resource conservation, sustainability and technical change', in R. Kerry Turner (ed.), *Sustainable Environmental Management: Principles and Practice*, Belhaven Press, London and Westview, Boulder, 1988. The markets for and availability of **advanced materials** have been surveyed by S. Fraser, A. Barsotti and D. Rogich,

'Sorting out materials issues', *Resources Policy* (March 1988). The complexities surrounding the actual non-renewable resources supply process are well explained (without the use of mathematical models) in Douglas Bohi and Michael Toman, 'Understanding non-renewable resource supply behaviour', *Science* (February 1983). **Economic resource scarcity indexes** were first analysed fully in a now famous book, Harold Barnett and Chandler Morse, *Scarcity and Growth: The Economics of Natural Resource Availability*, Johns Hopkins University Press, Baltimore, 1963. H. Barnett updated the analysis for the 1970s and came up with the same optimistic conclusion – no real resource scarcity evidence was present – see, 'Scarcity and growth revisited', in V. Kerry Smith (ed.) *Scarcity and Growth Reconsidered*, Johns Hopkins University Press, Baltimore, 1979. These basic findings have been challenged more recently in D. C. Hall and J. V. Hall, 'Concepts and measures of natural resource scarcity, with a summary of recent trends', *Journal of Environmental Economics and Management* (September 1984). The **geochemical theory** of resource availability is explained by D. P. Harris and B. J. Skinner, 'The assessment of long-term supplies of minerals', in V. Kerry Smith and John Krutilla (eds), *Explorations in Natural Resources Economics*, Johns Hopkins University Press, Baltimore, 1982. The economics of **recycling** are covered in David Pearce and Ingo Walter, *Resource Conservation: The Social and Economic Dimensions of Recycling*, Longman, London and New York University Press, New York, 1977. Some empirical evidence relating to **material substitution** possibilities can be found in John Tilton, *Material Substitution: Lessons from the Tin-Using Industries*, Resources for the Future, Washington, 1983.

CHAPTER 20

The **Krutilla–Fisher alogorithm** is developed in John Krutilla and Anthony Fisher, *The Economics of Natural Environments*, Resources for the Future, Washington DC, 1985. A fairly large literature surrounds the Krutilla–Fisher approach but it is largely brought together in this volume. The best exposition, one which develops some lines of thought, and which we have used as the basis of our presentation, is Richard Porter, 'The new approach to wilderness preservation through benefit cost analysis', *Journal of Environmental Economics and Management*, **9** (1982). The SMS approach is most clearly explained by Richard Bishop, 'Endangered species and uncertainty: the economics of a safe minimum standard', in *American Journal of Agricultural Economics* (February 1978).

CHAPTER 21

The value of **wetland ecosystems** and the causes and consequences of their destruction are analysed by R. Kerry Turner, 'Wetland conservation: economics and ethics', in David Collard, David Pearce and David Ulph (eds) *Economics, Growth and Sustainable Environments*, Macmillan, London, 1988. The situation in the USA has been surveyed in US Office of Technology Assessment, 'Wetlands, their use and regulation', OTA, Washington, 1984. The following studies have attempted to value in monetary terms various wetland functions and services:

1. G. D. Lynne *et al.*, 'Economic valuation of marsh areas for marine production processes', *Journal of Environmental Management* (April, 1981); **2**. S. Batie and J. Wilson, 'Economic values attributable to Virginia's coastal wetlands and inputs in oyster production', *Research Division Bulletin*, 150, Department of Agricultural Economics, Virginia Polytechnic Institute and State University, Blacksburg, Virginia, 1979; **3**. T. R. Gupta and J. H. Foster, 'Economic criteria for freshwater wetland policy in Massachusetts', *American Journal of Agricultural Economics* (January 1975); **4**. R. Bishop and T. A. Heberlein, 'Simulated markets, hypothetical markets and travel cost analysis: alternative methods of estimating outdoor recreation demand', Department of Agricultural Economics, Staff Paper, 187, University of Wisconsin, 1980; **5**. G. M. Brown and H. O. Pollakowski, 'Economic valuation of shoreline', *Review of Economics and Statistics*, **59** (April 1977); **6**. R. Constanza, S. Farber and J. Maxwell, 'The valuation and management of wetland ecosystems', paper presented at the Vienna Center Conference on Integrating Ecology and Economics, Barcelona, September, 1987; **7**. W. R. Fritz and S. C. Helle, 'Cypress wetlands for tertiary treatment', in USEPA, *Aquaculture Systems for Wastewater Treatment: Seminar Proceedings and Engineering Assessment*, USEPA, Office of Water Programme Operations, Series Water 430/9–80–006, Washington, 1979; **8**. J. G. Gosselink, E. P. Odum and R. M. Pope, 'The value of the tidal marsh', Centre for Wetland Resources, Louisiana State University, Baton Rouge, 1974; **9**. J. R. Kahn and W. M. Kemp, 'Economic losses associated with the degradation of an ecosystem: the case of submerged aquatic vegetation in Chesapeake Bay', *Journal of Environmental Economics and Management*, **12** (3), September, 1985; **10**. I. A. Mendelssohn, R. E. Turner and K. L. McKee, 'Louisiana's eroding coastal zone: management alternatives, *Journal of the Limological Society of South Africa*, **9**(2) (April, 1983); **11**. R. W. Nelson and W. Logan, 'Policy on wetland impact mitigation', *Environment International*, **10** (1), 1984; **12**. W. M. Park and S. S. Batie, 'Methodological issues associated with estimation of the economic value of coastal wetlands in improving water quality', Sea Grant Project Paper VP1-SG-79-09, Department of Agricultural Economics, Virginia Polytechnic Institute and State University, Blacksburg, Virginia, 1979; **13**. C. Seller *et al.*, 'Validation of empirical measures of welfare change: a comparison of non-market techniques', *Land Economics*, **61** (2), 1985; **14**. L. Shabman, S. Batie and C. Mabbs-Zeno, 'The economics of wetland preservation in Virginia', *Journal of the North Eastern Agricultural Economics Council*, **8** (2), 1979; **15**. L. Shabman and

M. Bertelsen, 'The use of development value estimates for coastal wetland permit decisions', *Land Economics*, **55** (2), 1979. **16**. G. Tchobanoglous and G. L. Culp, 'Wetland systems for wastewater treatment: an engineering assessment', in USEPA, *Aquaculture Systems for Wastewater Treatment*, Office of Water Programme Operations, Series Water EPA 430/9-80-007, Washington, 1986; **17**. F. R. Thibodeau and B. D. Ostro, 'An economic analysis of wetland protection', *Journal of Environmental Management*, **12** (1) (January 1981); **18**. T. C. Williams, 'Wetlands irrigation aids man and nature', *Water and Wastes Engineering*, (November 1980, pp. 28–31). **19.** R. Costanza & S. C. Faber, 'Valuation and Management of wetland ecosystems', Ecological Economics, 1 (4) pp. 335–361.

For an application of the **opportunity cost approach** see R. Kerry Turner, David Dent and Richard Hey, 'Valuation of the environmental impact of wetland flood protection and drainage schemes', *Environment and Planning* A. (December 1983), and R. Kerry Turner and Jan Brooke, 'Management and valuation of an environmentally sensitive area: Norfolk Broadland case study', *Environmental Management* (September 1988). For a US application see S. Batie and C. Mabbs-Zeno, 'Opportunity costs of preserving coastal wetlands: A case study of a recreational housing development', *Land Economics* (January 1985).

CHAPTER 22

Because of the relatively recent nature of the research there are few suitable overviews of the application of environmental economics to **developing economies**. However, Robert Repetto's *World Enough and Time: Successful Strategies for Resource Management*, Yale University Press, New Haven, 1986, is a useful starting point. An annual publication that assembles resource, environmental and energy data is *World Resources Report*, published jointly by the International Institute for Environment and Development and the World Resources Institute, Washington DC and London. An extensive text on the subject is being prepared by David Pearce and Jerry Warford. Ken Newcombe's paper on Ethiopia is 'The economic justification for rural afforestation: The case of Ethiopia', Energy Department Paper 16, World Bank, Washington DC, 1984. A. Sfeir-Younis's study of sedimentation is 'Soil conservation in developing countries', *mimeo*, Agriculture Department, World Bank, Washington DC, 1985.

INDEX